Parasitology is an immensely important aspect of biological science. This manual presents 50 easy-to-follow laboratory exercises for student practical (lab) classes. All the exercises are tried and tested by the authors and are used in a wide variety of university undergraduate teaching departments. They range from relatively simple observational exercises, using local materials and requiring little in the way of equipment, to more technically demanding experiments in physiology and molecular parasitology.

Each exercise includes a list of necessary equipment, consumables and sources of parasite material, instructions for staff and students, including aspects of safety, expected results and some analysis provided by questions; there are ideas for further exploration and information on similar exercises, and lists of selected further reading.

This book should be an essential purchase for all teachers of parasitology at the university undergraduate level and for students taking laboratory practical classes in the subject.

Practical Exercises in Parasitology

Edited by   **David W. Halton**
Queen's University of Belfast

Jerzy M. Behnke
University of Nottingham

Ian Marshall
Liverpool School of Tropical Medicine

CAMBRIDGE
UNIVERSITY PRESS

PUBLISHED BY THE PRESS SYNDICATE OF THE UNIVERSITY OF CAMBRIDGE
The Pitt Building, Trumpington Street, Cambridge, United Kingdom

CAMBRIDGE UNIVERSITY PRESS
The Edinburgh Building, Cambridge CB2 2RU, UK
40 West 20th Street, New York, NY 10011-4211, USA
10 Stamford Road, Oakleigh, VIC 3166, Australia
Ruiz de Alarcón 13, 28014 Madrid, Spain
Dock House, The Waterfront, Cape Town 8001, South Africa

http://www.cambridge.org

First published 2001

Printed in the United Kingdom at the University Press, Cambridge

Typeface Swift 10/13pt.   *System* QuarkXPress™   [SE]

A catalogue record for this book is available from the British Library

ISBN 0 521 79104 9 hardback

CONTENTS

CONTRIBUTORS

C. Arme
Department of Biological
Sciences
Centre for Applied Entomology
and Parasitology
Keele University
Keele
Staffordshire ST5 5BG
UK

I. Barber
Edward Llwyd Building
Institute of Biological Sciences
University of Wales Aberystwyth
Aberystwyth
Ceredigion
SY23 3DA
Wales, UK

P. A. Bates
Division of Molecular Biology and
Immunology
Liverpool School of Tropical
Medicine
Pembroke Place
Liverpool L3 5QA
UK

J. M. Behnke
School of Life and Environmental
Sciences
University of Nottingham
Nottingham NG7 2RD
UK

C. E. Bennett
School of Biological Sciences
University of Southampton
Biomedical Sciences Building
Bassett Crescent East
Southampton SO16 7PX
UK

J. W. Bowman
Animal Health Discovery
Research
Pharmacia and Upjohn Co
301 Henrietta Street
Kalamazoo
MI 49001
USA

D. Britten
Department of Infectious and
Tropical Diseases
London School of Hygiene and
Tropical Medicine
Keppel Street
London WC1E 7HT
UK

A. F. Brown
Uplands Team
English Nature
Northminster House
Peterborough PE1 1UA
UK

M. L. Chance
Division of Molecular Biology and
Immunology
Liverpool School of Tropical
Medicine
Pembroke Place
Liverpool L3 5QA
UK

L. H. Chappell
Department of Zoology
University of Aberdeen
Tillydrone Avenue
Aberdeen AB9 2TN
Scotland, UK

J. Chernin
School of Biological Sciences
University of Portsmouth
King Henry I Street
Portsmouth PO1 2DY
UK

R. L. Coop
Moredon Research Institute
Pentlands Science Park
Bush Loan
Penicuik
Midlothian EH26 0PZ
Scotland, UK

J. M. Crampton
School of Biological Sciences
Life Sciences Building
Crown Street
Liverpool L69 7ZB
UK

M. J. Doenhoff
School of Biological Sciences
University of Wales
Bangor
Gwynedd LL57 2UW
Wales, UK

B. Fried
Department of Biology
Lafayette College
Easton
Pennsylvania 18042
USA

D. W. Halton
School of Biology and Bio-
chemistry
Medical Biology Centre
Queen's University of Belfast
Belfast BT9 7BL
Northern Ireland, UK

R. E. B. Hanna
Veterinary Sciences Division
Department of Agriculture N. I.
Stoney Road
Belfast BT4 3SD
Northern Ireland, UK

P. A. Heuch
National Veterinary Institute
Fish Health Section
PO Box 8156 Dep.
N-0033 Oslo
Norway

W. M. Hominick
CABI Bioscience
Bakeham Lane
Egham
Surrey TW20 9TY
UK

M. Hommel
Division of Molecular Biology and
Immunology
Liverpool School of Tropical
Medicine
Pembroke Place
Liverpool L3 5QA
UK

H. Hurd
Department of Biological
Sciences
Centre for Applied Entomology
and Parasitology
Keele University
Keele
Staffordshire ST5 5BG
UK

G. A. Ingram
Department of Biological
Sciences
University of Salford
Salford M5 4WT
UK

S. W. B. Irwin
School of Applied Biological and
Chemical Sciences
University of Ulster at
Jordanstown
Shore Road, Newtownabbey
Co. Antrim, BT37 0QB
Northern Ireland, UK

E. Jackson
Moredon Research Institute
Pentlands Science Park
Bush Loan
Penicuik
Midlothian EH26 0PZ
Scotland, UK

F. Jackson
Moredon Research Institute
Pentlands Science Park
Bush Loan
Penicuik
Midlothian EH26 0PZ
Scotland, UK

D. A. Johnson
School of Biological Sciences,
University of Nottingham,
Nottingham NG7 2RD
UK

J. T. Jones
Department of Zoology
Scottish Crop Research Institute
Invergowrie
Dundee DD2 5DA
Scotland, UK

C. R. Kennedy
Department of Biological
Sciences
Hatherly Laboratories
University of Exeter
Exeter EX4 4PS
UK

T. Knapp
The Royal Castle International
Centre for Lung Cancer Research
200 London Road
Liverpool L3 9TA
UK

D. L. Lee
Shildon Cottage
Blanchland
Nr Consett
Co. Durham DH8 95U
UK

I. Marshall
Division of Parasite and Vector
Biology
Liverpool School of Tropical
Medicine
Pembroke Place
Liverpool L3 5QA
UK

N. J. Marks
School of Biology and Bio-
chemistry
Medical Biology Centre
Queen's University of Belfast
Belfast BT9 7BL
Northern Ireland, UK

R. J. Martin
Department of Biological
Sciences
College of Veterinary Medicine
Iowa State University
Ames
Iowa 50011-1250
USA

A. G. Maule
School of Biology and Bio-
chemistry
Medical Biology Centre
Queen's University of Belfast
Belfast BT9 7BL
Northern Ireland, UK

C. McGuire,
University of Southampton
Dermatopharmacology Unit
Southampton General Hospital
Southampton SO16 6YD
UK

A. W. Pike
Department of Zoology
The University
Tillydrone Avenue
Aberdeen AB9 2TN
Scotland, UK

D. I. de Pomerai
School of Life and Environmental
Sciences
University of Nottingham
Nottingham NG7 2RD
UK

J. G. Rea
School of Applied Biological and
Chemical Sciences
University of Ulster at
Jordanstown
Shore Road, Newtownabbey
Co. Antrim, BT37 0QB
Northern Ireland, UK

T. A. Schram
Department of Biology
Section of Marine Zoology and
Marine Chemistry
University of Oslo
Norway

M. W. Shirley
Institute for Animal Health
Compton Laboratory
Compton
Nr. Newbury
Berkshire RG16 0NN
UK

R. E. Sinden
Department of Biology
Biomedical Sciences Building
Imperial College of Science and
Technology
Imperial College Road
London SW7 2AZ
UK

J. E. Smith
Department of Pure and Applied
Biology
The University
Leeds LS2 9JT
UK

D. B. A. Thompson
Scottish Natural Heritage
2 Anderson Place
Edinburgh EH6 5NP
Scotland, UK

R. C. Tinsley
School of Biological Sciences
University of Bristol
Woodland Road
Bristol BS8 1UG
UK

C. M. R. Turner
Parasitology Laboratory
IBLS
Joseph Black Building
University of Glasgow
Glasgow G12 8QQ
Scotland, UK

D. Wakelin
School of Life and Environmental
Sciences
University of Nottingham
University Park
Nottingham NG7 2RD
UK

D. C. Warhurst
Department of Infectious and
Tropical Diseases
London School of Hygiene and
Tropical Medicine
Keppel Street
London WC1E 7HT
UK

J. E. Williams
Department of Infectious and
Tropical Diseases
London School of Hygiene and
Tropical Medicine
Keppel Street
London WC1E 7HT
UK

PREFACE

This manual describes a range of well-tried practical exercises, drawn largely from the membership of the British Society for Parasitology, and currently used in the teaching of parasitology at undergraduate level. The primary aim is to promote and, hopefully, stimulate practical teaching in parasitology in institutions where levels of experience and resources devoted to the subject vary from substantial to none. For this reason, the exercises selected range from the simple, requiring little in the way of sophisticated equipment, and focusing on locally-available materials, to the more elaborate, in the fields of molecular biology and immunology; it is not intended to provide comprehensive coverage of practical parasitology. Use of the manual outside of the UK may require alternative materials and/or sources of materials.

Although the seven sections presented comprise information and exercises in different aspects of the subject, it is recognised that merging of a number of the exercises cited, or other modifications, may suit the local situation. For example, the species used in Section 1 could be substituted depending on availability; these and other adaptations are to be encouraged and, in many instances, alternative sources of material are suggested by authors. The decision on how best to produce a bench-top handout is a personal one, but readers are welcome to use the text and ideas presented here.

It is important to note that while every consideration to health and safety has been given by the authors, editors and the Society, no responsibility can be accepted if things go wrong. In the UK, we advise each laboratory co-ordinator to carry out a hazard assessment for any of the exercises used so as to ensure that the Control of Substances Hazardous to Health Regulations (COSHH, 1986), part of the UK Health and Safety at Work Act, 1974, are fully met. Note that these now include the Categorisation of Biological

Agents, according to hazard and categories of containment, prepared by the Advisory Committee on Dangerous Pathogens, 1995.

A number of exercises include the use of laboratory animals. In the UK, it is essential that Home Office regulations are followed and the local laboratory co-ordinator and/or department must take responsibility for this. The transfer of animals from one laboratory to another also requires Home Office approval.

The editors hope that the ideas presented here will contribute to securing the future of teaching in parasitology, and welcome any suggestions or comments for possible improvement and further development of the content.

Finally, we are indebted to the authors of the exercises that have been selected for this publication, but equally grateful to those who submitted exercises which, for a variety of reasons, we have been unable to include. We would also like to record our thanks and appreciation to many other colleagues for their encouragement, criticisms and comments, and especially the Council of the British Society for Parasitology for their continued support.

David W. Halton, Jerzy M. Behnke, Ian Marshall
January 2000

General advice

This section provides information which may prove useful for those running practical classes for the first time, and which could serve as handouts for students.

The information is organised into four parts:

- General instructions on *health and safety* in the laboratory.
- A summary of the general principles of *good laboratory practice*.
- Guidance on *how to write up exercises* in the form of conventional scientific papers.
- Criteria which may be used in *assessing course work*, together with an example of a form that could be used to provide feedback to the students.

1. Health and safety

PLEASE READ CAREFULLY

The following section relates to aspects of safety during practical classes. By U.K. law, you are obliged to adhere to the regulations outlined and to seek additional advice if you are aware of any special circumstances in which your health or that of others may be at risk during the course of the work.

Listen carefully to the instructions given before the start of, or during, each class

1.1 Laboratory coats must be worn at all times when working in laboratories. If you do not have a laboratory coat you will be requested to leave the class.

1.2 All students must be in possession of safety glasses and you must wear these during any laboratory procedures which risk damage to your eyes.

You should take great care when handling both dead and living animal matter

1.3 Although the parasites which are the subject of study are mostly not infective to humans, you should treat **all** material as potentially hazardous.

1.4 If you contaminate yourself, or any laboratory equipment or bench space you are using, report it immediately to a member of staff.

1.5 **Do not** eat, drink, smoke or apply make-up in the laboratory.

1.6 If you have reason to believe that you are allergic to contact with rodents (mice, hamsters, rats), insects or any other animals (e.g. pets such as dogs, cats), please inform those running the class before undertaking any work. It may be possible for you to continue subject to appropriate advice.

1.7 If you need to wear gloves during practical work, you will be informed at the beginning of the class. **Remember, contaminated gloves must not be used to operate laboratory equipment** and should be removed and appropriately disposed of before leaving the laboratory.

1.8 Always wash your hands thoroughly during a break in the practical class and before you leave the laboratory for any purpose.

1.9 If you have any cuts or abrasions but wish to participate in the class, please ask for advice first and if permitted to continue, wear gloves throughout all laboratory procedures (see 1.7).

1.10 If you cut or in any other way injure yourself in the laboratory, please report at once to one of the laboratory staff.

1.11 If any chemicals employed in the work are hazardous to health you will be given precise instructions about aspects of safety on the day of the class. Please ensure that you follow these carefully. If you think you are allergic to any of the chemicals to be used, please inform those running the practical class concerned.

FIRE PRECAUTIONS

If you hear the fire alarm, leave the laboratory immediately and go to your assembly point as instructed. Make sure that you are familiar with all of the exits from the building.

2. Good laboratory practice

At all times aim to keep your laboratory environment neat and clean, as well as hazard-free for the benefit of others as well as yourself

2.1 Leave your personal belongings outside the laboratory, using the facilities provided. If bags are brought into the laboratory, they should be placed well under the benches or at the ends of the alley-ways between benches in order not to create a hazard.

2.2 Keep your immediate working environment on the benches neat and tidy.

2.3 Keep all fluids well away from electric cables, contacts and plugs. Report any unusual functioning of instruments immediately; there may be an electrical fault.

2.4 If you notice broken glassware, dispose of it safely.

2.5 Do not leave glass slides or any sharp instruments (e.g. scissors, forceps) in places which may cause injury.

Please help to make the work of technicians who set up the classes and clear the laboratories afterwards as hazard-free as possible. In order to minimise the work involved in clearing the laboratories for other classes:

2.6 Please ensure that your work area is tidy before you leave the laboratory.

2.7 Switch off all electric instruments, such as microscopes, bench incubators, lamps, etc.

2.8 Place all dirty glassware in the labelled containers provided.

2.9 Place all animal matter, including tissues, in appropriately-labelled containers.

Take great care with containers containing bleach!
If in doubt
ASK FOR HELP!

3. How to write up your practical work

3.1 You should keep accurate records of your work carried out during a practical, as a matter of routine.

3.2 Work carried out in specific practical classes may be written up in the form of separate, short scientific papers. No submission should be longer than eight sides of A4 paper, excluding illustrations.

3.3 Simple line drawings can be used to reduce descriptive text; they do not have to be works of art!

3.4 Your report should comprise the following sections:

- *Abstract or summary.* Concise but detailed summary of the experiments, results and main conclusions. Essentially, this should contain information on what you did, how you did it, and what you found.

- *Introduction.* Brief review of the literature, pointing out what contribution to knowledge you hope to make. You should end with a statement of the hypothesis you intend to test.

- *Materials and methods.* Full description of the methods actually employed on the day (a rewritten schedule will not do). You must give sufficient information to enable the experiment to be repeated exactly. Include information on experimental design and statistical analysis employed to evaluate the results, where appropriate.

- *Results.* Written account of the experimental plan and results supplemented by Tables and Figures (graphs and/or histograms as appropriate) showing the data obtained. If relevant, this section should also be supported by statistical analysis. Where there is no experiment, but biological material to describe, provide drawings (all in pencil). Provide a title. Use a full page for each drawing or series of related drawings and label clearly with horizontal indicator lines where possible. Include a scale bar on the drawing; alternatively, include a magnification factor taken from your microscope multiplying the objective lens magnification by the eyepiece magnification. Also indicate the amount of the field of view. Use a sampling technique to draw high power detail of cellular structure in an organism (e.g. a sector from the centre of a microscope slide specimen). On a facing page, provide an explanation of your material, life cycle or interpretation of cytology. Use material provided to you in the class or gleaned from books during the write-up period. A practical write-up in this way may contain several sets of drawings with accompanying pages.

- *Discussion.* This section should emphasise the results obtained, pointing out any sources of error and any reasons for reservation about the data. You should explain how your work has contributed to available knowledge in the literature. The following questions should also be considered:
    - Are your results as anticipated and do they meet the predictions of relevant hypotheses?
    - How do they compare to similar published data sets?
    - Are there any inconsistencies and, if so, how may these be explained?
    - What do you consider to be the most important finding?
    - How can the work be extended, i.e. what are the next questions which should be tackled?
- *References.* You should give full details of all the papers which you have consulted and used in earlier sections of the text. Only those actually used and referenced in the text should be given.

3.5 General points:

You should use the past tense throughout, as appropriate, and all key statements of knowledge should be supported by references to suitable literature. The mark awarded for your report will be based on an independent assessment of how you contributed to each of the headings given above as well as the overall structure of your report.

REFERENCE

Pechenik, J. & Lamb, B. (1994). *How to Write About Biology.* Harper Collins.

4. Assessment of reports

The following criteria may be taken into consideration in the marking of assessed course work. They may act as guidelines to examiners and students of the skills expected at each level of the degree classification scheme from First Class to Third Class. The final assessment mark will be a synthesis of the scores from the different areas:

4.1 Accuracy and omissions in factual components and principles.

4.2 Conceptual and critical ability.

4.3 Use of material gained from reading beyond lectures.
4.4 Presentation (text and diagrammatic).

The range of normally expected skills within these areas is:

4.1 Accuracy

I: High, with no major omissions or errors.
Ili: Thorough, comprehensive answer, few errors or omissions.
Iliii: Satisfactory answer but with some errors or omissions.
III: Basic knowledge with errors and omissions.

4.2 Conceptual and critical ability

I: High, with originality and inter-disciplinary thought.
Ili: Considerable, displaying clear insight and understanding.
Iliii: Reasonable appreciation though limited critical ability.
III: Basic ability with little evidence of criticism.

4.3 Evidence of reading beyond lectures

I: Extensive, well selected, researched and interpolated.
Ili: Clear use of additional reading matter.
Iliii: Some evidence of additional reading.
III: Very limited or no evidence of additional reading.

4.4 Presentation: text and diagrammatic

I: Excellent concise logic; well chosen diagrams/tables.
Ili: Good clear presentation, well supported by diagrams/
tables.
Iliii: Adequate structure with some supporting elements.
III: Evidence of training though with limited structure and
support.

Example of a feedback form
(Based on the *Hymenolepis* oncosphere hatching – Exercise 3.1)
The scores relate to the major skills to be expected in a good practical write-up. They are offered without prejudice to the final
mark provided, which is a synthesis of these elements. A score of
5 indicates a high standard or strong evidence of the skill. A
score of 1 indicates that the skill was of a weak standard or
lacking.
Note: when assessing quality or style of writing, due consideration must be given to students who are not using their first

language, or who may have a medically recognised condition such as dyslexia.

General

1 Was there good writing style? 5 4 3 2 1
2 Was presentation neat and well ordered? 5 4 3 2 1

Text

3 Were the descriptive elements of a high standard? 5 4 3 2 1
4 Was there sufficient descriptive material? 5 4 3 2 1

Drawings/Results

5 Was there a full range of possible drawings? 5 4 3 2 1
6 Were all drawn elements, graphs, tables or diagrams appropriate/helpful? 5 4 3 2 1
7 Were all drawn elements, graphs, tables or diagrams of a high standard (neat)? 5 4 3 2 1
8 Was labelling complete/accurate with titles to each item? 5 4 3 2 1

Discussion/Conclusions

9 Was there a useful discussion? 5 4 3 2 1
10 Were there specific and adequate conclusions or a critique? 5 4 3 2 1

References

11 Was there evidence that the references provided as background to the practical had been consulted? 5 4 3 2 1
12 Was there evidence that additional references had been found and incorporated in the write-up? 5 4 3 2 1

Other comments:

Section 1
Observational Exercises on Parasites

A. Local wild and domestic hosts as sources of parasites

1.1   Parasites of the earthworm: *Monocystis* (Protozoa) and *Rhabditis* (Nematoda)
D.Wakelin, D. I. de Pomerai & J. M. Behnke

1.2   Parasites of marine molluscs (*Littorina*)
R. E. B. Hanna & D. W. Halton

1.3   Parasites of fish:
(a) Whiting and *Diclidophora merlangi* (Monogenea)
D. W. Halton

1.4   Parasites of fish:
(b) Plaice/flounder and *Lepeophtheirus pectoralis* (Copepoda)
P. A. Heuch & T. A. Schram

1.5   Parasites of domestic livestock:
(a) Pig and *Ascaris suum* (Nematoda)
J. M. Behnke

1.6   Parasites of domestic livestock:
(b) Sheep and *Fasciola hepatica* (Trematoda)
C. E. Bennett

1.7   Parasites of crops:
Potato cyst nematode (PCN) *Globodera pallida* (Nematoda)
J. T. Jones

B. Laboratory maintained species

1.8   Protozoan parasites of the intestinal tract of the cockroach, *Periplaneta americana*
J. E. Williams & D. C. Warhurst

1.9   Protozoan parasites of the mouse intestinal tract
J. E. Williams & D. C. Warhurst

1.10 Rodent malaria
J. E. Smith

1.11 Malaria: an example of a vector transmitted parasite
H. Hurd & R. E. Sinden

1.12 Larval and adult *Echinostoma* spp. (Trematoda)
B. Fried

1.1 Parasites of the earthworm: *Monocystis* (Protozoa) and *Rhabditis* (Nematoda)

D. WAKELIN, D. I. DE POMERAI & J. M. BEHNKE

Aims and objectives

This exercise is designed to demonstrate:

1. The general morphology of the protozoan parasite *Monocystis*.
2. The general morphology of the parasitic and free-living stages of the nematode parasite *Rhabditis*.

Introduction

Earthworms are common terrestrial invertebrates from the phylum Annelida. They are exploited by a number of protozoan and nematode parasites and act as intermediate hosts for many parasites of birds and mammals. Two common parasites are *Monocystis* (Protozoa, Sporozoa) and *Rhabditis* (Nematoda); these infect earthworms only.

Monocystis is the commonest protozoan genus to infect earthworms, but another nine genera have been recorded in Britain. The growing form (trophozoite) and reproductive forms occur in the seminal vesicles of the earthworm; the reproductive forms (sporocysts – contained within larger cysts or spores) also enter the body cavity.

The nematode parasite *Rhabditis* also infects earthworms but only in its larval phase. The parasitic third-stage larvae occur free in nephridia of the earthworm, encysted in the body wall, or encapsulated in the coelom. Adult nematodes develop only when the worm dies, when the larvae begin feeding on the bacteria that break down the tissues. Adult *Rhabditis* is therefore a free-living organism. Several species of *Rhabditis* are involved, of which the most common is *R. maupassi*, but they are difficult to differentiate.

Earthworms have a body cavity (a true coelom), a well-developed blood system and the capacity to defend themselves against some types of invaders. The fluid in the coelom contains **11** free cells (amoebocytes), which help to recognise foreign materials. Amoebocytes can phagocytose material as well as encapsulate objects that are too large to ingest. Aggregations of amoebocytes around foreign bodies form the large 'brown bodies' that accumulate in the tail-end of the coelom.

Laboratory equipment and consumables
(per student or group)

Equipment

Microscope
Dissecting dish
50 ml beaker

Consumables

Pasteur pipettes and bulbs   *Rhabditis* cultures
Slides and coverslips   1 freshly killed earthworm
Pins (for dissection)   Filter paper
Lens tissue   Chloroform
Paper towel   Petri dishes
Invertebrate saline (0.7% NaCl)   Disposable gloves

Sources of parasite material
Earthworms can be found in most types of soil and the larger the specimen the better for this exercise. Compost heaps are a good source but the species in decomposing matter tend to be relatively small. Earthworms may be available in some shops specialising in tackle for fishermen. In the USA, earthworms can be purchased from Carolina Biological Supply Co., Burlington, NC 27215. Information on the extraction of *Pellioditis* (*Rhabditis*) *pellio* from *Lumbricus* spp. earthworms can be found in the publication by Eyster & Fried (1999).

Safety

There are few health hazards associated with this exercise if good laboratory practice is followed. As far as is known, neither the parasites in question nor any others carried by earthworms are likely to cause medical/health problems in humans. However, some soil micro-organisms, particularly bacteria (adhering to or ingested by the earthworms), may be considered a health hazard, and hence disposable gloves may be necessary, e.g. when there are cuts on fingers. Staff should also take care when handling chloroform.

Instructions for staff

The earthworms should be collected a few days or weeks before the class and kept in suitable containers with soil and organic material (e.g. leaves) for food, and ventilation (do not seal in air-tight jars). They should be killed in a chloroform jar about 15–20 min before the class commences (a shorter exposure to chloroform may be insufficient to kill all the worms, particularly the larger specimens, and they may show limited movement during subsequent dissection).

Rhabditis cultures should be prepared about 7–10 days before the class. Freshly killed earthworms should be cut into 1 cm pieces, which are then opened out by cutting along the length and placed onto filter paper moistened with saline in a Petri dish. These Petri dishes should be kept at room temperature (20 °C) for about 7–10 days before the class, but must be kept damp throughout (not waterlogged)

Instructions for students

You are provided with freshly killed earthworms.
  1. Dissect the anterior end of a worm from the dorsal side (Fig. 1.1.1).
  2. Remove the seminal vesicles and place in saline. Note that at certain seasons of the year, the seminal vesicles regress and may be very small or even undetectable is some individuals; even in winter, the majority of larger earthworms (>10 cm in length) should possess easily distinguishable seminal vesicles.
  3. Take small pieces of seminal vesicle, squash under a coverslip, and look for the different stages of Monocystis (Fig. 1.1.2).
  4. Make drawings of each stage found and construct an annotated life cycle.
  5. Record (i) stages of parasite found and (ii) intensity of infection.
  6. Take nephridia (scrapings of body wall) to look for Rhabditis larvae.
  7. Examine the Rhabditis cultures to find adults, eggs and larvae (Fig. 1.1.3).

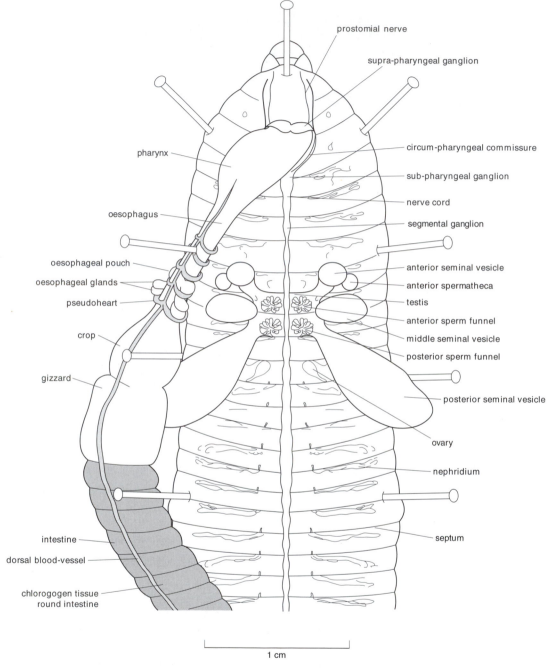

prostomial nerve

supra-pharyngeal ganglion

pharynx

circum-pharyngeal commissure

sub-pharyngeal ganglion

nerve cord

oesophagus

segmental ganglion

oesophageal pouch

anterior seminal vesicle

oesophageal glands

anterior spermatheca

pseudoheart

testis

anterior sperm funnel

crop

middle seminal vesicle

posterior sperm funnel

gizzard

posterior seminal vesicle

ovary

nephridium

intestine

dorsal blood-vessel

septum

chlorogogen tissue
round intestine

1 cm

Fig. 1.1.1 Dissection of an
earthworm to show the
arrangements of the organs and
the location of the seminal
vesicles. Adapted from Rowett
(1963).

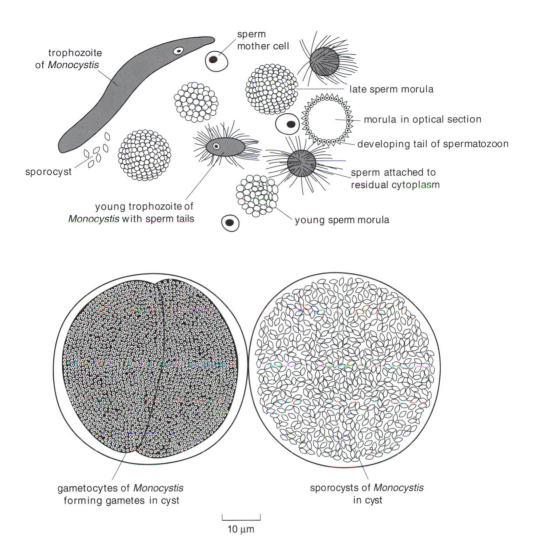

trophozoite
of *Monocystis*

sperm
mother cell

late sperm morula

morula in optical section

developing tail of spermatozoon

sperm attached to
residual cytoplasm

sporocyst

young trophozoite of
Monocystis with sperm tails

young sperm morula

gametocytes of *Monocystis*
forming gametes in cyst

sporocysts of *Monocystis*
in cyst

10 μm

Fig. 1.1.2 *Monocystis:* stages of the life cycle encountered in the seminal vesicles of the earthworm, drawn to approximate scale. Adapted from Rowett (1963).

8. Make drawings of each stage found and construct an annotated life cycle. Record the number of infected worms in the class.

9. Dissect the posterior end of the worm to find the 'brown bodies', spherical/oval bodies formed by the accumulation of amoebocytes around foreign bodies in the body cavity. Remove and place in saline.

(*Note:* Care is needed in dissecting the posterior end of the worm, since if the gut becomes punctured, soil particles and other debris make brown bodies difficult to distinguish. Chaetae (of self or foreign origin?) are very commonly found within brown bodies, along with *Monocystis* and encysted *Rhabditis* larvae. Flushing out the posterior

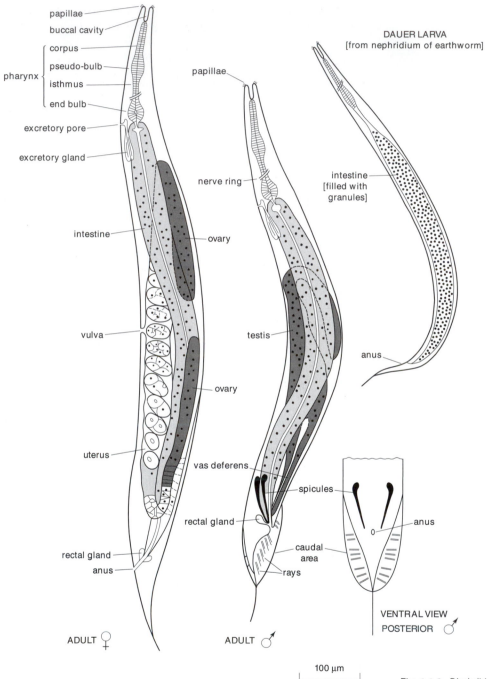

papillae

buccal cavity

pharynx {
corpus
pseudo-bulb
isthmus
end bulb
}

excretory pore

excretory gland

intestine

vulva

uterus

rectal gland

anus

ADULT ♀

papillae

nerve ring

testis

vas deferens

spicules

rectal gland

caudal area

rays

ADULT ♂

DAUER LARVA
[from nephridium of earthworm]

intestine
[filled with granules]

anus

ovary

ovary

anus

0

anus

VENTRAL VIEW
POSTERIOR ♂

100 μm

Fig. 1.1.3 *Rhabditis maupassi:*
morphology of free-living male
and female worms from laboratory
cultures and a dauer larva from an
earthworm. Adapted from Smyth

end of the earthworm, with saline from a Pasteur pipette, is a useful procedure; examine the liquid for dislodged brown bodies.)

10. Select good-sized brown bodies, measure against graph paper, squash each under a coverslip and examine. Examine at least 3 or 4 specimens. Identify, record and draw the contents of the brown bodies. Assess the numbers of each object per body.

Ideas for further exploration

- Does the presence of *Monocystis* or *Rhabditis* impair the fitness of worms? How would you set about determining whether the presence of these parasites has an effect on host fitness?
- Can either parasite survive without earthworms as hosts?
- Do earthworms respond to the presence of the parasites? Can earthworms protect themselves in any way from being invaded by *Monocystis* and *Rhabditis*?
- What signs of the host's response did you detect and how efficient do you think it might be in regulating parasite numbers?

Additional information

The smears from the seminal vesicles of the earthworms can be fixed on glass slides with Bouin's fixative and stained with haematoxylin and eosin. This will enable further details to be seen and a permanent record to be kept of the species and developmental stages of the parasites. Alternatively, a drop of dilute methylene blue can be added to the smears to help enhance the detail.

This exercise can also generate quantitative data on the distribution of parasite populations in their hosts. Provided that sufficient numbers of students participate and each counts the numbers of brown bodies as accurately as possible, the data can be analysed mathematically as explained in the Ecology section (See Exercise 2.4)

REFERENCES

Cox, F. E. G. (1968). Parasites of British earthworms. *Journal of Biological Education* **2**, 151–164.

Eyster, L. S. & Fried, B. (2000). Easy extraction of roundworms from earthworm hosts. *American Biology Teacher* **62**, 370–373.

Mackinnon, D. L. & Hawes, R. S. J. (1961). *An Introduction to the Study of Protozoa*. Oxford, Clarendon Press.

Rowett, H. G. Q. (1963). *Dissection Guides. V. Invertebrates.* London, John Murray.

Smyth, J. D. (1994). *Introduction to Animal Parasitology*. 3rd edition. Cambridge, Cambridge University Press

1.2 Parasites of marine molluscs (*Littorina*)

R. E. B. HANNA & D. W. HALTON

Aims and objectives

This exercise is designed to demonstrate:

1. A range of digenean parasites of *Littorina* spp.
2. The general morphology of the intermediate stages of Digenea.

Introduction

The larval stages of digenetic trematodes develop in intermediate hosts, almost all of which belong to the phylum Mollusca. The miracidium hatches from the egg and penetrates the molluscan host giving rise to the next larval stage (the first intramolluscan stage), the primary or mother sporocyst. In some trematodes, development of cercariae occurs directly within the secondary or daughter sporocysts; in others, cercariae develop within rediae. In both cases, development usually takes place in the gonad or digestive gland of the molluscan host.

The digestive glands of gastropod molluscs exhibit profound physico-chemical changes as a result of infection with larval trematodes. The observable manifestations of these changes are apparent in alterations in the structure, histochemistry and biochemistry of the molluscan digestive gland. Excellent reviews of the subject are provided by Wright (1971) and Smyth and Halton (1983).

During this exercise, you will be examining specimens of the marine mollusc, *Littorina* (the common or edible periwinkle) for infection with a variety of digenean parasites. Other species may be common locally and could be examined additionally or substituted if necessary.

Laboratory equipment and consumables
(per student or group)

Equipment

Microscope
Dissecting dish
50 ml beaker
Pliers

Consumables

Pasteur pipettes and bulbs	Sea water
Slides and coverslips	Filter paper
Pins (for dissection)	Petri dishes
Lens tissue	Disposable gloves
Paper towel	

Sources of parasite material

Periwinkles (*Littorina* spp.) can be collected easily at low tide on most rocky shores in the UK. The snails are found on the middle shore and below on rocks and weeds and the species is very widely distributed. The examples described in this exercise were collected from the shore off Portavogie, County Down, UK.

Safety

There are few health hazards associated with this exercise if good laboratory practice is followed. As far as can be ascertained, neither the parasites in question nor any others carried by *Littorina* are likely to cause medical/health problems in humans. However, some marine micro-organisms (adhering to or ingested by the periwinkles) may be considered a health hazard, particularly bacteria, and hence use of disposable gloves is a sensible precaution.

Instructions for staff

The periwinkles should be collected a few days before the class and kept in suitable containers with cold aerated sea water and organic material (e.g. seaweed) for food, and with ventilation (do not seal in air-tight jars).

Instructions for students

Materials

You are provided with living periwinkles of the species *Littorina littorea*, *L. littoralis* and *L. saxatilis* collected from a local sea shore.

Methods

Crush the shell gently with pliers (or blunt forceps for *L. saxatilis*) and remove the broken pieces. Find samples of the digestive gland, gonad and foot and transfer each separately to a glass slide in a drop of sea water for examination. Squash with a second slide and examine with the microscope.

When an infection is found, observe, draw and make notes on the size, colour, movements, etc., of the various larval stages present. Use the key to help determine the digenean cercariae present. Keep a record of the number of molluscs infected and the sites of infection.

Key to the digenean cercariae found in *Littorina* spp. from Portavogie, Co. Down

1. Cercariae with eyespots ....*Cryptocotyle lingua*
    Cercariae lacking eye spots 2
2. Tail at least as long as the body 3
    Tail vestigial or absent...................................5
3. Cercariae with stylet (Xiphidiocercaria) 4
    Cercariae without stylet; with spines on anterior collar
    ....*Himasthla* sp.
4. Cercaria monostome (one sucker) ....*Microphallus similis*
    Cercaria distome (two suckers)....................*Cercaria emasculans*
5. Xiphidiocercaria...*Cercaria littorina saxatilis* V

    Cercaria lacking stylet....................................6
6. Cercaria distome, metacercariae retained within sporocyst
    ....*Microphallus pygmaeus*

    Cercaria monostome with greatly enlarged oral sucker, metacercariae encysted in digestive gland
    ....*Parapronocephalum symmetricum*

Cryptocotyle lingua

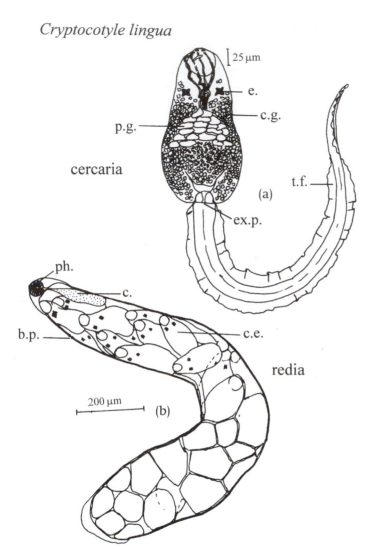

cercaria

redia

Fig. 1.2.1 *Cryptocotyle lingua* (a) cercaria and (b) redia. b.p., birth pore; c., caecum; c.e., cercaria; c.g., cystogenic gland cell; e., eyespot; ex.p., excretory pore; p.g., penetration gland; ph., pharynx; t.f., tail fin.

Cryptocotyle lingua
(Hosts, *Littorina littorea*, *L. saxatilis*, *L. littoralis*)
Cercaria (Fig. 1.2.1a) 200–250 μm in length, aspinose, tail longer than body bearing prominent fin, penetration glands well-developed opening around oral sucker, two conspicuous eyespots anteriorly.
Redia (Fig. 1.2.1b) 1–2 mm in length, grey, well-developed pharynx, short gut caecum, birth pore anterior, most mature intra-redial cercariae occur anteriorly.

Fig. 1.2.2 *Microphallus similis* (a) cercaria and (b) sporocyst. a.p.g., anterior penetration gland cell; c.e., cercaria; ex.b., excretory bladder; ex.p., excretory pore; o.s., oral sucker; p.p.g., posterior penetration gland; s., stylet.

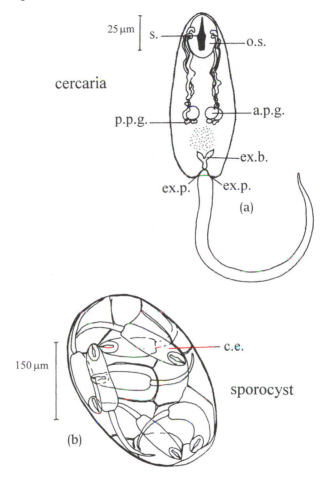

Microphallus similis

cercaria

25 μm

s.

o.s.

p.p.g.

a.p.g.

ex.b.

ex.p.   ex.p.

(a)

c.e.

150 μm

sporocyst

(b)

Microphallus similis
(Hosts, *L. littoralis, L. saxatilis*)
Cercaria (Fig. 1.2.2a) body 100–200 μm in length, tail longer than body, body contractile, spinose, stylet characteristic. Sporocyst (Fig. 1.2.2b) 300–400 μm in length, white or colourless, contains 4–30 cercariae.

Microphallus pygmaeus

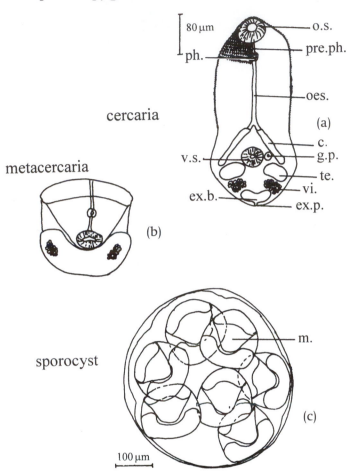

cercaria

metacercaria

sporocyst

Fig. 1.2.3 *Microphallus pygmaeus* (a) cercaria, (b) metacercaria, (c) sporocyst. c., caecum; ex.b., excretory bladder; ex.p., excretory pore; g.p., gonopore; m., metacercaria; oes., oesophagus; o.s., oral sucker; ph., pharynx; pre.ph., prepharynx; te., testis; vi., vitellaria; v.s., ventral sucker.

Microphallus pygmaeus
(Host, *L. saxatilis*)
Cercaria (Fig. 1.2.3a) 350–400 μm in length, tail vestigial, spinose, distome, no stylet or penetration glands, cercariae are not liberated but form metacercariae which remain within the sporocyst (Fig. 1.2.3c), metacercariae folded and lacking tail (Fig. 1.2.3b).

Fig. 1.2.4 *Cercaria littorina saxatilis V* (a) cercaria, (b) sporocyst. a.p.g., anterior penetration gland cell; c.e., cercaria; ex.b., excretory bladder; ex.p., excretory pore; o.s., oral sucker; p.p.g., posterior penetration gland; s., stylet.

Cercaria littorina saxatilis V

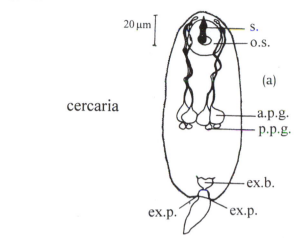

cercaria

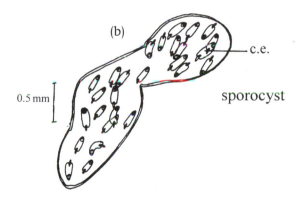

sporocyst

Cercaria littorina saxatilis V
(Host, *L. saxatilis*)
Cercaria (Fig. 1.2.4a) 120 μm in length, tail vestigial, spinose body, monostome, stylet is characteristic. Sporocyst (Fig. 1.2.4b) 2 mm approximately, irregular, yellowish.

Cercaria emasculans

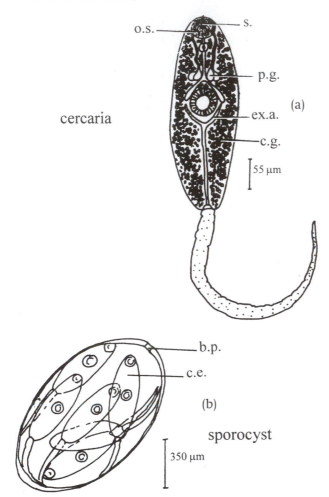

cercaria

o.s. ——

s.

p.g.

(a)

ex.a.

c.g.

55 μm

b.p.

c.e.

(b)

sporocyst

350 μm

Fig. 1.2.5 *Cercaria emasculans* (a) cercaria, (b) sporocyst. b.p., birth pore; c.e., cercaria; ex.a., arm of excretory vesicle; c.g., cystogenic gland cell; o.s., oral sucker; s., stylet; p.g., penetration gland.

Cercaria emasculans
(Hosts, *L. littorea, L. saxatilis*)
Cercaria (Fig. 1.2.5a) 325–400 μm in length, tail longer than body, prominent Y-shaped excretory duct reaches ventral sucker, body packed with cystogenic cells, ventral sucker larger than oral sucker, stylet present. Sporocyst (Fig. 1.2.5b) 1 mm approximately, immobile, colourless, terminal birth pore.

Fig. 1.2.6 *Himasthla* spp. (a) cercaria, (b) redia. c., caecum; c.e., cercaria; c.g., cystogenic gland cell; co., collar; co.sp., collar spines; ex.b., excretory bladder; p.g., penetration gland; ph., pharynx.

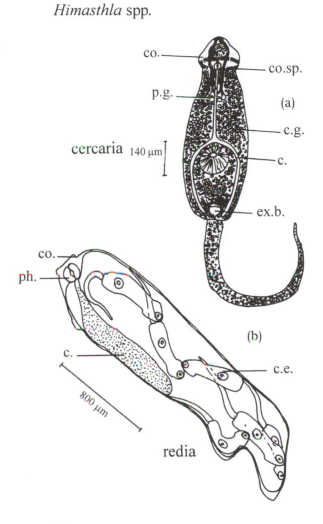

Himasthla spp.

cercaria 140 μm

redia 800 μm

Himasthla spp.
(Host, *L. littorea*)
Cercaria (Fig. 1.2.6a) 700–900 μm in length, aspinose, tail as long as body, collar bearing row of spines around oral sucker, large ventral sucker, prominent bifid gut with caeca reaching posterior of body, penetration glands inconspicuous.
Redia (Fig. 1.2.6b) 2.5 mm approximately, orange yellow or colourless, caecum reaches halfway along body, collar around well-developed pharynx, lateral projections posteriorly.

Parapronocephalum symmetricum

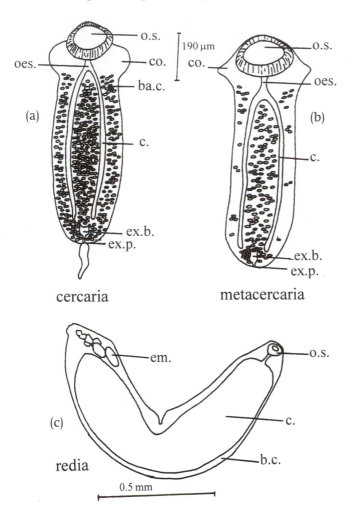

cercaria

metacercaria

redia

0.5 mm

Fig. 1.2.7 *Parapronocephalum symmetricum* (a) cercaria, (b) metacercaria, (c) redia. ba.c., batonnet cystogenic gland cell; b.c., body cavity; c., caecum; co., collar; em., embryo; ex.b., excretory bladder; ex.p., excretory pore; oes., oesophagus; o.s., oral sucker.

Parapronocephalum symmetricum
(Host, *L. saxatilis*)

Cercaria (Fig. 1.2.7a) 900–950 μm in length, tail vestigial, body aspinose bearing numerous cystogenic cells, monostome with enormous oral sucker, gut with caeca reaching almost to posterior end of body, pronounced collar round oral sucker.

Metacercariae (Fig. 1.2.7b) encysted in digestive gland, lack tail, fewer cystogenic cells than cercariae. Redia (Fig. 1.2.7c) 1.5–2 mm in length, gut caecum occupies most of the body space, 2–3 daughter rediae or cercariae at posterior end.

Ideas for further exploration

- Does the presence of the digenean parasites you have observed impair the fitness of the mollusc? How would you set about determining whether the presence of these parasites has an effect on host fitness?
- Do the periwinkles respond to the presence of the parasites? Can they protect themselves in any way from being invaded?
- What signs of the host's response did you detect and how efficient do you think it might be in regulating parasite numbers?
- Did you observe any effects on the digestive gland, gonad or foot? Compare these organs in infected and non-infected periwinkles.
- Can the cercariae move? How do they move? Did the various species you have examined differ in their movements?
- How long do you think that cercariae will live once they have left their host? How do they obtain energy to enable them to continue swimming?
- If you are able to examine different species of *Littorina*, what comments can you make about the host specificity of the parasites you found.

Additional information

Many different species of marine molluscs can be used for practicals of this nature. Fresh water snails also provide good hosts for digenean parasites. If these are collected several days before the practical, they can be isolated in beakers and those releasing cercariae can be identified before the class. A particularly good source of live trematode larvae in the USA is the California hornshell snail, *Cerithidea californica*, which can be found on mud beaches of estuaries in southern California. These snails can also be purchased from Jones Biomedicals & Laboratory, Inc. who supply fresh naturally infected snails at a reasonable price. A key to the cercariae that develop in *C. californica* can be found in Martin (1972). Carolina Biological supplies physid snails and Marine Biological (Woods Hole, MA) supplies *Ilyanassa obsoletus* and *Thais lapillus*; in both cases, the snails carry larval digeneans. Preserved trematodes can be obtained from Ward's Biological. Other suppliers include: American Type Culture Collection, Grand Isle Biological Company (GIBCO), Presque Isle Culture, and Sea Life Supply.

REFERENCES

Dawes, B. (1968). *The Trematoda.* Cambridge, Cambridge University Press.

James, B. L. (1969). The Digenea of the intertidal prosobranch, *Littorina saxatilis* (Olivi). *Zeitschrift für Systematik und Evolutionsforschung* **7**, 273–316.

James, B. L. (1969). The distribution and keys of species in the family Littorinidae and of their digenean parasites, in the region of Dale, Pembrokeshire. *Field Studies* **2**, 614–650.

Martin, W. E. (1972). An annotated key to the cercariae that develop in the snail *Cerithidea californica. Bulletin of the Southern California Academy of Sciences* **71**, 39–43.

Smyth, J. D. & Halton, D. W. (1983). *The Physiology of Trematodes.* Cambridge, Cambridge University Press.

Wright, C. A. (1971). *Flukes and Snails.* London, George Allen & Unwin.

1.3 Parasites of fish: (a) Whiting and *Diclidophora merlangi* (Monogenea)

D. W. HALTON

Aims and objectives

This exercise is designed to demonstrate:

1. The external morphology of the monogenean parasite *Diclidophora merlangi*, with particular emphasis on the organs of attachment.
2. The arrangement of internal organs including the reproductive system and intestine as a model for understanding the general body plan and organisation of the Monogenea.
3. Site specificity and distribution of parasites among hosts differing in sex and age.

Introduction

Monogenean parasites are generally ectoparasitic on aquatic hosts and are frequently encountered on both marine and freshwater fish. Some have invaded body openings and have become specialised for sites such as the gills and the bladder of vertebrate hosts, and in so doing have developed specialised attachment organs that cannot be used for attachment to other sites on their host. One such species is *Diclidophora merlangi*, a common gill parasite of whiting (*Merlangius merlangus*) (Fig. 1.3.1). The adult worms can grow up to 1 cm in length and can be easily located by eye without need for microscopy.

Laboratory equipment and consumables
(per student or group)

Equipment

Microscope
Dissecting microscope
Hand lens
Dissecting dish

50 ml beaker
Dissecting instruments,
   including fine forceps
   and sharp scissors
Stained, whole-mount
   preparations of *D. merlangi.*

Consumables

Pasteur pipettes and bulbs	Ruler
Petri dishes	Disposable gloves
Slides and coverslips	Lens tissue
Paper towel	Sea water (see Exercise 1.2)
Pins (for dissection)	

Sources of parasite material

Whiting are best acquired fresh from a local fishing port, but can be expensive. However, once filleted, the fish 'heads' (head and entrails) are a cheap and excellent source of parasites and can usually be acquired from local fish processing establishments through appropriate arrangement, often for little or no cost. It is important, however, that the heads are examined within 24 h of the fish being caught and that they have been kept cold in the intervening period.

Safety

There are few health hazards associated with this exercise providing good laboratory practice is followed. As far as is known, neither the parasites in question nor any others carried by fresh fish are likely to cause medical/health problems in human beings through contact during handling and subsequent dissection. No materials should be consumed. However, some marine environmental micro-organisms (adhering to or ingested by the fish) may be considered a health hazard, particularly bacteria, and wearing disposable gloves while handling fish is recommended.

Instructions for staff

This exercise is best carried out with freshly obtained material, less than 24 h from capture of fish. The fish or fish heads can be kept for up to a day in the refrigerator but if there is a longer period between collection and use in the class the worms may deteriorate and the fish are unpleasant to work with. The class should also be provided with stained (e.g. borax carmine) whole-mount preparations of the worms for examination of the internal organisation of the parasites. Students should be provided with detailed diagrams of the internal organisation of the parasites and these can be found in many standard parasitology text books.

Instructions for students

Materials
Fresh whole whiting *Merlangius merlangus*, or whiting heads.

Methods
If whole whiting are available, record as many parameters as you can about the fish that you are about to examine. Weigh the fish. Measure its length from the tip of the snout to the posterior end of the fleshy lobe of the tail. Identify the sex of the fish. Make a ventral incision in the body wall and, depending on the season in the year, determine the sex of the fish by looking for the presence of an ovary (orange/yellow) or testis (creamy colour). When the fish is ripe, gentle pressure on the sides of its body cavity will expose drops of white 'milt' (sperm) through the urino-genital pore.

Expose the gill cavity of the fish by removing the operculum with scissors. The four gill arches face outwards from the main axis of the fish, and can be separated from each other by gentle handling with forceps or, better still, fingers. Each gill arch can be removed by cutting at the top and bottom. Begin with the most anterior arch and remove the arches in order, carefully label each so that its identity is known. In numbering the gill arches, the rudimentary arch is ignored and the remaining four arches are designated I–IV in anterior–posterior succession. The arches can then be placed in a Petri dish with some sea water and the parasites can be observed easily on the gills. Repeat the procedure on the gills on the other side of the fish. Note the adhesive attitude of the parasite on the gills (see Fig. 1.3.1).

Generally, some 10–30% of whiting caught in the Irish Sea harbour *D. merlangi* and it is often the only trematode parasite on whiting gills, although the copepod parasite, *Lernaeocera branchialis* may also be encountered (see below). Estimate by

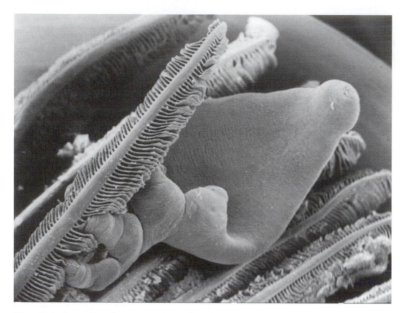

Fig. 1.3.1. Scanning electron micrograph of *Diclidophora merlangi* on the gills of whiting (*Merlangius merlangus*). Note that the pincer-like clamps of the posterior haptor secure the parasite to the secondary gill lamellae of the fish so that the forebody of the worm lies downstream relative to the gill-ventilating current.

direct visual examination the number of *D. merlangi* on each gill arch; use the dissecting microscope to confirm that none has been missed. The results from all the students should be pooled for analysis.

Carefully remove one of the worms from the gill arch into sea water and observe its behaviour, using the dissecting microscope. Note the extensile forebody of the worm and the opening and closing activity of the eight pincer-like clamps. Place a specimen onto a microscope slide in some sea water and examine it more closely. Pay particular attention to the structure of the haptor and its clamps, which are used for attachment to the gills. Examine also the anterior of the worm and locate the main body openings.

Examine the stained specimens of *D. merlangi* and identify the reproductive system and the intestine. Inspect also the structure of the clamps on the haptor.

Expected results

Infected whiting generally harbour only a few parasites, commonly one or two worms, with fewer than 20% of infected fish having more than three worms. The majority of *D. merlangi* are encountered on gill arch I, which will generally harbour about 60–80% of all the parasites observed. The percentage of parasites on the remaining arches then declines, with gill arch II accounting for some 20–30% and the other two <5%. This can be checked by comparing class data.

Ideas for further exploration

- How many parasites did you record on the various gill arches? Did each gill arch have the same number of parasites? Was the distribution along each gill arch even or clumped? What determines where the parasites will attach? Were there different numbers on the arches of the left and right sides of the fish?
- How do the clamps keep the parasite fixed to the gills? Try to work out the role of each individual component of the clamp.
- Is attachment permanent or can the worms move about on their fish host? How are such complex organs of attachment likely to have evolved?
- Having examined the haptor and the clamps, do you think there are any other sites in the fish to which this organism can attach?
- Looking at the pooled class results, can you detect any relationship between the size of the fish (length and/or weight) and the total number of parasites?
- Is there any difference in the prevalence of *D. merlangi* between male and female fish?
- On what does the worm feed and by what mechanism?

Additional information

There are a large number of possibilities for running similar classes on other species of fish and their respective parasites (see Exercise 2.2). The quantitative data from this exercise can also be analysed to determine whether the parasites are aggregated on particular fish.

Whiting are a good source of other metazoan parasites. In addition to *D. merlangi*, the gills may also be occupied by the bulky worm-like body of *Lernaeocera branchialis,* a blood-feeding female copepod the head of which is armed with horns and a protrusible proboscis that extends to the heart region of the fish and serves to procure blood from the ventral aorta. Most specimens of whiting caught in the Irish Sea are infected with the metacercarial stage of the gasterostome trematode, *Bucephaloides gracilescens* (as cysts in nasal tissue and cranial fluid around the brain) and the plerocercoid of the trypanorhynch tapeworm, *Grillotia erinaceus* (encysted on viscera).

In the USA, it may be possible to obtain various species of Monogenea from commercial marine aquaria and fish hatcheries. Alternatively, industrial fishing centres and fish ports located at marine or freshwater locations may be good sources of fresh material. It may be possible to trace potential suppliers through recent publications in one of the biological publications databases.

REFERENCES

Arme, C. & Halton, D. W. (1972). Observations on the occurrence of *Diclidophora merlangi* (Trematoda: Monogenea) on the gills of whiting, *Gadus merlangus. Journal of Fish Biology* **4**, 27–32.

Smyth, J. D. (1994). *Introduction to Animal Parasitology*. Cambridge, Cambridge University Press.

1.4 Parasites of fish: (b) Plaice/flounder and *Lepeophtheirus pectoralis* (Copepoda)

P. A. HEUCH & T. A. SCHRAM

Aims and objectives

This exercise is designed to demonstrate:

1. The external morphology of the parasitic copepod, *Lepeophtheirus pectoralis*.
2. The anatomy of *L. pectoralis* as a model for understanding the general body plan and organisation of parasitic copepods of the order Siphonostomatoida.

Introduction

Lepeophtheirus pectoralis belongs to the family Caligida, which includes many parasites of common fishes. It is found on the flounder, *Platichthys flesus* and on plaice, *Pleuronectes platessa* in the North Atlantic Ocean. Adult, egg-producing females are located under the pectoral fins of the host. Like most parasitic copepods, this species has only one host in its life cycle. The eggs, which hatch while still attached to the female, give rise to two free-swimming nauplius stages, and subsequently one free-swimming copepoid stage. The latter infects the host and anchors itself by means of a chitinous thread to the host skin before moulting to the first chalimus stage. *L. pectoralis* has four sessile chalimus stages and two mobile preadult stages preceeding the adult. The pre-gravid adult female is about 4 mm long, but expansion of the genital complex adds another 1–2 mm during egg production. The general morphology and anatomy of the parasite is therefore easily studied with the aid of a dissecting microscope. (Note: in the US, *L. salmonis* would be a close alternative.)

Laboratory equipment and consumables
(per student or group)

Equipment

Dissecting microscope	Access to compound microscope
Petri dishes	Dissecting instruments, including fine scalpel, needles and fine forceps

Consumables

Filtered sea water (at 20–30 g/l NaCl – or commercial substitutes from aquarist suppliers)	Paper towel Lactic acid Glass slides and coverslips

Sources of parasite material

Material may be obtained from any fishmonger. Mature females are found under the pectoral fins of plaice and flounder. Chalimi, pre-adults and adult males may be found all over the body surface. Flounder are easily caught in gill nets set in muddy bottom estuaries and bays, and can be transported live to the laboratory in sea water-filled buckets topped with oxygen. If the fish are not required, the parasites may be removed from the host with fine forceps, and kept alive in a sea water-filled thermos bottle lined with a polythene bag. In the laboratory, the bag is pulled out of the bottle and may be turned inside out to gain access to the parasites. If the water is cold and clean, the parasites may stay alive for about 12 h in the bottle. A little crushed ice may be added to the bottle before inserting the bag to keep the temperature down. Alternatively, the parasites may be fixed in 4% formalin in sea water, in which case the material should be rinsed thoroughly in sea water before commencing the exercise.

Safety

No safety precautions are required for this exercise as *L. pectoralis* does not parasitise humans. If working with parasites fixed in formalin, the normal safety measures for this chemical should be observed.

Instructions for staff

This exercise is best carried out with freshly caught material, as the parasites fall apart relatively quickly when not fixed.

Instructions for students

Materials
You are provided with fresh (or fixed) adult female *Lepeophtheirus pectoralis*.

Methods
Observe the following external features of the parasite:

1. The body of *L. pectoralis* (Fig. 1.4.1) is divided into two main regions, the cephalothorax and the genital complex. In mature, gravid and post-gravid females, these regions are about the same length.
2. The free thoracic segment (the fifth thoracic segment) articulates anteriorly with the cephalothorax, and posteriorly with the genital complex.
3. The abdomen is anterior to the genital complex.
4. The cephalothorax comprises the cephalic (head) segments and four thoracic segments. The latter segments bear the maxillipeds (first thoracic) and three pairs of swimming legs (second to fourth thoracic).
5. The whole region is covered by a carapace, the margin of which is provided with a finely striated membrane.
6. The eyes are situated either side of the midline, anterior to the midpoint of the carapace.
7. On the very anterior margin of the carapace, but slightly ventral, you will find the antennules, which project outward on both sides from the frontal plate.
8. The basal segment of the antennules provides a marginal extension of the cephalothorax.
9. Facing anteriorly in the notch between the antennules is the filament gland, which secretes the chitinous thread that anchors the parasite through the chalimus stages and moults. Try to find it by tilting the carapace upwards.
10. On the posterior margin there are two deep notches, which are called sinuses. These are equipped with folds that function as valves.
11. Turn the animal around so it lies on its dorsal surface. You will see that the cephalothorax is concave ventrally and

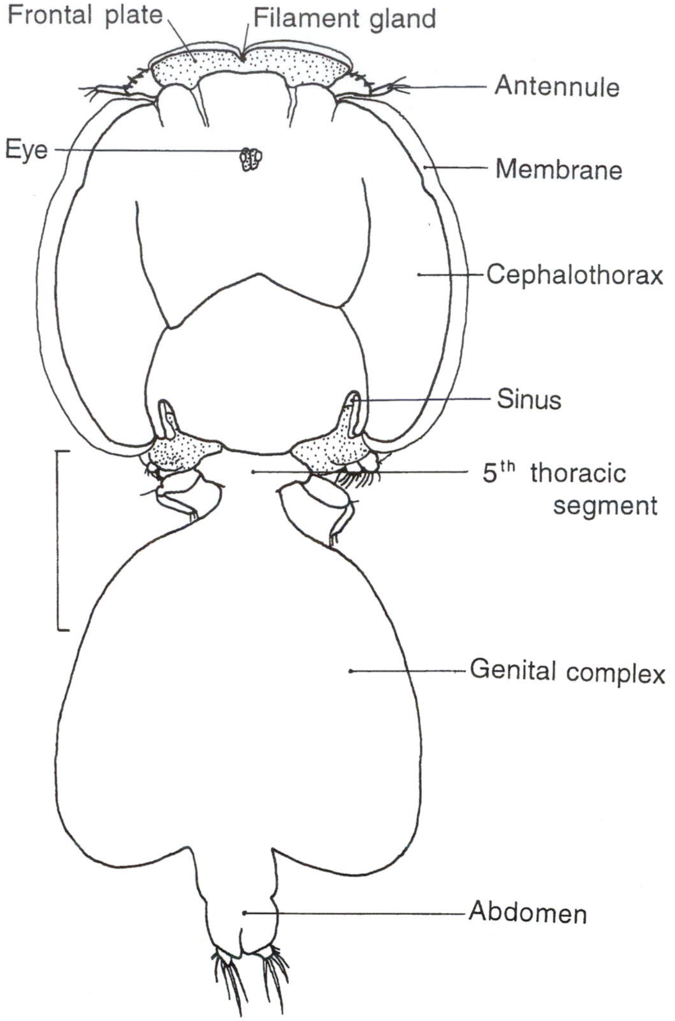
Frontal plate
Filament gland
Antennule
Eye
Membrane
Cephalothorax
Sinus
5th thoracic segment
Genital complex
Abdomen

Fig. 1.4.1 Schematic dorsal aspect of *L. pectoralis*. Scale bar = 1 mm.

convex dorsally so that the whole structure, including the third pair of legs, works as a suction cup.

12. The antennae have three segments, but only the joint between that which is closest to the body (proximal) and the second segment is functional (Fig. 1.4.2).
13. Attachment is initiated by movement of the proximal segments towards the body (Fig. 1.4.2A), which forces the 'hooks' into the fish skin, and reduces the distance between them. The body is then pulled towards the host by flexion in the functional joint (Fig. 1.4.2B). Surplus water is forced out and suction is produced (Fig. 1.4.2C).

Pressure from the water currents surrounding the fish will

Fig. 1.4.2 Carapace of Caligid copepod, showing how the antennae produce suction. A. Initial attachment by hooking into the skin. B. Creation of suction by flexion of the antennal joint. C. Copepod attached by suction and gripping. Scale bar = 1 mm.

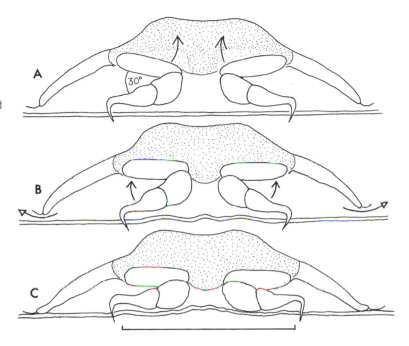

also act to depress the middle of the carapace and so increase suction. This can be observed by trying to wash the animal away with water (currents), such as from the tap or a pipette. The valves in the sinuses control the pressure under the carapace.

14. The animal is fairly translucent, and the gut should be visible through the cuticula. In live specimens, are there any movements in the gut? Is it branched, or just a tube?

15. Posterior to the antennules, you will see the antennae, and between these, the mouth cone (Fig. 1.4.3). The mandibles are inserted into the cone through slits in the sides. Posterior and lateral to the cone are the maxillules, which are rather short and stout, and the long and thin maxillae. On living animals, the latter frequently perform long sweeping strokes. All the above appendages belong to the cephalic region of the copepod.

16. The maxillipeds and the swimming legs follow posteriorly to the maxilla. Between the former appendages is a fork-shaped structure called the sternal furca, the function of which is uncertain. The third pair of swimming legs is flattened and seals the carapace posteriorly. If you work with a live animal, you may notice that the right and left swimming legs always move together. Can you see why? Between

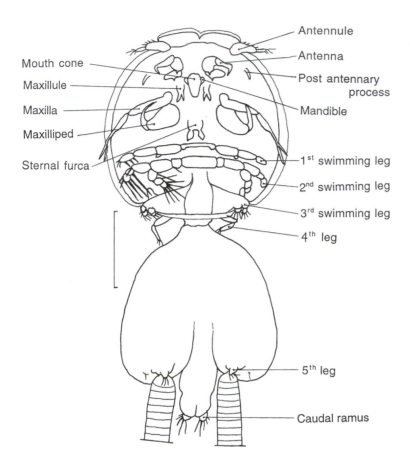

Mouth cone

Maxillule

Maxilla

Maxilliped

Sternal furca

Antennule

Antenna

Post antennary process

Mandible

1st swimming leg

2nd swimming leg

3rd swimming leg

4th leg

5th leg

Caudal ramus

Fig. 1.4.3 Schematic ventral aspect of *L. pectoralis.* Scale bar = 1 mm.

the legs is a bar (the inter-podal bar), which links them together. This is one of the distinguishing features of the class Copepoda. The fourth pair of legs sits on the free thoracic segment.

17. Examine the genital complex. The fifth pair of legs is found in the posterior section of this, just anterior to the lobes terminating the complex. In *L. pectoralis*, the egg strings are extruded from separate gonopores situated underneath the fifth legs. Does your specimen have two oval vesicles attached? These are the spermatophores, which contain the sperm from the male, and are "glued" onto the female with a cement-like substance (Fig. 1.4.4). A tube grows out from each of these, terminating in the genital apertures (copulatory pores) between the legs.

18. The abdomen terminates in two blade-like appendages, the caudal rami. Between these is the anus.

19. Cut off the mouth cone, place it in a drop of water on a slide,

Fig. 1.4.4 *L. pectoralis* female. Genital complex and abdomen, showing the internal anatomy. From Huys, R. & Boxshall, G.A. (1991). *Copepod Evolution*. London: The Ray Society (reprinted by kind permission). Scale bar = 0.5 mm.

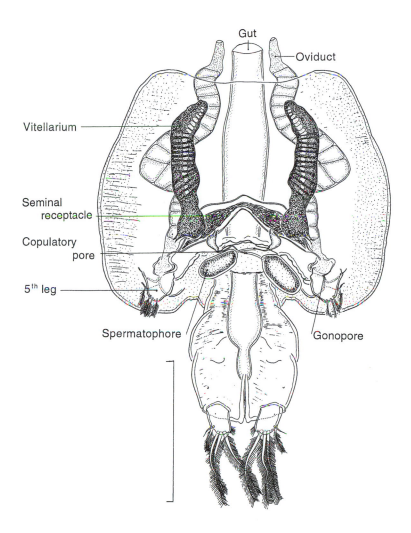

and cover it with a coverslip. Examine the preparation under a microscope to see the toothed rim and the mandibles. These gnaw and cut epithelial cells from the host.

20. If your specimen has egg strings, study the arrangement of eggs within the string. Black pigment in eggs means the nauplii are fairly developed. You may try to liberate nauplii by making a shallow incision along the string, and gently pressing it down.

Ideas for further exploration

- How does *L. pectoralis* move? Which appendages may be used for walking, grasping, etc?

- How does it feed? How is the food transported from the rim of the mouth cone to the mouth?
- If the animal is fixed in formalin (4% (v/v) in sea water), you can see the eggs being formed in the genital complex and the sperm stored in the seminal receptacle (Fig. 1.4.4). Can you see the tubes growing out of the spermatophores crossing on their way to the receptacles?

Additional information

L. pectoralis is a convenient parasite for this exercise, but it could be based on *L. salmonis*, the salmon louse, if this is available. As it is bigger, it would in fact be preferable to *L. pectoralis*. Salmon lice can be collected from farmed salmon, so a call at a fish farm or processing plant would be necessary.

At the end of the class a discussion could be held on the observations and findings. How different is *L. pectoralis* from free-living copepods? Which parts of the parasite are modified for a parasitic way of life?

In the USA, it may be possible to obtain various species of Copepoda from commercial marine aquaria and fish hatcheries. Alternatively, industrial fishing centres and fish ports located at marine or freshwater locations, may be good sources of fresh material. It may be possible to trace potential suppliers through recent publications in one of the biological publications databases.

REFERENCES

Boxshall, G. A. (1974a). *Lepeophtheirus pectoralis* (O.F. Müller 1776); a description, a review and some comparisons with the genus *Caligus* Müller, 1785. *Journal of Natural History* **8**, 445–468.

Boxshall, G. A. (1974b). The developmental stages of *Lepeophtheirus pectoralis* (Müller, 1776) (Copepoda: Caligidae). *Journal of Natural History* **8**, 681–700.

Kabata, Z. (1974). Mouth and mode of feeding of Caligidae (Copepoda), parasites of fishes, as determined by light and scanning microscopy. *Journal of the Fisheries Research Board of Canada* **31**, 1583–1588.

Kabata, Z. (1979). *Parasitic Copepoda of British Fishes*. London, The Ray Society.

Kabata, Z. & Hewitt, G. C. (1971). Locomotory mechanisms in Caligida (Crustacea: Copepoda). *Journal of the Fisheries Research Board of Canada* **28**, 1143–1151.

1.5 Parasites of domestic livestock: (a) Pig and *Ascaris suum* (Nematoda)

J.M. BEHNKE

Aims and objectives

This exercise is designed to demonstrate:

1. The external morphology of the nematode parasite *Ascaris suum*.
2. The anatomy of *A. suum* as a model for understanding the general body plan and organisation of the nematodes.

Introduction

Ascaris lumbricoides (also known as the large round worm) is one of the most common of the nematode parasites affecting humans. It has a close relative, *A. suum*, a common parasite of domestic pigs, which is usually readily available from local abattoirs for research and teaching. These two species are virtually indistinguishable on morphological features, although recent work has shown that DNA techniques can be used to separate them. It is generally considered that *A. suum* is not capable of developing to the adult stage in the human host, although embryonated eggs may well hatch in the human intestine after oral ingestion and the larvae may begin their migration with consequent local damage to the liver and lungs depending on the number of eggs consumed.

Ascaris are relatively large worms, with males up to 30 cm or more in length and females up to 35 cm. The worms can therefore be dissected with relative ease to show the internal body organisation of the worm as a general model of the arrangement of internal organs in members of the phylum Nematoda.

Laboratory equipment and consumables
(per student or group)

Equipment

Compound microscope	Dissecting dish
Dissecting microscope	50 ml beaker
Hand lens	Dissecting instruments

Consumables

Pasteur pipettes and bulbs	Pins (for dissection)
Petri dishes	Hanks' saline
Slides and coverslips	Paper towel
Disposable gloves	Lens tissue

Sources of parasite material

A. suum can be obtained from local abattoirs in regions where pig farming is common. The adult worms live in the small intestine and can be picked out when the intestine is removed and its contents vacated. Abattoir workers will often be prepared to collect worms as long as a suitable container is provided. The best plan is to deliver a plastic bucket, with a volume of about 5 litres and a screw-on lid, to the abattoir manager a day before a pig consignment is expected and then collect the worms in the afternoon of the day during which the pigs have been slaughtered or on the following day.

Safety

Freshly obtained worms do not have infective stages directly infectious to humans, but the adult worms have extremely potent secretions/body fluids, which can induce allergic reactions in some individuals. **Care should be taken in puncturing the body cavity because of the high internal pressure and likelihood that allergens will escape and may squirt onto the face and hands.** It is most unlikely that anyone will react severely if exposed to worms for the first time. However, the pseudocoelomic fluid in particular can sensitise individuals and a severe reaction may follow if the subject concerned is re-exposed to the worms on a subsequent occasion. **In very sensitive subjects these allergic reactions can be life-threatening.** Advice on appropriate precautions should be sought from the health and safety officials (see also Exercise 3.5 for further details). Among some of the pre-cautions that should be considered are: wearing plastic dispos-able gloves when handling the worms, respirators, access to fume cupboards, availability of anti-histamine injections.

If worms have been fixed in formalin or other preservatives in dilute concentration for a few days/weeks (depending on tem-perature) it is possible that eggs within the female worms may have developed to the infective stage. Fully embryonated and infective eggs pose a threat to human health, because if ingested the larvae may hatch and migrate through the liver and lungs, as well as to other body organs and cause local pathology.

Instructions for staff

This exercise is best carried out with freshly obtained material. The worms can be kept for a day or so in the refrigerator but if there is a longer period between collection and use in the class they should be preserved in a suitable fixative. The standard fixa-tives are 5% formalin or 70% ethanol, although worms should be placed in hot alcohol so as to ensure penetration through the cuticle. Long-term preservation is best in 70% ethanol with 5% glycerine. Students should be provided with detailed diagrams of the internal organisation of the parasites and these can be found in many standard parasitology text books.

Instructions for students

You are provided with fresh adult male or female *Ascaris suum* (see Fig. 1.5.1).

External features of the worm

1. Observe the external features of the worm and measure its length. Note that the body is cyclindrical in shape and is long, with tapering anterior and posterior ends.
2. Examine the anterior end of the worm and note in particular the lips. You may need to use a hand lens to help you see detail of the three lips, one dorsally located and two in the lateral–ventral position.
3. About 2 mm behind the mouth opening you should find the external opening of the excretory system, an excretory pore. The location of this opening marks the ventral aspect of the worm.

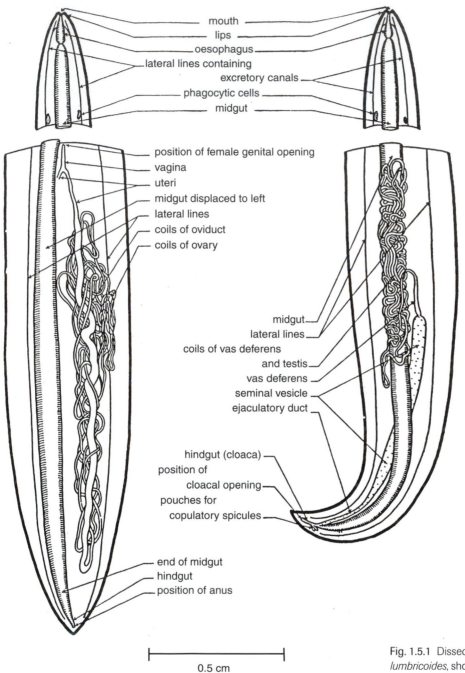

mouth
lips
oesophagus
lateral lines containing
excretory canals
phagocytic cells
midgut

position of female genital opening
vagina
uteri
midgut displaced to left
lateral lines
coils of oviduct
coils of ovary

midgut
lateral lines
coils of vas deferens
and testis
vas deferens
seminal vesicle
ejaculatory duct

hindgut (cloaca)
position of
cloacal opening
pouches for
copulatory spicules

end of midgut
hindgut
position of anus

0.5 cm

Fig. 1.5.1 Dissection of *Ascaris lumbricoides*, showing (left) female worm with the reproductive organs from the left side omitted, and (right) male worm. Adapted from Bullough (1966).

4. Examine the posterior of the worm. The anal opening should be located a few mm from the end and again identifies the ventral aspect of the worm.
5. Males are generally smaller than females but size depends on the age of the organisms. The tail of the male worm curls ventrally, whereas that of the female is straight. In males the posterior opening is a common opening for both the reproductive system and the gut and is therefore a cloaca. By careful examination it should be possible to see the tips of the spicules within the cloacal opening. Note also the small papillae around the cloaca. These are sensory structures.
6. In females there is a separate genital opening, which is located about one-third of the way from the anterior tip of the worm. It is often marked by a constriction running right around the parasite.
7. Note the four prominent lines that appear to run the length of the worm. The ventral and dorsal lines are thin, whilst the two lateral lines should be more prominent and obvious. These lines demarcate the positions of the nerve cords (dorsal and ventral) and the lateral excretory canals.
8. Distinguish the anterior and posterior ends by reference to the diagrams provided and decide the sex of your worm.
9. Examine briefly a worm of the opposite sex from one of your neighbours before proceeding further.

Internal features of the worm

1. Cover your worm with Hanks' saline.
2. Use sharp fine scissors to make the first incision; take care when doing so because the body cavity (pseudocoelom) is under relatively high hydrostatic pressure and some of the internal fluid may squirt out. You should avoid contact with this if possible (see instructions under Safety).
3. Cut along the length of the worm taking care not to penetrate too deeply with the scissors. The body cavity of male worms should be opened along the dorso-lateral side from the anterior to the posterior and the walls pinned down in the dissecting tray. Female worms should be opened dorsally and the body walls likewise pinned down.
4. Note the following:
   (a) In both sexes, the gut, the lateral lines.
   (b) In males, the testis, vas deferens, seminal vesicle, copulatory spicules.

(c) In the female, the two ovaries, coils of the oviduct, uterus, vagina.

5. Cut a small section of the uterus, and remove some of the eggs. Place these into a drop of saline on a glass microscope slide and cover with a coverslip. Examine the shape and the organisation of the egg under the microscope.

Ideas for further exploration

- How does *Ascaris* move? Given that the worm has only longitudinal somatic muscles how does it straighten out again after contraction? What mechanism is used to antagonise the contraction of the longitudinal muscles?
- On what does *Ascaris* feed?
- Why is the intestine of the parasite so thin and flacid? How are ingesta moved along its length?
- How does the male worm fertilise eggs?
- What constraints are there likely to be for movement in organisms with similar construction to *Ascaris*?
- Did you observe anything unusual when the body cavity of your worm was penetrated for the first time? Internally the worm is under high hydrostatic pressure. How then does it manage to feed?
- What devices does it have to ensure that food is not immediately regurgitated? How does it eliminate undigested materials from its gut? How does it ensure that gametes are only liberated when required to do so?
- Comment on the possible effects of large worm burdens.

Additional information

A. suum is a convenient parasite for this exercise but there are other large nematode worms that can also be dissected (e.g. *Toxocara*, *Ascaridia*), although some of these might entail microdissection with the aid of a microscope. The internal organisation can also be demonstrated on smaller nematodes on stained slide preparations.

If sufficient numbers of worms have been examined, quantitative analysis could include frequency distribution plots of the length of male and female worms separately.

At the end of the exercise a discussion could be held on the observations and findings. How does the structure of *Ascaris* relate to the rest of the phylum? What constraints follow from the structure and how are appropriate solutions found to problems such as feeding, excreting, insemination, given that the internal body cavity, the pseudocoelom, is under high hydrostatic pressure?

In the USA, embryonated eggs of *Ascaris lumbricoides* can be purchased from Carolina Biological Supply Co., Burlington, NC 27215. When fed to young rats, the eggs will hatch and undergo the early migratory phase of the life cycle in mammalian hosts. However, they are unlikely to develop into mature worms. The same company will supply preserved adult *Ascaris* for teaching purposes. As in Europe, *A. suum* can be obtained from local abattoirs in the USA in districts where pig husbandry is practised and, again, it is recommended that the management is approached well in advance of any class to enable a mutually convenient arrangement to be made.

REFERENCES

Bullough, W. S. (1966). *Practical Invertebrate Anatomy*. Second edition. London, Macmillan & Co. Ltd.

Cox, F. E. G., Dales, R. P., Green, J., Morton, J. E., Nichols, D. & Wakelin, D. (1969). *Practical Invertebrate Zoology, A Laboratory Manual*. London, Sidgwick & Jackson.

Smyth, J. D. (1994). *Introduction to Animal Parasitology*. 3rd edition. Cambridge, Cambridge University Press.

1.6 Parasites of domestic livestock: (b) Sheep and *Fasciola hepatica* (Trematoda)

C. E. BENNETT

Aims and objectives

This practical is designed to demonstrate:

1. Part of the life cycle of a parasite of medical and veterinary significance, the common liver fluke *Fasciola hepatica*, from infection to maturation.
2. The general body plan of an acoelomate flatworm and its adaptations for parasitism with respect to the surface tegument, digestive, excretory and reproductive systems.

Introduction

F. hepatica, the common liver fluke, is a platyhelminth digenean trematode. The life cycle of the liver fluke (Fig. 1.6.1) is indirect and includes freshwater snails as intermediate hosts; in the UK this is the amphibious mud snail, *Lymnaea truncatula*. *Fasciola* is an example of a 'zoonosis'; i.e. an organism that is fully infectious to humans but which is maintained in the ecosystem by a range of other animals, including rabbits, acting as definitive hosts. *F. hepatica* occurs widely in Eurpoe. It causes fasciolosis or 'liver rot', mainly in cattle, sheep and goats. It is not usually an important parasite of humans, with the exception, for example, of parts of Latin America. The related liver fluke, *F. gigantica*, is an important zoonosis and is found in humans and domestic and wild ruminants in Africa and the orient.

Liver fluke causes serious veterinary problems in livestock where the disease varies from year to year depending largely on climatic conditions, such as increased temperature and rainfall in June and July. Ecological studies in the UK have enabled a valuable forecasting system to be developed to warn farmers when to expect outbreaks of the disease (Ollerenshaw & Rollands, 1959). In the 1980s and 1990s, *F. hepatica* was estimated to cause as much as £300M of economic loss per annum in the UK, measured **53**

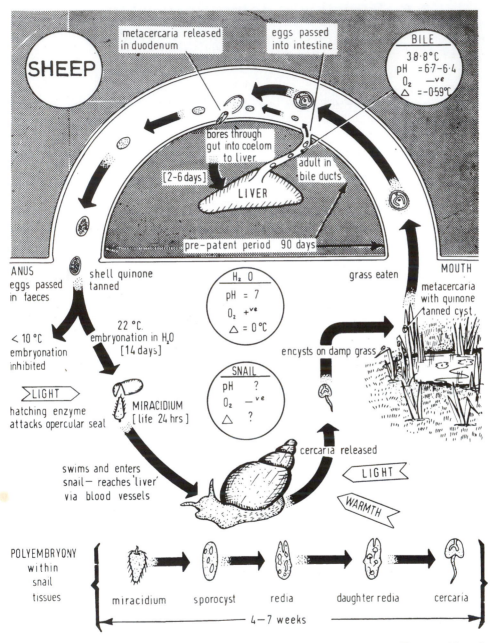

Fig. 1.6.1 Life cycle of *Fasciola hepatica* (after Smyth, 1994).

largely from losses resulting from reduced weight gain and yields by sheep and cattle. Diseased livers, which are condemned at the abattoir, represent only a small proportion of the total losses to the farmer.

Infection of sheep, cattle and goats begins with the ingestion of herbage carrying metacercarial cysts of *F. hepatica*. These infective stages result from the encystment of the free-swimming cercariae released from *L. truncatula*. Excystment occurs in the small intestine and the juvenile fluke migrates through the intestinal wall and reaches the peritoneal cavity within 12 h. After random wanderings in the body cavity, the young worms eventually penetrate the liver capsule and burrow through the liver parenchyma to become established in the biliary system. This later phase may take 6 weeks in sheep and up to 12 weeks in cattle.

Maturation and development of the migrating flukes has been recorded by light microscopy (Dawes, 1962). The spines, which protrude through the tegument, are vital to the migratory movement of the juveniles and serve the adult in abrasion of the bile-duct epithelium (Bennett, 1975a,b). Movement of both the infective juvenile stage and the adult fluke is by a vermiform peristalsis.

Note that *F. hepatica* provides an example of a metazoan animal of bilateral symmetry with an acoelomate, triploblastic body plan. As such, the parasite allows demonstration of movement via a hydrostatic skeleton and a vermiform peristalsis. A basket of circular, longitudinal and diagonal muscles operates on the tissue-filled mesoderm. This peristalsis is also demonstrated in the emergence of the first larval stage, the miracidium, from the egg (see Exercise 7.1).

Cellular organisation of adult flukes

The tegument is composed of a surface layer of anucleate cytoplasm, which is connected at intervals by tubular elements to the nucleated regions that lie some distance below the surface layer.

The digestive tract in *F. hepatica* comprises paired blind-ending caeca. These are composed of a cellular gastrodermis whose luminal surface is amplified by sheet-like lamellae rather than microvilli.

The excretory system is a network of tubules leading from flame cells to the major ducts and excretory bladder. The mesodermal parenchyma is also in contact with the excretory ducts and bladder and allows for the circulation of nutrients and passage of excretory waste.

The vitelline glands or vitellaria comprise cells that produce the so-called yolk of the egg and shell material; the shell of the egg is 'tanned' or hardened in the presence of oxygen, producing a resistant, translucent yellow-brown egg.

Laboratory equipment and consumables
(per student or group)

Dissecting microscope for examination of whole-mount preparations.
Compound microscope for examination of transverse sections.
Immersion oil for detailed examination of transverse sections.

Sources of parasite material

Stained whole-mount preparations and transverse sections are commercially available (e.g. Philip Harris Education).

Instructions to staff

The laboratory may be organised with a set of two or three slides per group of students. If slides are limited, the students may move around a circuit examining prepared demonstrations of slides and sets of diagrams and photographs created from the publications listed. Laser photocopies provide excellent reproduction of photographs from journals and can be laid out under glass. Microscopes may be set up with specimens already focused, requiring students to use only fine focus adjustment. Labelled diagrams may be placed next to the microscopes as a guide to what may be observed.

Instructions to students

F. hepatica adult: whole mounts
1. You are provided with prepared slides of adult worms (Fig. 1.6.2) (stained with haemotoxylin and eosin, triple-stained with Mallory or Masson stains, or stained by the catechol technique [this involves use of phenolase to demonstrate shell protein and hence its presence in the reproductive system: vitellaria, vitelline ducts and eggs in the uterus]).
2. Using the dissecting microscope for initial observation of

Fig. 1.6.2 Schema of the alimentary and reproductive systems of *F. hepatica* (after Smyth, 1994).

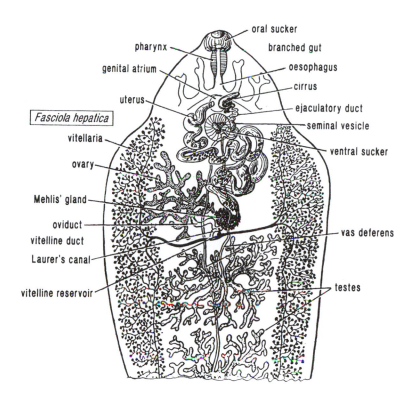

slides, and the lowest power of your compound microscope (×4) CARE! YOU MUST USE ONLY LOW MAGNIFICATION OBJECTIVES BECAUSE OF THE THICKNESS OF THE WHOLE-WORM PREPARATIONS, draw and label the preparations; include a scale-bar or magnification and a title. Note the position and shape of the digestive caeca; these are most easily seen near to the anterior 'shoulders' of the worms where they are not obscured by the vitellaria and testes.

F. hepatica adult: transverse sections

1. Using Fig. 1.6.3 as a guide, and with the aid of a low power objective (×10), draw accurately and label only what you see on your slide.
2. Look at the structure of the tegument, digestive caeca and any portions of the excretory and reproductive systems that are present. During or after the exercise, make notes to describe the functioning of these organ systems. Make representative high power diagrams (×40 or ×100) of each of the organ systems. Compare the gastrodermal cells with Fig. 1.6.5A,B. In your transverse section, compare the percentage

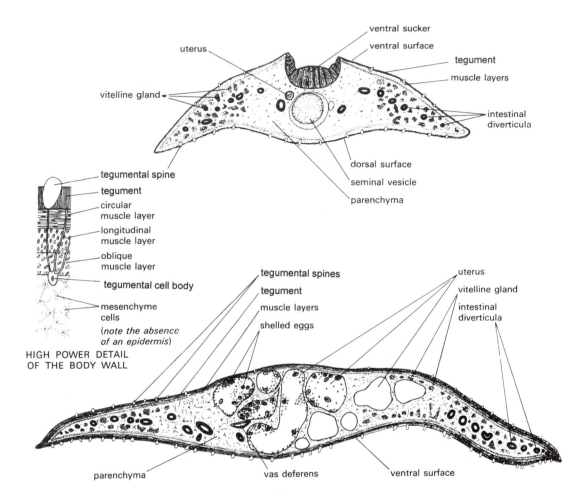

vitelline gland

tegumental spine
tegument
circular
muscle layer
longitudinal
muscle layer
oblique
muscle layer
tegumental cell body
mesenchyme
cells
(*note the absence
of an epidermis*)

HIGH POWER DETAIL
OF THE BODY WALL

of cells that are in a secretory phase (cells with large numbers of secretory granules) in the main caeca with cells in the lateral caeca (diverticula). Do the main and lateral caeca have cells with different morphologies? Observe the vitelline follicles in the lateral margins of the worm and consider the route that the shell material takes from synthesis to egg release. Look for all signs of musculature in your transverse sections. In addition to the main basket of circular and longitudinal muscles, where are muscles found or not found?

Fig. 1.6.3 Transverse sections (through ventral sucker and through main body) of *F. hepatica* (after Freeman & Bracegirdle, 1971).

Brief account of expected results

1. There will be differences between the structure of the tegument you have drawn using light microscopy with that in the diagram (Fig. 1.6.4), which is based on electron microscopy.
2. All spines are attached to the basal lamina. Transverse sec-

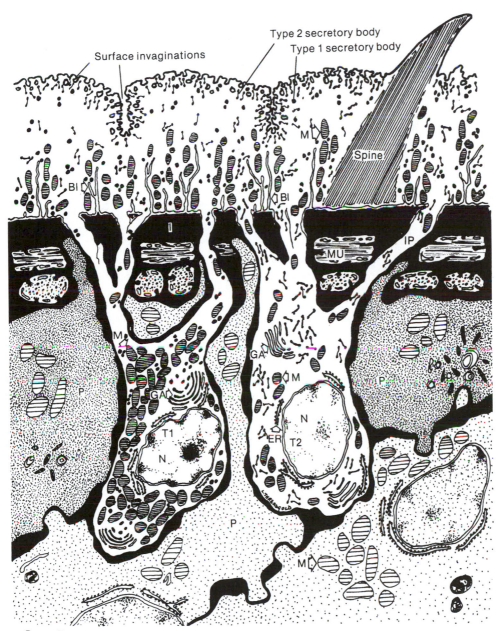

Surface invaginations

Type 2 secretory body
Type 1 secretory body

Spine

P = Parenchymal cell
T1 = Type 1 tegumentary cyton
T2 = Type 2 tegumentary cyton
GA = Golgi complex
I = Interstitial material (connective tissue)
IP = Internuncial process

MU = Muscle
BI = Basal invagination
N = Nucleus
ER = Granular endoplasmic reticulum
M = Mitochondria

Fig. 1.6.4 Schema of the fine
structure of the tegument of
F.hepatica (after Threadgold,
1963).

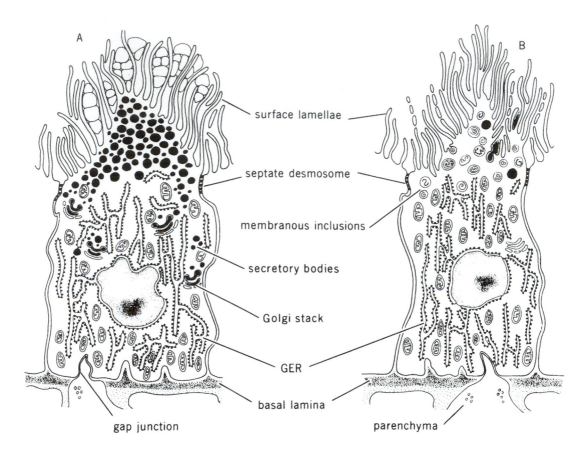

surface lamellae

septate desmosome

membranous inclusions

secretory bodies

Golgi stack

GER

basal lamina

gap junction

parenchyma

tions of backwardly pointing spines may appear to 'float' in the tegumental syncytium.

3. You should be able to identify the shell material in the vitelline follicles and in any eggs you observe in the uterus. Further details on the structure of the vitellaria in *Fasciola* are given by Irwin & Threadgold (1970).

4. There will be little muscle evident around the main digestive caeca/lateral diverticula. Dorso-ventral muscle bands cross the body through the paraenchyma and often near to the ventral sucker. Further details of the muscle system in *F. hepatica* are given by Mair *et al.* (1998).

Fig. 1.6.5 Schema of the fine structure of the gut cells of *F. hepatica*, showing left, in secretory phase; right, in absorptive phase (after Smyth & Halton, 1983).

Ideas for further exploration

• What generalised platyhelminth features do you consider to be preadaptive to a parasitic way of life: the tegument, the gut, the excretory system, the egg?

- If you can prepare transverse sections of the oral sucker and pharynx how do the muscles in these organs work by comparison with the musculature in the ventral sucker (see diagram based on Bennett 1973 in Smyth & Halton 1983)?
- Perform catechol staining of whole adult worms for egg shell protein. This method uses phenolase to demonstrate shell protein and hence its presence in the reproductive system: vitellaria, vitelline ducts and eggs in the uterus. Catechol-stained slides may be obtained through contacts within the British Society for Parasitology.

Alternative sources of material

Tapeworm slides, both whole-mount preparations and transverse sections of *Taenia,* are also available from Philip Harris Education.

Additional Information

In the USA, it may be possible to obtain *Fasciola* life cycle stages (cysts are most convenient for transportation) from workers who maintain and work with this parasite. It should be possible to trace potential suppliers through recent publications in one of the biological publications databases. As in the UK, *Fasciola* may be obtained through local abattoirs.

REFERENCES

Bennett, C. E. (1975a). Surface features sensory structures and movement of newly excysted juvenile *Fasciola hepatica* L. *Journal of Parasitology* **61**, 886–891.

Bennett, C. E. (1975b). Scanning electron microscopy of *Fasciola hepatica* L. during growth and maturation in the mouse. *Journal of Parasitology* **61**, 892–898.

Dawes, B. (1962). On the growth and maturation of *Fasciola hepatica* L. in the mouse. *Journal of Helminthology* **36**, 11–38.

Freeman W. H. & Bracegirdle, B. (1971). *An Atlas of Invertebrate Structure.* London, Heinemann.

Irwin, S. W. B. & Threadgold, L. T. (1970). Electron microscope studies of *Fasciola hepatica.* VIII *Experimental Parasitology* **28**, 399–411.

Mair, G. R., Maule, A. G., Shaw, C., Johnston, C. F. & Halton, D. W. (1998). Gross anatomy of the muscle systems of *Fasciola hepatica* as visualised by phalloidin-fluorescence and confocal microscopy. *Parasitology* **117**, 75–82.

Ollerenshaw, C. B. & Rollands, W. T. (1959). A method of forecasting the incidence of fascioliasis in Anglesey. *Veterinary Record* **71**, 591–598.

Smyth, J. D. (1994). *Introduction to Animal Parasitology*. 3rd edition. Cambridge, Cambridge University Press.

Smyth, J. D. & Halton, D. W. (1983). *The Physiology of Trematodes*. 2nd edition. Cambridge, Cambridge University Press.

Threadgold, L. T. (1963). The tegument and associated structures of *Fasciola hepatica*. *Quarterly Journal of Microscopical Science* **104**, 505–512.

1.7 Parasites of crops: Potato cyst nematode (PCN) *Globodera pallida* (Nematoda)

J. T. JONES

Aims and objectives

This exercise is designed to demonstrate:

1. The biology of plant parasitic nematodes.
2. The complexity of the host–parasite relationship in these organisms.

Introduction

Plant parasitic nematodes are serious pests of agricultural crops throughout the world. The nematodes cause damage directly to the crop plant and may also act as vectors for plant viruses. The chemicals used to control plant parasitic nematodes are amongst the most toxic applied to the soil but, as yet, other ways of controlling these parasites are not commercially viable. Some plant parasitic nematodes are classified as migratory ectoparasites; these are soil dwellers which simply feed on roots with which they come into contact. Others, including the most economically important species (the root knot and cyst forming nematodes), are sedentary endoparasites and have far more complex life cycles.

An example of a sedentary endoparasite is the potato cyst nematode (PCN) *Globodera pallida*. This nematode has an intimate relationship with its host. Cysts, each of which contains 2–300 eggs, lie dormant in the soil until the second stage juveniles (J2) within the eggs (the moult from J1 to J2 occurs in the egg) are stimulated to hatch by the presence of potato root diffusates. After hatch, the J2 locates and enters the growing host root, usually just behind the zone of elongation. It then migrates intracellularly until it reaches the differentiating vascular cylinder tissue, where it sets up a feeding site. The feeding site is a large, multinucleate, metabolically active cell, the formation of which involves massive changes in plant gene expression **63** thought to be induced by nematode secretions. Once the feeding site is established the nematode becomes sedentary and over the following weeks feeds and undergoes three moults to the sexually mature adult stage. Adult males revert to the vermiform body shape, leave the root and locate and fertilise adult females. Adult females become swollen to such an extent that their bodies stick out of the root, allowing the migratory males to mate with them. After mating, the female dies and her body wall forms the cyst that encloses the eggs, protecting them until the next host plant is infected.

An excellent video showing the invasion process of plant parasitic nematodes and discussing the induction of feeding sites is available for a small fee from Dr Florian Grundler, Department of Phytopathologie, Universität Kiel, Rodewald Strasse 9, 2300 Kiel, Germany. This video, and others in the same series, serve as an excellent introduction to plant nematology and may be suitable for those teaching departments unable to set up the exercise described here.

Laboratory equipment and consumables
(per student or group)

Equipment

Dissecting microscope	Pasteur pipettes and bulbs
Slides and coverslips	Petri dishes
Standard light microscope	Dissection kit
Solid glass watch-glasses	Bucket

Consumables

Small (Eppendorf) tubes	Muslin bag
Acid fuschin	Distilled water
Lactic acid	Tissues
Glycerol	Ice

Sources of parasite material

Careful planning will be needed to ensure all stages of the parasite are available for examination. Samples will need to be prepared several weeks in advance. Cysts of *Globodera pallida* and tubers of a suitable potato variety can be provided by Dr John Jones, Nematology Department, SCRI, Invergowrie, Dundee, DD2 5DA.

Storage
Dried cysts can be stored in a small jar for many years on a shelf or in a drawer.

Preparing potato root diffusate (PRD)
Cut the roots from a growing potato plant and leave in distilled water for a few hours. Filter and store the resulting solution in the refrigerator until required. PRD will keep for about 3 months at 4 °C.

Recovering hatched J2s for observation
Place about 30 cysts in a small container with a few millilitres of distilled water and cover to prevent from drying out. Leave for 5 days at room temperature, changing the water when necessary to prevent fungal growth. Replace the water with PRD and leave at room temperature. J2s will start hatching within a few days and will continue to hatch for 2–3 weeks. Once hatched, the J2s are active for about 1 week.

Infecting plants
Cut a sprout from a sprouting tuber of a susceptible potato culti-var (e.g. Désirée) along with a small piece of the tuber and leave to allow the surface to dry for 2–3 h. Inoculate a small pot of soil with 10 cysts approximately 2 inches (5 cm) below the surface and plant the dried plant material in the soil. The plant can be used to observe either the developing juveniles in the roots after 4 weeks or adult females after 7 weeks.

Observing developing juveniles within roots
Carefully remove the pot and suspend the soil and root system in a bucket of water. Massage the roots to remove as much of the soil as possible. Chop the roots into small pieces and place in a muslin bag. Boil for 3 min in a mixture of 1:1:1 lactic acid/glyce-rol/0.05% acid fuschin in distilled water. Do this in a fume hood. Rinse the roots repeatedly in water. Clear the roots for 24 h in 1:1 glycerol/distilled water.

Observing adult females
This can usually be done by simply removing the pot from the soil and examining the roots on the surface of the soil/rootball. Adult females appear as small, round, white growths, slightly larger than sand grains on the surface of the roots. The females may appear white, golden or brown in colour depending on their age.

Safety

There are few health hazards associated with this exercise as long as good laboratory practice is maintained. However, since the chemicals and procedures used to stain developing nematodes in roots are hazardous (see above), it is suggested that this is carried out in advance by staff rather than by students and that students are asked simply to observe the various parasite stages.

To avoid the risk of infecting arable land with nematodes, great care should be taken to ensure all nematodes are dead before disposal. Plant tissue and soil containing live nematodes should be autoclaved before disposal. Nematode stages not in soil or plants should be placed in a dilute bleach solution before washing away through drainage outlets.

Instructions for staff

The various stages of the parasite to be examined should be prepared in advance. Each student should be given some cysts soaked in tap water for 48 h (this will allow them to be dissected under water – when dried they tend to float), hatched J2s, a small quantity of stained infected roots containing developing juveniles and a pot containing an infected plant that is likely to have adult females present (see above).

Instructions for students

Materials
You are provided with stages of the potato cyst nematode, *Globodera pallida*, some of which are within the roots of their host plant (the potato, *Solanum tuberosum*).

Methods
Start with the cysts. Observe the cysts under the dissection microscope. Using a scalpel or a needle, break open a cyst to reveal the eggs. Note the approximate number of eggs within a cyst and consequently the small size of each individual nematode.

1. Why should the invasive stage juveniles be so small?
    Each egg contains a second stage juvenile (J2) in a dormant state. The moult from J1 to J2 occurs in the egg before dormancy starts.

Break open a few eggs with a needle. Observe the released J2. If 10 J2s are released from eggs how many show signs of movement?

Take a small volume of liquid containing hatched J2s. Place half in a tube and put on ice to cool the nematodes. Observe the rest at the highest power allowed by the dissection microscope. Contrast the movement and appearance of the hatched J2s with the lethargy of those artificially released from the eggs. The stimulus to hatch and become active is found in diffusates from the roots of the host plants.

2. What advantage does the nematode gain from this dormancy and the way it breaks it?

Take the now cooled nematodes from the ice and mount on a microscope slide. Cooling the nematodes slows them down and allows them to be observed at higher magnification. Focus on the anterior end of the nematode and identify the structures shown in Fig. 1.7.1.

The stylet is used by the nematode to pierce a chosen root cell and to inject secretions from the two gland cells. These secretions induce massive changes in plant gene expression and root morphology. The stylet is then used to withdraw nutrients from the root cell, which has now become a multinucleate feeding site.

3. Make a drawing of the anterior end of a second stage juvenile. Use the fine focus of the microscope to examine the surface structure of the cuticle. Indicate on your diagram some of the features of the cuticle surface.

Next move to the developing juveniles. Take a small amount of stained root tissue and observe under the dissection microscope. The developing nematode will be stained pink. The nematode becomes sedentary at the feeding site and undergoes three moults to the sexually mature adult stage. Observe the nematode carefully, dissecting away the plant tissue surrounding it if necessary. Note the intimate relationship between host and parasite and how the nematode manages to avoid triggering any of the normal plant defense mechanisms.

4. How does the body of the nematode in the root differ from that of the invasive stage J2? Do all the nematodes in the root look the same? Nematodes use many ways of avoiding detection by their hosts – speculate as to what some of these may be.

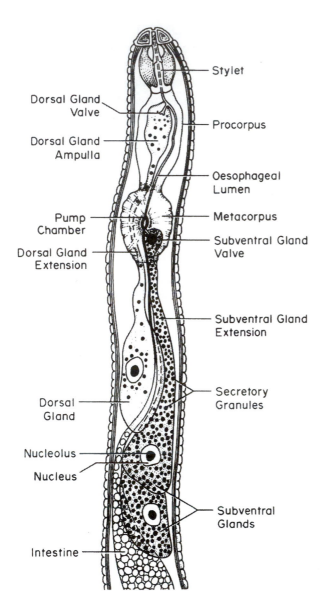

Stylet

Dorsal Gland Valve

Dorsal Gland Ampulla

Procorpus

Oesophageal Lumen

Pump Chamber

Metacorpus

Subventral Gland Valve

Dorsal Gland Extension

Subventral Gland Extension

Dorsal Gland

Secretory Granules

Nucleolus

Nucleus

Subventral Glands

Intestine

Fig. 1.7.1 Diagram of the anterior end of a plant parasitic nematode. Structures visible under a light microscope are labelled.

Now observe the adult females. Take a pot with a growing plant. Tap sharply and remove pot from the soil and rootball, which should stay in one piece. Look carefully at the roots on the surface of the soil for adult females. Note that in contrast to the developing juveniles, the adult female body has grown so much that it sticks out of the root surface. The females may be white or brown in colour. Using a pair of fine forceps, pick off a couple of adult females and observe them under the microscope. Are any of the structures observed in the J2s present in the females?

5. Pick off some more adult females – preferably a range of sizes and colours. Crush them or cut them open on a microscope slide. What do you observe under the microscope?

6. You have now observed all stages of the plant parasitic nematode *G. pallida* except for the adult male. The adult male develops in the root in exactly the same way as the adult female but when moulting to the sexually mature adult stage it reverts to the worm-like shape of the J2 and leaves the root in order to find the females to mate. Given this information, and what you have observed in the practical, draw a diagram summarising the life cycle of *G. pallida*. Include brief notes on the most important events in the life cycle.

Expected results

This exercise is mainly an observational one although 'problems' have been included in the Instructions for students. These are designed to encourage the students to complete the exercise diligently and to persuade them to think about what they are seeing. The final 'problem' is designed to make them draw together what they have observed. The expected answers are:

1. Invasive-stage juveniles are small to allow them to enter the roots of the host plant and remain within the roots to establish feeding sites. *G. pallida* is an r-strategist, producing many (consequently small) offspring to maximise the chances of at least one being able to reach sexual maturity successfully.

2. The nematode remains dormant in the soil until stimulated to hatch by the presence of host root diffusates. This allows it to hatch only when a potential host is nearby, maximising the parasite's chance of successful invasion. Such mechanisms are common in parasites of both plants and animals.

3. The drawing should show most of the features in Fig. 1.7.1 although the gland cells may not be as obvious. The cuticle has an annulated appearance and lateral ridges run the length of the J2, but this question is designed to encourage the students to examine the nematode carefully rather than to tax them intellectually.

4. The body of the nematode in the root becomes much more swollen as it grows, taking on the appearance of a thick sausage. Some students may observe two forms of nematodes in the roots, some very swollen, others shaped more like the J2s. These represent the developing females and males respectively. Nematodes use a range of methods to avoid detection

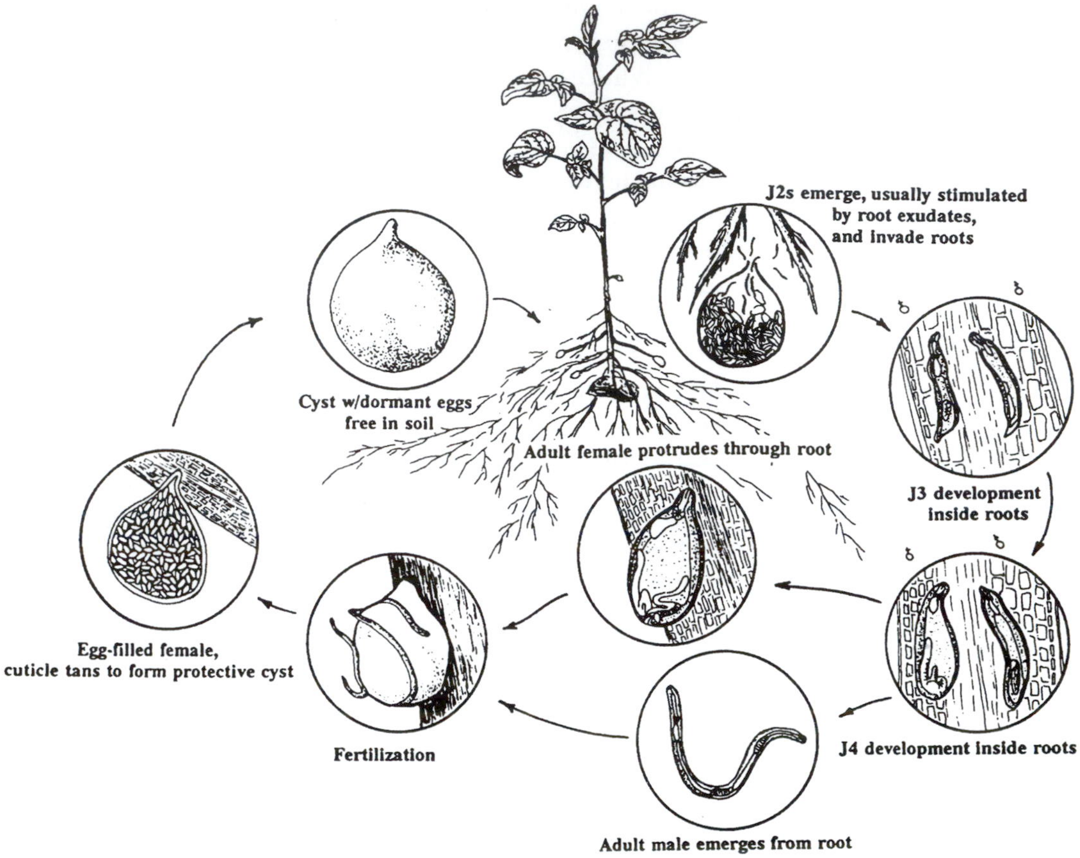

Fig. 1.7.2 Life cycle of the potato cyst nematode.

by the host plant; these may include moving and living between root cells rather than within them, secreting compounds that mask their presence and by binding plant molecules onto their surfaces.

5. Adult females should contain developing eggs. The eggs may contain juvenile nematodes or, depending on the age of the females, developing embryos.

6. A life cycle diagram is shown in Fig. 1.7.2.

Ideas for further exploration

• Experiments comparing hatch of J2s from wet cysts placed in water or dilute PRD may be possible but take a long time. Students could set up two sets of cysts one week and then count the emerged J2s the following week.

• It may also be possible to relate fecundity of adult females with density on roots. Plant parasitic nematodes are extremely awkward experimental animals and it is probably best to stick to an observational exercise.

Similar exercises

There are many plant parasitic nematodes with varying life cycles infecting a wide range of plant species. Some, like *G. pallida*, are obligate endoparasites whereas others live outside the roots. Although similar practical exercises using different nematodes may be possible, many plant parasites are subject to quarantine regulations and many are almost impossible to culture in large numbers. Therefore, *G. pallida* is probably the nematode of choice for this exercise.

Functional studies on plant parasitic nematodes are difficult, again due to the limitations imposed by the difficulties in obtaining large numbers of nematodes for analysis. Such studies are probably best carried out with other nematodes.

Additional Information

In the USA, it may be possible to trace potential suppliers through recent publications in one of the biological publication databases, or through contact with crop research stations locally.

REFERENCES

Gommers, F. J. & Maas, P. W. T. H. (1992). *Nematology from Molecule to Ecosystem*. Invergowrie, Scotland, European Society of Nematologists Inc.

Perry, R. N. (1989). Dormancy and hatching of nematode eggs. *Parasitology Today* **5**, 377–383.

Sijmons, P. C., Atkinson, H. J. & Wyss, U. (1994). Parasitic strategies of root-knot nematodes and associated host-cell responses. *Annual Review of Phytopathology* **32**, 235–259.

1.8 Protozoan parasites of the intestinal tract of the cockroach, *Periplaneta americana*

J. E. WILLIAMS & D. C. WARHURST

Aims and objectives

This exercise is designed to:

1. Examine the protozoan parasites found in the various regions of the cockroach intestine and to draw and attempt to identify those found.
2. Make detailed drawings of the internal and external arrangement of each cell type.
3. Identify the organisms.

Introduction

A large number of Protozoa are found in the intestinal tract of almost all species of animals; most of those present in the cockroach intestine are flagellates but amoebae and ciliates also occur.

Laboratory equipment and consumables
(per student or group)

Equipment

Compound microscope
    ×10, ×40, ×100 objectives
Wax-filled dissection dish

Sharp-pointed scissors
Fine-point forceps

Consumables

Pasteur pipettes and bulbs
Glass microscope slides
Lugol's iodine (or other
    bacteriological iodine
    solution)*

22 × 22 mm coverslips
Physiological saline (0.9% NaCl)

*Dissolve 2 g potassium iodide in approximately 30 ml distilled water. Add 1 g iodine crystals, dissolve and make up to 100 ml.

Sources of parasite material

Freshly killed laboratory-reared cockroaches, *Periplaneta americana* (the bigger the better!) are the usual source, but other species, e.g. *P. blatta*, may be used.

Safety

Other than normal laboratory safety procedures, no special precautions are required. The only aspect worth pointing out is that some students may be allergic to insects and appropriate precautions should be taken for such persons.

Instructions for staff

Cockroaches should be killed in a jar containing ethyl acetate or chloroform and fixed ventral side up in a small dish of wax. This is best done for the students as there is a potential for numerous cockroach 'escapees' in the class.

Instructions for students

1. Before beginning to examine the cockroach intestine, ensure that the microscopes are set up in bright-field illumination and that the condenser iris is closed down to a very small aperture. Doing this will increase the contrast of the image obtained and allow more of the structures of the various parasites to be observed.
2. Open up the abdomen using fine scissors and forceps and squeeze a drop of rectal contents onto a slide; mix with physiological saline. Make sure that the gut contents do not dry during preparation of the smears, as the organisms are very susceptible to desiccation. Examine using wet film technique.
3. Pull out the intestine and Malpighian tubules and examine the gut contents at various points along the length, making wet preparations with saline to help visualise the parasites. Examine for amoebae, flagellates and ciliates. Examples of some of the protozoans you may encounter are illustrated in Fig. 1.8.1.
4. Make a note of which part of the gut the different organisms are found. Using the diagrams, try to identify what you find.
5. Large numbers of motile bacteria will be found in the preparations and these should not be confused with protozoan flag-

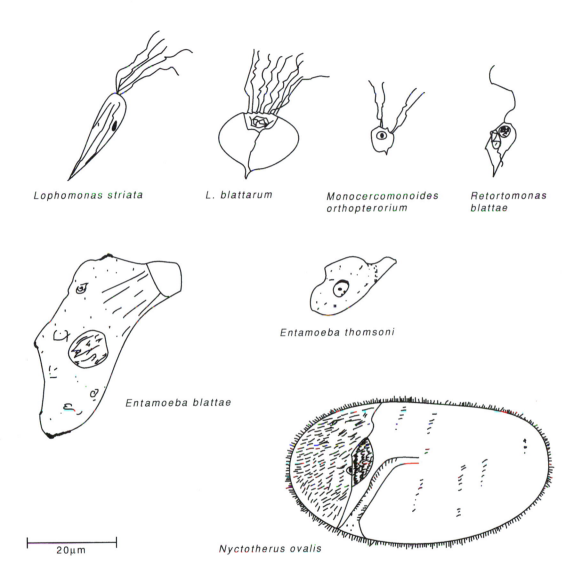

Lophomonas striata

L. blattarum

Monocercomonoides orthopterorium

Retortomonas blattae

Entamoeba thomsoni

Entamoeba blattae

20μm

Nyctotherus ovalis

Fig. 1.8.1 Examples of protozoans likely to be found in the cockroach intestine.

ellates. Some of the bacteria are very large, rod-shaped with fibrils or flagella at both ends of the body and these can be differentiated from smaller protozoan flagellates by addition of Lugol's iodine to the preparation, which will kill any of the organisms present and stain their internal and external structures.

Ideas for further exploration

- This practical can be expanded to include permanently stained preparations of the various organisms found in the intestine.

- Gut contents can be placed in sodium acetate–acetic acid–for-malin (SAF) fixative and smears made after fixation, which can then be stained using an iron haematoxylin stain to show morphology.
- Alternatively, thin smears can be made direct from the gut contents and fixed with methanol and stained using Giemsa; this is not so good for amoebae but does work reasonably well with flagellates.

Additional information

In the USA, live cockroaches (*Periplaneta* and *Blatta*) are supplied commercially by Carolina, Wards and other commercial suppliers. Woodroaches and termites (from Carolina) are also a good source of *Trichonympha* and related hypertrichs. You should request a catalogue from Carolina, Wards or your local supplier for recent information on availability before ordering.

REFERENCES

Mackinnon, D. L. & Hawes, R. S. J. (1961). *An Introduction to the Study of Protozoa*. Oxford, Clarendon Press.

1.9 Protozoan parasites of the mouse intestinal tract

J. E. WILLIAMS & D. C. WARHURST

Aims and objectives

This exercise is designed to:

1. Examine the protozoan parasites found in the various regions of the mouse intestine.
2. Make detailed drawings of the internal and external arrangement of each cell type.
3. Identify the organisms.

Introduction

A large number of Protozoa are found in the intestinal tract of almost all species of animals; most of those present in mammals are members of the order Amoebida ('*Amoeba*'), although flagellates are also found. The amoebae mainly inhabit the large intestine, whereas the flagellates are found in both the large and small intestine.

Laboratory equipment and consumables
(per student or group)

Equipment

Compound microscope with
  × 10, × 40 and
  × 100 objectives

Cork board or wax-filled
  dissection dish
Sharp-pointed scissors

Consumables

Pasteur pipettes
Glass microscope slides
22 × 22 mm coverslips
Disposable gloves

Physiological saline (0.9% NaCl
  solution)
Lugol's iodine (or other
  bacteriological iodine
  solution)*

*Dissolve 2 g potassium iodide in approximately 30 ml distilled water. Add 1 g iodine crystals, dissolve and make up to 100 ml.

Sources of parasite material

Freshly killed laboratory mice should be used. These should not be SPF or other high specification animals.

Safety

Other than normal laboratory safety procedures, no special safety precautions are required. Animal material and sharps must be collected and disposed of in the appropriate manner.

Instructions for staff

If it is preferred not to give whole mice to the class, then the intestinal tract should be removed from a freshly killed mouse immediately before the practical. The tract should be cut at the point where the ileum leaves the stomach and as low down the large intestine as possible so that the rectum is included. Although not vital for this practical, a phase-contrast microscope is useful to allow students to see the organisms they isolate in better detail.

Instructions for students

1. Before beginning to examine the intestine, ensure that your microscope is set up in bright-field illumination and that the condenser iris is closed down to a very small aperture. Doing this will increase the contrast of the image obtained and allow more of the structures of the various parasites to be observed.
2. The intestine should be gently uncoiled and the ileum, caecum, ileo-caecal junction, colon and rectum identified (Fig. 1.9.1).
3. Open each region of the gut in turn, starting with the rectum, and prepare a wet film of the contents. The physiological saline can be used to dilute and emulsify the contents of the gut, which is then used to prepare a thin smear under a coverslip. A small drop is placed on a microscope slide, emulsified if necessary with an applicator stick or similar, such that the material spreads evenly under the coverslip. Avoid excess fluid as this may lead to contamination of the microscope.

Fig. 1.9.1 Sections of mouse intestine to be examined.

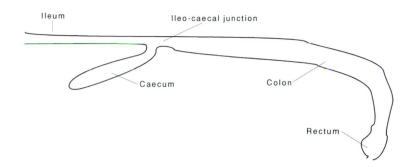

Ileum   Ileo-caecal junction   Caecum   Colon   Rectum

4. Look for parasites in each region of the intestine and record the approximate numbers (e.g. few, many, swarming) and the types found. Lugol's iodine can be added to the preparation to kill the organisms and help make nuclei and flagella more visible.

5. Large numbers of motile bacteria will be found in the preparations and these should not be confused with protozoan flagellates. The bacteria are generally very large and rod-shaped.

Figure 1.9.2 gives an indication of the common species and the general morphology of the parasitic protozoa occurring naturally in the mouse gut.

Ideas for further exploration

- This practical can be expanded to include permanent stained preparations of the various organisms found in the intestine. Gut contents can be placed in sodium acetate–acetic acid–formalin (SAF) fixative and smears made after fixation, which can then be stained using an iron haematoxylin stain to show morphology.

- Alternatively, thin smears can be made direct from the gut contents and fixed with methanol and stained using Giemsa; this is not so good for amoebae but does work reasonably well with flagellates.

- Material can also be cultured using a general faecal culture system, such as Robinson's medium, which will grow all amoebae and gut trichomonads, but not *Giardia*.

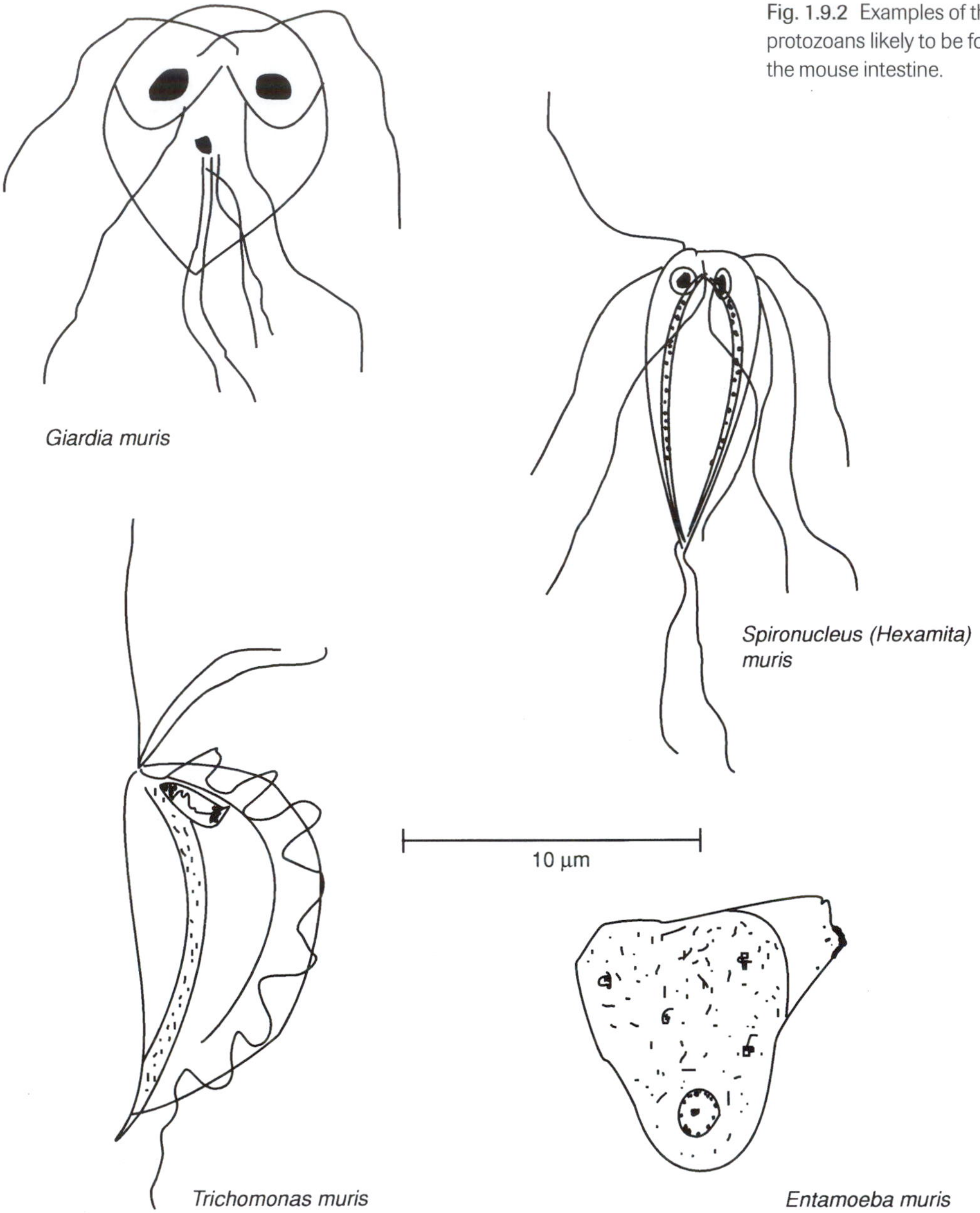

Fig. 1.9.2 Examples of the protozoans likely to be found in the mouse intestine.

Giardia muris

Spironucleus (Hexamita) muris

Trichomonas muris

10 μm

Entamoeba muris

REFERENCES

Cox, F. E. G. (1970). Parasitic Protozoa of British Wild Mammals. *Mammal Review* **1**, 1–28.

Mackinnon, D. L. & Hawes, R. S. J. (1961). *An Introduction to a Study of Protozoa*. Oxford, Claredon Press.

1.10 Rodent malaria

J. E. SMITH

Aims and objectives

This exercise is designed to demonstrate:

1. The morphology of the asexual blood stages of malaria parasites.
2. The relationship between parasitaemia and disease pathology.

Introduction

Plasmodium yoelii is one of a group of rodent malaria parasites that provides a model of the important human disease. In the wild the parasite is transmitted by mosquitoes and undergoes a cycle of development in the liver before releasing merozoites, which infect red blood cells. Once in the red blood cell the parasite multiplies rapidly, causing host cell lysis and releasing large amounts of parasite antigen and cell debris into the bloodstream. Most of the symptoms of malaria are related to this phase of the parasite life cycle; lysis of red blood cells causes anaemia, while infected and damaged red blood cells are taken up by macrophages in liver and spleen.

This exercise aims to demonstrate the relationship between bloodstream parasiteamia and pathology, following direct infection with bloodstream forms of the parasite. *P. yoelii* causes a self-curing infection that lasts for approximately two weeks in some strains of mice.

Laboratory equipment and consumables
(per student or group)

Equipment

Dissecting board
Dissecting kit

Compound microscope, $\times 100$
oil objective

81

Coplin jar containing 70%
  alcohol

Coplin jar containing 20%
  Giemsa in buffer

Access to tap water

Tally counters

Weighing balance

Haematocrit centrifuge

Timer

Consumables

Infected and control
  uninfected mice

Disposable gloves

Foil or weighing boats

Petri dishes

Pasteur pipettes and bulbs

Microscope slides

Haematocrit or blood
  centrifuge

Capillary tubes for haematocrit
  centrifuge

Plasticine

Immersion oil

Cinbins

Marker pens/labels

Sources of parasite material

Non-lethal strains of rodent malaria parasite, such as *Plasmodium yoelii yoelii* 17XNL can be obtained from schools of tropical medicine and from major universities that teach parasitology; it must be maintained under Home Office licence. In the USA, it may be possible to trace potential suppliers through recent publications in one of the biological publications databases or via contact with medical schools and biology departments of universities where research into tropical diseases is routinely conducted. Subject to quarantine regulations and safety legislation, it may be possible to obtain infected mosquitoes, infected mice, or frozen blood samples.

Safety

There is no human health risk associated with the murine malaria, but laboratory animals may carry other pathogens and it is therefore advisable that students carrying out dissection wear disposable gloves. Animal material and sharps must be collected and disposed of in the appropriate manner.

Instructions for staff

P. yoelii is one of several malaria parasites that infect laboratory mice. The parasite is routinely maintained by weekly serial passage in adult female outbred mice, under appropriate Home Office licence. Infection may be evaluated by making and examining a blood smear from a single drop of tail blood. The parasite may also be cryopreserved and stored in liquid nitrogen.

For storage, blood should be taken from animals at 3–7 day post-infection. Animals should be killed by Schedule 1 procedure and blood taken by cardiac puncture into a heparinised syringe. Ten percent v/v cryopreservation agent (dimethylsulphoxide [DMSO] or glycerol) should be added drop-wise and the blood transferred to storage vials and cooled at $-70\,^{\circ}$C for 1 h prior to storage in liquid nitrogen. To retrieve frozen samples, remove vial from liquid nitrogen, thaw in water at $37\,^{\circ}$C and inject 0.1 ml blood intraperitoneally into adult outbred mice.

This exercise works best if the students are not aware of the infection status of the mice. Groups of animals should be infected 15, 12, 9, 6 and 3 days prior to the practical class. Immediately before the practical, infected and control mice should be killed by Schedule 1 method and blood removed by cardiac puncture into a heparinised syringe and transferred to Eppendorf tubes. Each mouse, together with its blood sample, should then be placed in a plastic bag and given a code letter.

It is also a good idea to make and stain some spare blood-smear slides from known infected mice. These can be given out to enable students/groups, who have been allocated control animals, to observe the different stages of parasites.

Instructions for students

Materials

You are provided with a mouse, which may be infected with the rodent malaria parasite, *P. yoelii*. Using the methods below assess the level of parasitaemia in the blood and describe the pathology of infection.

Methods

Organ weights

Weigh and sex the mouse, take a blood sample then dissect out and weigh the liver and spleen. Calculate the liver and spleen weights as a proportion of the total body weight and enter into the class results table. Place your tissue samples in the Petri dishes provided and label with the code number of the mouse. Set up a class demonstration of these organs and make notes on their appearance.

Group/student name(s) _____

Mouse ID # _____   Sex _____ Mouse wt _____ g

Liver wt _____ g   Liver (% total body wt) _____

Spleen wt _____ g   Spleen (% total body wt) _____

Haematocrit _____ % RBC

Parasitaemia

Fig. 1.10.1 Individual data sheet.

Shake blood sample to re-suspend and transfer one small drop of blood near to one end of a clean microscope slide using a Pasteur pipette. It is essential that the clean slide is handled by its edges only. Take a second clean slide and, holding it at an angle of 45°, spread the blood sample with a single smooth action. Allow the sample to air dry, label, fix for 1 min in 70% alchohol and then stain for 8 min in 20% Giemsa. Differentiate the staining by placing the slide under running tap water for 2 min then air dry and examine under the microscope using oil immersion.

Giemsa stains nuclear material red and cytoplasm blue. In the first instance find and make a scale drawing of normal red and white blood cells. Once you have done this look for and draw the different stages of malaria parasites inside the red blood cells (ring forms, other trophozoites, schizonts, gametocytes). If your mouse has a low or unapparent infection ask for a demonstration slide to make these drawings.

To assess the level of parasitaemia use tally counters to score the number of infected red blood cells in a sample of 1000. Repeat this estimate twice to check accuracy and enter your mean result into the class results sheet.

Haematocrit

Gently shake the blood sample to re-suspend and take up a sample by capillary action into haematocrit tubes. Plug the end of the tube with plasticine then spin in a haematocrit centrifuge, to separate blood cells and plasma. Read off the percentage of red blood cells and enter into the class results sheet.

Records

Record individual data (Fig. 1.10.1) and combine in a class results sheet (Fig. 1.10.2). Also make notes and annotated diagrams (appearance of tissues, infected and uninfected blood cells).

Mouse No.	Days PI	Parasitaemia	Liver wt	Spleen wt	Haematocrit

Fig. 1.10.2 Class results sheet.

Expected results

Morphology of malaria parasites

The malaria parasite invades the red blood cell as a tiny ($1 \mu m$) uninucleate merozoite. The parasite forms a vacuole in the red blood cell and digests haemoglobin to provide nutrients. These nutrients are taken across the vacuole membrane to enable the parasite to grow. Growth and division is by a process known as schizogony. Three stages of the parasite can be defined during this asexual cycle and, in addition, some merozoites may enter the sexual cycle and form micro- and macro-gametocytes (Figs. 1.10.3, 1.10.4).

Rings:
The earliest stage of infection in which a small parasite can be seen within the vacuole.

Older trophozoites:
Growing stage in which the parasites make large amounts of protein and DNA.

Schizonts:
Final stage in which the DNA is packaged into 8–16 daughter merozoites.

Gametocytes:
Sexual stages which circulate in the bloodstream and produce gametes when taken into the mosquito gut. (Note that these are very rare in old laboratory strains).

Disease pathology

Although individual results may show some variation, means and standard deviations calculated from pooled class results will give a clear picture of disease pathology.

Parasitaemia

The parasitaemia will rise to a peak of around 30% of red blood cells infected at day 10–12 of infection and will then fall to zero by day 20.

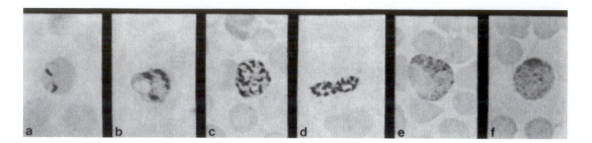

Haematocrit

This will fall from around 40% to 30% by day 9 to 12 of infection, due partially to lysis and partially to removal of infected red blood cells by phagocytosis. Recovery should be seen by day 15.

Organ weights

There is a small increase in the weight of liver and a large increase in spleen weights throughout the period of study. Part of the reason for this is that macrophages in these two organs phagocytose infected red cells and debris. Evidence of this can be seen clearly by looking at the demonstration material. Liver and spleen become progressively darker due to the presence of undigested crystalline haem pigment found in infected cells. The massive increase in spleen weight is also due to stimulation of the immune response, which causes lymphocyte proliferation in this organ.

Fig. 1.10.3 Comparative morphology of the blood stages of rodent malaria (*Plasmodium yoelii*) in the mouse (Giemsa's stain). Note surrounding red and white blood cells. After Landau & Boulard (1978). Left to right: a, early ring stage; b, young trophozoite; c, late trophozoite; d, mature schizont; e, microgametocyte; f, macrogametocyte.

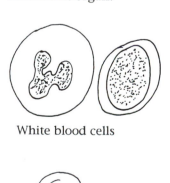

| White blood cells | Ring form | Trophozoite |
| Red blood cell | Schizont | Gametocyte |

Fig. 1.10.4 Drawing to illustrate the major cell types in a blood smear from a mouse infected with *P. yoelii*.

REFERENCES

Cox, F. E. G. (1993). *Modern Parasitology.* Second Edition. Oxford, Blackwells Scientific Publications.

Cox, F. E. G. (1993). Plasmodia of rodents. In: *Parasitic Protozoa*, 2nd edn, vol. 5 (ed. J. P. Kreider), pp. 49–104. San Diego, Academic Press.

Landau, I. & Boulard, Y. (1978). Life cycles and morphology. In: *Rodent Malaria* (ed. R. Killick-Kendrick & W. Peters), pp. 53–84. London, Academic Press.

Schmidt, G. D. & Roberts, L. S. (2000). *Foundations of Parasitology*, 6th edn. The McGraw-Hill Companies, Inc.

Smyth, J. D. (1994). *Introduction to Animal Parasitology.* Cambridge, Cambridge University Press.

1.11 Malaria: an example of a vector transmitted parasite

H. HURD & R. E. SINDEN

Aims and objectives

This exercise is designed to demonstrate:

1. The different stages in the life cycle of *Plasmodium* spp.
2. How to recognise infection in a mammal and a mosquito.

Introduction

Over 40% of the world's population live at risk from infection with malaria, a parasitic protozoan of the genus *Plasmodium*. The malaria parasite is transmitted by its mosquito vector that feeds on blood. The typical symptoms of malaria include short episodes of severe fever, occurring every two or three days, anaemia and enlargement of the spleen. Between 300–500 million clinical cases of malaria occur each year and 1.5–2.7 million people die of the disease annually. Ninety percent of cases are in tropical Africa and half of the deaths are in children under five years. A more detailed account of malaria can be found in Knell (1991).

The life cycle of the malaria parasite (Fig. 1.11.1) has four phases: three asexual and one sexual. **Sporozoites** from the mosquito are injected into the blood and immediately invade the liver. Here they become **trophozoites** and grow and multiply to form **schizonts** containing many **merozoites** (asexual phase 1). When the liver cells burst, merozoites are released into the blood and invade red blood cells. **Erythrocytic trophozoites** grow and divide to form **schizonts** containing new merozoites. These are released when the infected red cell bursts and they invade new red cells (asexual phase 2). Within the red cells some merozoites develop into male or female **gametocytes** rather than trophozoites. If gametocytes are taken up by a mosquito during feeding they pass into the midgut of the vector to initiate the sexual phase. Male gametes are produced and released during the process of **exflagellation**. They swim through the **89**

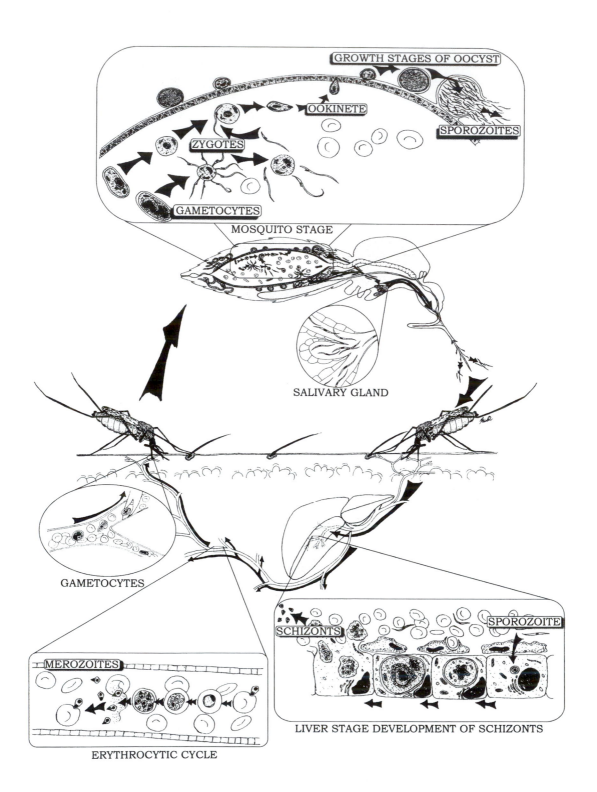

GROWTH STAGES OF OOCYST

OOKINETE

SPOROZOITES

ZYGOTES

GAMETOCYTES

MOSQUITO STAGE

SALIVARY GLAND

GAMETOCYTES

SCHIZONTS

SPOROZOITE

MEROZOITES

LIVER STAGE DEVELOPMENT OF SCHIZONTS

ERYTHROCYTIC CYCLE

Fig. 1.11.1 (opposite) The life cycle of malaria. The liver stage: sporozoites of *Plasmodium vivax* may form hypnozoites and remain dormant in the liver cells for months or years before developing. The erythrocytic cycle: after invasion, a merozoite develops from the signet ring stage to a feeding trophozoite. It forms schizonts, which divide asexually by schizogony to form more merozoites. Some merozoites may develop into male or female gametocytes. The mosquito midgut: within minutes gametocytes are released from erythrocytes. Male gametocytes form flagella and are released by exflagellation. Fertilisation occurs in the midgut and the zygote develops into a motile ookinete, which invades the midgut wall. It transforms into an oocyst, which divides to form sporozoites. The mosquito salivary gland: sporozoites that invade the salivary glands are injected into the blood of a new host when the mosquito bites and rapidly invade the liver cells. (Original drawing by Derric Nimmo, Keele University.)

blood meal to fertilize a female gamete. The resultant zygote develops into a mobile **ookinete** and invades the mosquito midgut, passing through or between the epithelial cells to the basement membrane. Here it rounds up and forms an **oocyst**. Within the oocyst, division takes place and sporozoites develop (asexual phase 3). After two weeks the oocyst bursts and sporozoites are released into the haemocoel of the mosquito. Sporozoites travel through the haemocoel and invade the salivary glands, from which they are injected into the blood when the mosquito next feeds.

Infection can be detected in the vertebrate by examining blood cells for the presence of *Plasmodium*. Infected mosquitoes can be recognised by the presence of oocysts on the midgut or sporozoites in the salivary glands.

Laboratory equipment and consumables
(per student or group)

Equipment

Compound microscope with oil immersion lens
Dissecting microscope
Staining rack

Pair of mounted needles and a pair of forceps
Preserved or freshly anaesthetised male and female mosquito
Blood film from uninfected and infected mouse

Consumables

Microscope slide
Prepared mount of ookinetes
Dropping bottles with methyl alcohol
Dropping bottles with phosphate-buffered saline (PBS)
Lens tissues

Prepared mount of an infected midgut
Prepared mount of infected salivary glands
Dropping bottles with Giemsa's stain solution Gurr® BDH*
Immersion oil
Paper towels

*Stock diluted x 5 with buffered water (0.7 g KH_2PO_4, 1.0 g Na_2HPO_4, per litre, pH 7.2).

Sources of parasite material

This practical exercise could be performed using any *Plasmodium* species. Examples of those containing rodent malaria, such as *P. yoelii nigeriensis* or *P. berghei*, or avian malaria, such as *P. gallinaceum*, are maintained in several institutions. Thin blood films can be made and stored dry for several years. Infected mosquito midguts can be dissected after a short period of storage at $-20\,°C$, but salivary glands need to be dissected from fresh, chilled or anaesthetised mosquitoes.

Safety

Provided good laboratory practice is followed, there should be few health hazards associated with this exercise. However, both methanol and Giemsa's stain are toxic and highly flammable. Never leave bottles open. Ensure that students dispose of slides in a container provided.

Instructions for staff

If the exercise is confined to preserved material there are no special instructions once this material has been obtained. For *P. y. nigeriensis* infections, live mosquitoes will have to be fed on a malaria-infected animal eight days before collection for dissection of oocysts or about 14 days before salivary gland dissections. Details of parasite and mosquito maintenance can be obtained from the manual of Crampton *et al.* (1997). Live mosquitoes should be kept on ice until used or anaesthetised with ethyl acetate. If available, it is helpful to demonstrate salivary gland and midgut dissection using a microscope attached to a video system.

Instructions for students

1. Diagnosis of malaria infection in the blood
The traditional way to diagnose malaria infection is to identify parasite stages in the erythrocytes. Skilled workers can distinguish between different species this way. However, this is difficult for non-skilled personnel and, in the case of falciparum malaria, infected cells may be sequestered in the organs and not circulating. Modern methods use molecular techniques incorporated in a 'dip stick' test, but these may be too expensive. Here, the traditional staining technique is used to determine which of two mouse blood samples is infected with a rodent

malaria, *Plasmodium yoelii nigeriensis*. A thin smear of blood will be made for you. A stain based on acid and basic dyes is used. Giemsa's stain uses a combination of acid and basic dyes to produce three colours. Nuclei are stained purple, cytoplasm is stained blue and erythrocytes are stained red.

1. Fix the smear by covering with methyl alcohol for 60 sec.
2. Drain and air-dry completely.
3. Cover with Giemsa's stain (ready diluted 1:5 with buffered water (0.7 g KH_2PO_4, 1.0 g Na_2HPO_4)) for 20–30 min. Do not allow to dry.
4. Drain, wash vigorously under tap water and gently blot dry.
5. Initially observe the blood film under low power. Select an area with a thin, even covering of cells and observe under oil immersion.
6. Draw erythrocytes, white cells and platelets.
7. Identify the infected slide. Identify and draw the following.
   (a) very young trophozoites (signet-ring stages), and developing trophozoites. How many parasites invade one cell?
   (b) mature schizonts
   (c) gametocytes

2. Malaria and the mosquito

Haematophagous insects require a blood meal to provide the protein necessary for egg production; thus, bloodfeeding, egg production and transmission of blood-borne diseases, such as malaria, are linked. Examine the insect head and how the generalised structure of the mouth parts is adapted for bloodsucking. For this, make notes of the differences in structure between male and female antennae and mouth parts. Examine the antennae of your specimen to check that you have a female mosquito. You will be provided with an anopheline mosquito and told the species name. If they are fresh, work quickly; once they revive they will fly away!

1. Using a mounted needle as instructed, gently separate the head of the mosquito from the thorax. The **salivary glands** will be left trailing from the head. They are tri-lobed and transparent.
2. Using the diagram provided (Fig. 1.11.2), identify the **labium**, which forms a sheath round the mouth parts, and ends in a hinged labella, which is covered in sensilla.
4. The remaining mouth parts form the **stylet**, which is inserted into the skin during feeding. Identify:

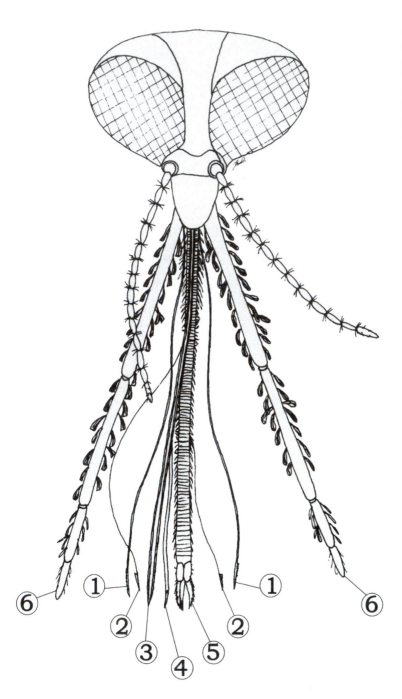

Fig. 1.11.2 Diagram of the head and mouthparts of *Anopheles stephensi*. 1. Maxilla with serrated tip; 2. mandible; 3. labrum; 4. hypopharynx; 5. labium with hinged labellum; 6. maxillary palps. (Original drawing by Derric Nimmo, Keele University.)

   (a) the **labrum** or upper lip, which forms a food channel up which blood can flow;
   (b) the **maxillae**, which have pointed, toothed ends and are used for piercing the skin;
   (c) the **maxillary** palps, which contain sensory organs;
   (d) the **mandibles**, whose function is unknown;
   (e) the **hypopharynx**, which forms the salivary channel.

5. Sugar meals are stored in the **crop** of the mosquito. The bloodmeal is taken directly into the mosquito **midgut** and stored there whilst digestion occurs. Locate these organs and also find the **forgut**, **hindgut** and **Malpighian tubules**.

6. Dissect the alimentary canal and reproductive system of the mosquito in a drop of PBS. Use one mounted needle pressed on the thorax to steady the insect. With the other needle, tear the cuticle between abdominal segments six and seven. Press down on the seventh and eighth segment and pull away from the thorax, bringing the gut, **ovaries** and **oviduct** out. Look for the dark brown, spherical **spermatheca** (a sperm store) in the segments that remain attached to the gut. Make annotated drawings of your dissection.

7. You are provided with three slides of the stages of the malaria parasite found in the mosquito, **ookinetes**, **oocysts** and **sporozoites**. Using the oil immersion lens, examine the ookinete preparation.

8. Using the low power of your compound microscope examine the permanent mounts of infected midguts that have been stained to make the oocysts more visible.

9. Using the oil immersion lens examine the sporozoite preparation.

10. Make annotated drawings of these stages.

Remember to wipe the oil immersion lens with lens tissue when you have finished.

Ideas for further exploration

If infected material is available in the institution:

• Students could be provided with thick blood films to observe exflagellation and time the events as they occur.
• Mosquitoes could be observed blood feeding through membranes

- Dissections of infected mosquitoes could be performed. Prevalence and intensity of infection could be calculated as a class exercise.
- Practical work could be supplemented with the CD *Topics in International Health: Malaria*, Oxford, CAB International.

REFERENCES

Crampton, J. M., Beard, C. B. & Louise, C. (1997). *The Molecular Biology of Insect Disease Vectors. A Methods Manual.* London, Chapman & Hall.

Knell, A. J. (1991). *Malaria.* Oxford, Oxford University Press.

Clements, A. N. (1992). *The Biology of Mosquitoes, Volume 1; Development, Nutrition and Reproduction.* London, Chapman & Hall.

1.12 Larval and adult *Echinostoma* spp. (Trematoda)

B. FRIED

Aims and objectives

This exercise is designed to demonstrate:

1. The morphology of adult echinostomes.
2. The morphology of larval echinostomes, particularly the egg, miracidium, daughter redia, cercaria, and encysted metacercaria.
3. Development and hatching of miracidia from the eggs.
4. Infection of vertebrate hosts with encysted metacercariae.

Introduction

Echinostomes are ubiquitous intestinal trematodes with adult stages that live in various avian and mammalian hosts. Their first intermediate hosts are gastropod molluscs whereas the second intermediate hosts are various invertebrates, particularly gastropods and bivalves, but also cold-blooded vertebrates such as fishes and frogs (both adults and tadpoles). The life cycle of an echinostome is shown in Fig. 1.12.1. Numerous echinostome life cycles have been described, the best known being the 37-collar-spined echinostomes of the genus *Echinostoma*. Most information on echinostomes is based on species that use hosts in fresh-water habitats, but there is literature on echinostomes of marine hosts. This exercise focuses on echinostomes associated with fresh-water habitats. Although of minimal importance in human medicine, there are occasional reports of echinostome infections in man acquired as a food-borne trematodiasis. Echinostomes provide useful models for biological research at all levels of organisation, from the molecular to the organismic.

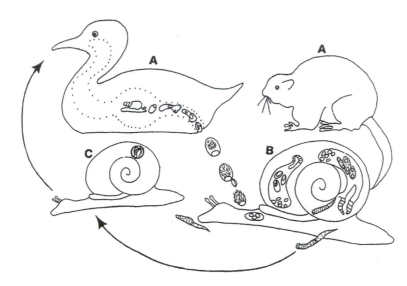

Fig. 1.12.1 The life cycle of an echinostome (based on the life cycle of *Echinostoma caproni*). A, definitive hosts; B, first intermediate host; C, second intermediate host. Reproduced from Fried & Huffman (1996) with the permission of Academic Press Limited.

Laboratory equipment and consumables

(per student or group)

Equipment

Compound microscope	Finger bowls of various sizes
Dissecting microscope	Beakers of various sizes
Hand lens	Hammer (for cracking open
Dissecting trays and dishes	snails)
	Dissecting instruments

Consumables

Pasteur pipettes and bulbs	Pins for dissection (optional)
Standard-sized Petri dishes	Tissue paper and paper towels
Physiological saline (0.9% NaCl)	Disposable gloves
Slides and coverslips	Aquarium water/conditioned tap water*

*water allowed to stand for about 24 h prior to use.

Sources of parasite material

A variety of waterfowl, e.g. ducks and geese, and aquatic mammals, e.g. muskrat and beaver in the USA, carry infections of adult echinostomes in their intestines and, if available freshly killed, provide a possible source of material. Gastropods of the following genera serve as first and second intermediate hosts of echinostomes: *Biomphalaria*, *Physa*, *Helisoma*, *Planorbis*, *Lymnaea*,

Viviparus and others. Locally available snails may be a good source, but this can vary from year to year. Some academic and government laboratories maintain the life cycles of several echinostome species, and suitable material may be available from these sources in the form of infected snails, infected vertebrate hosts, or encysted metacercariae stored in physiological saline at 4 °C. Subject to availability, metacercarial cysts or infected snails can be supplied by Dr Bernard Fried, Department of Biology, Lafayette College, Easton, PA 18042, and in Europe by the Danish Bilharziasis Laboratory, Jaegersborg Alle 1D, Charlottenlund, DK-2920.

Safety

As long as the material available for this exercise comprises animals infected with echinostomes, there are no reasons for additional precautions to those taken when running other classes based on the use of living animal material. A possible infective stage to humans is the metacercarial cyst (encysted metacercaria) but this stage would have to be inadvertently swallowed by the student to pose a risk. Students are advised not to put fingers or dissecting tools into their mouths. Appropriate care should also be taken when handling hosts or parasites in case of allergies to these organisms. Environmental pathogens associated with animals and animal waste products need to be considered as potential contaminants with some possible risk factor to human health.

Instructions for staff

Obtain echinostome material from naturally or experimentally infected hosts or from laboratory personnel able to supply the material. Infected snails provide suitable material for examining intramolluscan stages, such as cercariae, encysted metacercariae and rediae. Snails with patent cercarial infections can be isolated in water to determine if they are releasing cercariae. Snails can also be crushed lightly with a hammer or with forceps (shell is broken and then removed with forceps). Infected snails normally contain hundreds of rediae in their digestive gland–gonad (DGG) complex and encysted metacercariae in the kidney–pericardium region. Some basic knowledge of snail morphology is useful in finding the intramolluscan stages of these echinostomes.

It is assumed that staff members are licensed to infect appropriate vertebrate hosts with larval stages of parasites, and that necropsy procedures and the proper disposal of freshly killed vertebrate hosts should follow institutional guidelines.

Instructions for students

You are provided with snails naturally or experimentally infected with larval echinostomes.

1. Place a snail on a paper towel and, with a hammer or forceps, break its shell.
2. Transfer the snail to a Petri dish containing physiological saline and remove the rest of the shell with forceps.
3. Using needles and/or forceps, tease larval stages of the echinostomes from the DGG (mainly rediae and cercariae) or the kidney–pericardium region (mainly metacercarial cysts) of the snail. Recognise the following stages in the saline solution: rediae (Fig. 1.12.2A), cercariae (Fig. 1.12.2B,C), metacercarial cysts (Fig. 1.12.2D).
4. Make wet mounts of these stages on a microscope slide with a coverslip and examine under the compound microscope. Recognise the cercarial body (Fig. 1.12.2B,C) and the cercarial tail, which may contain fin folds (Fig. 1.12.2B). Note the collar with cephalic or collar spines (Fig. 1.12.2E) at the anterior end of the cercarial body from which it gets its name (echino = spiny; stoma = mouth).
5. Note the following structures in your wet-mount of the redia stage: pharynx; gut, often filled with orange, red or black material; ambulatory buds and other intramolluscan stages (typically cercariae) within the redia (Fig. 1.12.2A).

It is most likely that the redia you are examining is a daughter redia and contains cercarial bodies within. The encysted metacercariae average about 150 μm in diameter and are characterized by a transparent outer cyst wall and a denser inner cyst wall. Within the inner cyst will be the juvenile echinostome and you may see collar spines, suckers and excretory (calcareous) concretions in the body of the juvenile (Fig. 1.12.2D). Encysted metacercariae may be used to infect vertebrate hosts. If the species of echinostome is known, a susceptible vertebrate host can be chosen. For instance, if you are working with either European or North American *E. revolutum*, you must select avian hosts (domestic chicks or ducks) since mammalian hosts are not sus-

Fig. 1.12.2 Larval and adult stages of echinostomes based on several species of 37-collar-spined *Echinostoma*.
A. Daughter redia. Note the pharynx (p), gut (g), cercaria (c), and ambulatory bud (a). Scale bar = 100 μm.
B. Cercaria. Note the body (b), tail (t), and finfold (f). Scale bar = 100 μm.
C. Cercarial body. Note the oral sucker (o), collar (c), pharynx (p), oesophagus (e), excretory concretions (ex) in a protonephridial tube, acetabulum (a), intestinal caeca (i) and excretory bladder (b). Scale bar = 100 μm.
D. Encysted metacercaria (cyst). Note the outer cyst (o), the inner cyst (i), the collar spines (c), the acetabulum (a), the intestinal caeca (ic) and excretory concretions (ex). Scale bar = 100 μm.
E. Collar spines. These spines are present on the collar of the cercaria and the adult. To simplify the line drawings, collar spines have been omitted from the cercariae in B and C and from the adult in Fig. 1.12.3 Scale bar = 200 μm.

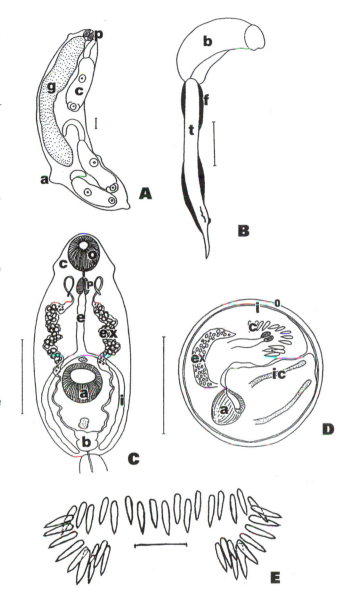

ceptible. Should you be using cysts of the North American species *E. trivolvis*, both hamsters and domestic chicks are good hosts. Should you be using the African echinostome, *E caproni,* mice are excellent definitive hosts, although hamsters and domestic chicks can also be used. Similarly, a well-adapted laboratory strain is *E. paraensei* and encysted metacercariae of this species infect both mice and hamsters. Mammalian hosts are best infected with 25–50 cysts via stomach tube and avian hosts with 50–100 encysted metacercariae by pipetting the cysts directly into the mouth of the bird. Hosts should be maintained according to standard laboratory protocols and necropsied 2 weeks post-infection (PI) at which time you should find ovigerous adults in the gut anywhere from the pyloric valve to the cloaca, including the caeca and even the bursa of Fabricius. Some knowledge of the topography of the intestinal tract of a vertebrate host is expected on the part of the student. Echinostome adults are usually tenaciously attached to the mucosal lining of the gut but will eventually lose their hold in cold physiological saline.

6. Examine segments of open gut tissue in Petri dishes half-filled with saline. The live worms will contract and extend from about 5–15 mm in length and there may be petechial haemorrhages in gut areas associated with the infection. In some infections, e.g. *E. caproni* in the mouse, the infected areas of the gut are dilated and about 3 times the diameter of the control uninfected gut areas.
7. Worms should be removed from the gut, transferred to fresh saline solution and cleaned of mucus and other debris by rapid transfer with a pipette.
8. The morphology of the live worm should be examined, with emphasis on the following structures: suckers, collar of spines, gonads, vitellaria, excretory system, and uterus filled with eggs (Fig. 1.12.3A). The uterus will be red due to the presence of haemoglobin. The bifid intestinal caeca may contain food material colored orange, red, tan, or white, reflecting the worm diet in the host. Worms feed on a mixture of host luminal contents and mucus from the host gut lining and also perhaps on host blood.

Adult worms can be used as a source of eggs to obtain miracidia. Most eggs of echinostomes *in utero* are fertilized, but poorly developed (Fig. 1.12.3B) and require incubation in aquarium water or conditioned tap water for about 10 days at 28 °C or 14–21 days at 22–24 °C to reach the stage where they contain fully developed miracidia (Fig. 1.12.3C).

9. Dissect eggs out of the adult worms, rinsing them in water and incubating them in Petri dishes half-filled with water for 10–21 days at the above-mentioned temperatures.
10. Premature hatching can be prevented by keeping the cultures in the dark (wrap them in aluminum foil).

Fig. 1.12.3A. Adult. Note the oral sucker on the collar (c), pharynx (p), oesophagus (e), intestinal caeca (ic), acetabulum (a), vitellaria (v), uterus (u) with eggs, ovary (ov), anterior testis (At), posterior testis (Pt), excretory bladder (B) and excretory pore (Ep). Scale bar = 1 mm.
B. Egg *in utero* showing embryo (E); operculum (O) and knob (K). Scale bar = 50 μm.
C. Fully developed egg containing the miracidium (m). Scale bar = 50 μm.
D. Newly hatched miracidium showing the apical papilla anterior to the eyespots (E); flame cells (F); cilia (C) and germinal cells (G). Scale bar = 50 μm.

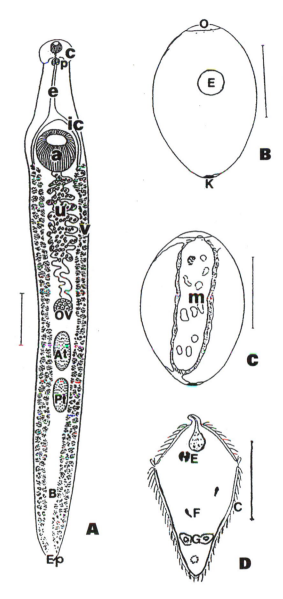

11. Expose cultures that contain eggs with fully developed miracidia to a source of bright incandescent light to obtain a synchronous miracidial hatch. The released miracidia (Fig. 1.12.3D) can be used for observational studies and even for snail infectivity work, should a susceptible first intermediate host be available.

Brief account of expected results

You will learn the basic morphology of a monoecious digenean intestinal trematode and will gain an appreciation of the intricacies of the adult worm. You will see the inherent contractility of the worm, its mode of sucker action, expulsion of eggs, peristaltic action of the intestinal caeca, and release of excretory products from the excretory bladder. You should also gain a sense of the diversity of forms that have evolved in the Digenea based on the various larval stages seen. What accounts for this interesting change from the encysted metacercaria to the sexually mature adult worm?

Ideas for further exploration

The problem of parasite development and maturation can be explored and some of the following questions answered:

- How does the metacercaria excyst?
- How does the excysted larva find its optimal habitat in the intestine?
- What is the growth rate of the parasite from the excysted metacercaria to the ovigerous adult?
- What is the association of the parasite with the gut?
- What does the parasite feed on in the gut?
- What about reproductive patterns?
- Is a single worm capable of self-copulation and then self-fertilisation?
- How do eggs develop and produce miracidia that are infective to snails?
- How do miracidia find their snail hosts?
- What happens during intramolluscan development from the time that miracidia infect snails until cercariae are released?
- What is the fate of the released cercariae, and how do they find the second intermediate host?
- What is the mechanism of cercarial encystment in the second intermediate host?
- How are metacercarial cysts transmitted to definitive hosts?

Ideas for staff on similar exercises

The following ideas can be explored further by staff:

- *In vitro* cultivation of echinostomes from excysted metacercariae to the ovigerous adults.
- *In ovo* cultivation of echinostomes from excysted metacercariae to ovigerous adults.
- Methods for the *in vitro* excystation of echinostome metacercariae.
- Chemical communication of larval and adult stages of echinostomes.
- Effects of larval echinostomes on their intermediate hosts.
- Pathophysiological and pathobiochemical effects of adult echinostomes on their definitive hosts.

Additional information

To the best of our knowledge, there are no commercial suppliers of *Echinostoma*. However, several laboratories maintain relevant species and material may be requested from them. In the USA, Dr Bernard Fried can supply metacercarial cysts or infected snails subject to availability. In Europe, the Danish Bilharziasis Laboratory maintains *E. caproni* and may supply material on request.

REFERENCES

Beaver, P. C. (1937). Experimental studies on *Echinostoma revolutum* (Froelich) from birds and mammals. *Illinois Biological Monographs* **15**, 1–96.

Fried, B. (1989). Cultivation of trematodes in chick embryos. *Parasitology Today* **5**, 3–5.

Fried, B. (1994). Metacercarial excystment of trematodes. *Advances in Parasitology* **33**, 91–144.

Fried, B. & Huffman, J. E. (1996). The biology of the intestinal trematode *Echinostoma caproni*. *Advances in Parasitology* **38**, 311–368.

Fried, B. & Stableford, L. T. (1991). Cultivation of helminths in chick embryos. *Advances in Parasitology* **30**, 107–165.

Huffman, J. E. & Fried, B. (1990). *Echinostoma* and echinostomiasis. *Advances in Parasitology* **29**, 215–269.

Idris, N. & Fried, B. (1996). Development, hatching, and infectivity of *Echinostoma caproni* (Trematoda) eggs, and histologic and histochemical observations on the miracidia. *Parasitology Research* **82**, 136–142.

1.13 *Schistosoma mansoni* (Trematoda)

M. J. DOENHOFF, L. H. CHAPPELL & J. M. BEHNKE

Aims and objectives

This exercise is designed to demonstrate:

1. The external and internal morphology of the adult schistosome.
2. The pairing behaviour of the adult worms.
3. Destination of eggs.
4. Hatching of miracidia from eggs.
5. Some gross effects of infection on the mouse host.

Introduction

Schistosomiasis is one of the most important public health problems in the tropics. Despite intensive research, there is still no easy cure for the disease and no simple way to control transmission. It is estimated that some 200 million people are infected with schistosomes.

The human schistosomes and many of the other species in mammals belong to the genus *Schistosoma*, in the family Schistosomatidae. Members of this family show morphological and physiological peculiarities that set them apart from all the other trematodes. Firstly, they are dioecious, the male carrying the female in a ventral canal, the gynaecophoric canal, and secondly, they live in the bloodstream of warm-blooded hosts (See Fig. 1.13.1 for details of the life cycle).

Five species of blood flukes may infect humans, causing schistosomiasis or Bilharzia. Three of these species are very common. *S. mansoni* is found in Africa and South America and affects chiefly the large intestine in humans; *S. haematobium* is found in Africa and the Middle East and affects the urino-genital system; *S. japonicum* is found in parts of Asia and affects mainly the large intestine. The remaining two species are *S. mekongi*, which is related to *S. japonicum* and is also found focally in Asia, and *S. intercalatum* in Africa.

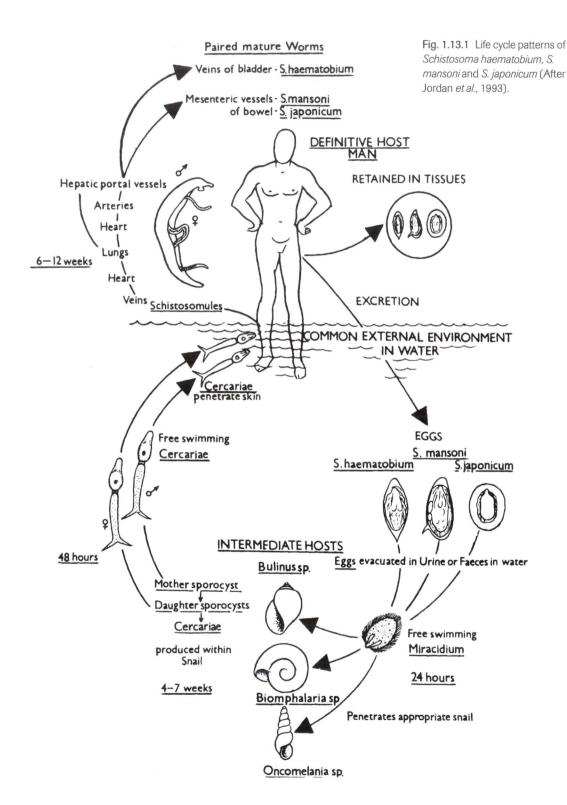

Paired mature Worms

Veins of bladder - S.haematobium

Mesenteric vessels - S.mansoni
of bowel - S. japonicum

Fig. 1.13.1 Life cycle patterns of *Schistosoma haematobium*, *S. mansoni* and *S. japonicum* (After Jordan *et al.*, 1993).

DEFINITIVE HOST
MAN

RETAINED IN TISSUES

Hepatic portal vessels
Arteries
Heart
Lungs
Heart
Veins Schistosomules

6—12 weeks

EXCRETION

COMMON EXTERNAL ENVIRONMENT
IN WATER

Cercariae
penetrate skin

Free swimming
Cercariae

EGGS

S. mansoni

S. haematobium   S. japonicum

48 hours

Mother sporocyst
Daughter sporocysts
Cercariae

produced within
Snail

4—7 weeks

INTERMEDIATE HOSTS

Bulinus sp.

Eggs evacuated in Urine or Faeces in water

Free swimming
Miracidium

24 hours

Biomphalaria sp

Penetrates appropriate snail

Oncomelania sp.

The laboratory mouse has proved to be a suitable host for *S. mansoni* and one can expect a reasonable percentage of cercariae to mature. Mice usually show a moderate number of eggs in their faeces, many eggs in the liver and a low death rate, although all of these features depend on the dose of cercariae administered, the parasite isolate used, and the strain of the mouse (see Instructions for Staff for further details).

This exercise is designed to demonstrate the location of adult worms in the vascular system of the host, and to enable close examination to be made of some aspects of the fascinating biology of the adult worms.

Laboratory equipment and consumables
(per student or group)

Equipment

Compound microscope
Dissecting microscope
Hand lens
Dissecting dish

Solid watch-glass or small
    dishes
50 ml beaker
Dissecting instruments

Consumables

Pasteur pipettes and bulbs
Petri dishes (standard)
Slides and coverslips
Disposable gloves
Pins (for dissection)
Hanks' saline

Paper towel
Lens tissue
Cotton blue-lactophenol
Citrate-saline
20 ml hypodermic syringe

Sources of parasite material

Infected mice are best obtained from a laboratory specialising in research on schistosomes and which maintains the entire life cycle. It is beyond the scope of this manual to describe procedures required to maintain the parasite throughout its life cycle.

Safety

As long as the material available for this class comprises mice infected with *S. mansoni*, there are no reasons for additional precautions to those taken when running other classes based on mouse material. **Note that schistosome cercariae are infective to humans and should therefore not be used in this practical**. One way of minimising the risk to students is to have a microscope linked to a television monitor via a video camera, so that students can see live cercariae without handling the material themselves. Appropriate care should be taken when handling mice, adult worms, eggs and miracidia. Some subjects may be allergic to mice and the parasites. Environmental pathogens associated with animals and animal faeces need to be considered as potential contaminants, with some associated risk factor to human health.

Instructions for staff

This exercise is best carried out with freshly obtained material. Mice should be killed by cervical dislocation a few minutes into the class so that the students receive warm animals for dissection. In order to demonstrate eggs in the faeces, moderate–heavy intensity infections will be required. In most cases, inocula of 50–100 cercariae result in good establishment of worms, with the majority of mice surviving for long enough to ensure that the worms have matured and eggs have begun to be shed in host faeces (5–7 weeks). However, different isolates and laboratory passaged strains of *S. mansoni* show varying levels of pathogenicity and you should be aware of the characteristics of the strain you intend to use. Moreover, different strains of mice also vary in their response to infection, some being more resistant than others.

Instructions for students

You are provided with mice infected with *Schistosoma mansoni*.

Observations on adult worms in the blood vessels of the mouse host

1. Place the mouse in a dissecting dish, open the abdominal cavity and examine the small intestine, blood vessels associated with the intestine, and the liver *in situ*. Note that in the mouse it is the small rather than the large intestine that is principally infected with *S. mansoni*. Is there any evidence indicating that the animal is infected? Compare the size and appearance of the liver from an infected mouse with that from an uninfected animal.

2. Observe the mesenteric veins with the stereomicroscope and look for small, greyish worms through the walls of these vessels. These are the adult worm pairs and they should still be alive and moving in the blood vessels. They can be removed from the blood vessels by careful dissection. If you do so, transfer them immediately to warm Hanks' saline. Note that some worms may have shifted to the hepatic portal system shortly after the death of the mouse.

3. Another way to extract the adult worms is to flush them through the liver with a citrate-saline buffer. For this procedure, after exposing the body cavity make a single incision from the lower end into the thorax. Cut away the rib-cage to expose the heart but take care not to cut the heart. Locate the hepatic portal vein and cut it close to the liver. Place the mouse in a large Petri dish then take a 20 ml hypodermic syringe full of citrate-saline and insert the top of the needle into the left ventricle. Slowly inject the citrate-saline into the heart to flush the worms from the blood vessels. Rinse the worms with citrate-saline.

4. Recover as many worm-pairs as possible and observe them with the binocular microscope. Many of the pairs will separate from one another on transfer and this will give you an opportunity to examine male and females separately. Identify the sexes and make detailed drawing of all you see. How do the sexes differ morphologically?

5. Place live male and female worms (Fig. 1.13.2) onto a glass slide in a drop of saline, cover with a cover slip and examine under the microscope. Try mounting in a drop of cotton blue-lactophenol. This time heat gently on a hot plate for a couple of minutes then cover with a coverslip. Draw a male and a female worm, identifying all the features, and highlighting those which distinguish the sexes. Observe the oral and ventral suckers, bifurcating gut containing haematin (blood pigment), the gynaecophoric canal and reproductive organs. How many eggs does the female worm have in the uterus? Now select a pair of worms *in copulo*. Observe them under the microscope, noting how the two are held together.

Schistosome eggs

1. Collect a soft faecal pellet from the rectum of your mouse and mix some of this with Hanks' saline on a slide.

2. Squash the faecal slurry with a coverslip and examine with the microscope for the presence of eggs.

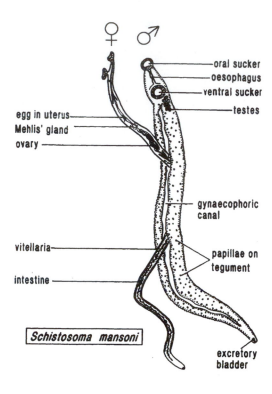

oral sucker
oesophagus
ventral sucker
testes

egg in uterus
Mehlis' gland
ovary

gynaecophoric canal

vitellaria

papillae on tegument

intestine

Schistosoma mansoni

excretory bladder

Fig. 1.13.2 Adult pair of *Schistosoma mansoni* (After Smyth, 1994). Male is approximately 1 cm in length; the female 1.5 cm in length.

3. Observe these carefully and make detailed drawings.
4. Depending on the number of worms carried by the mouse and the age of the infection, there may be too few eggs in faeces for them to be easily detected. In this case, cut a small section of the liver containing the white lesions, macerate it in water and observe under the microscope. What is unusual about the appearance of the eggs? Note the movement of the miracidium inside the egg after it has been in water for a while.

Liver eggs and lesions

1. Examine a piece of the liver of the mouse and account for its measly appearance by looking for small white lesions in squash preparations, using glass slides.
2. Look also for the presence of worms in this preparation as some worms may have migrated back into the liver after the mouse was killed.
3. Remove one of the lesions and examine its contents in more detail with your microscope. What are the inclusions? Similarly make squash preparations of the mouse intestine and look for similar lesions.

Hatching miracidia

1. Collect eggs from the faecal pellets of your mouse, or if the numbers are low, from a portion of macerated liver.
2. Transfer half of these to a small dish (solid watch-glass) containing 0.85% saline and the other half into a dish containing distilled water.
3. Label the dishes and place both close to light for 10–15 min.
4. Using the dissecting microscope, count the number of free-swimming miracidia recovered from each dish at 5 min intervals.
5. If time permits, add distilled water to the eggs in saline and repeat the experiment to see if the miracidia will hatch. What can you conclude from your observations?
6. Prepare different dilutions of saline and observe the effect of varying salinity levels on the rate of hatching and also the total hatch.

Ideas for further exploration

- How do the sexes differ?
- How do pairs maintain contact?
- Why do eggs not hatch while still within the host?
- How do you think your mouse was infected?
- Is it possible for you to become infected through contact with the infected mouse? Why?
- Would you expect all trematodes to have miracidial hatching inhibited by high salt concentrations? Why?
- How do you account for the lesions in the mouse tissues? What is the correct technical term for such lesions?
- Estimate the number of eggs in a gram of faeces of your mouse and calculate the number released by each female/24 h, bearing in mind that some eggs are retained in the tissues.
- What is the principal cause of pathology in schistosomiasis? Is it the adult worms or the eggs?

Additional information

Infections of mice with *S. mansoni* can be used to illustrate different aspects of the host–parasite biology. By examining mice exposed to different numbers of cercariae the relationship between the severity of infection and pathology can be quantified by measurements on liver/spleen size. Serum samples can be examined for antibodies. (see Section 4 Exercises). Histologi-cal slides, stained appropriately, could be provided to demonstrate the cellular reaction around eggs in the liver. Slides of miracidia and cercariae could also be provided to supplement the material for study.

In the USA, it may be possible to obtain *Schistosoma* life cycle stages (eggs or infected mice) from workers who maintain and work with this genus. It should be possible to trace potential suppliers through recent publications in one of the biological publication databases; the parasite may also be obtained from the Biomedical Research Institute, where various species and strains of *Schistosoma* are maintained (see Appendix 3).

REFERENCES

Jordan, P., Webbe, G. & Sturrock, R. F. (1993). *Human Schistosomiasis.* Wallingford, Oxford CABI.

Rollinson, D. & Simpson, A. J. G. (1987). *The Biology of Schistosomes. From Genes to Latrines.* London, Academic Press.

Smyth, J. D. (1994). *Introduction to Animal Parasitology*, 3rd edn. Cambridge, Cambridge University Press.

Wilson, R. A. & Coulson, P. S. (1986). *Schistosoma mansoni* : dynamics of migration through the vascular system of the mouse. *Parasitology* **92**, 83–100.

1.14 *Hymenolepis diminuta* (Cestoda)

J. M. BEHNKE

Aims and objectives

This exercise is designed to demonstrate:

1. The external morphology of the adult stage of the cestode parasite, *Hymenolepis diminuta*.
2. The morphology of the metacestode stage (cysticercoid) of *H. diminuta*.
3. The distribution of tapeworms in the intestine.

Introduction

Many parasites use more than a single host species during the course of their life cycle. Often the larval stages develop in one host and the adults in another. Thus, the parasite must be transmitted from one host to the next in order to complete its development and the most common relationship between hosts exploited by parasites for transmission, particularly those living in the intestine as adults, is that involved in the predator–prey food chain. For example, some tapeworms develop as larvae in small flour beetles, *Tribolium confusum*, but parasitise rodents such as mice and rats as adults. Rodents are essentially grain feeders, but will consume insects also and therefore by depending on two host species, both of which live in close association with cereals and grain, the tapeworms ensure that their transmission cycle is completed successfully.

This exercise is based on the tapeworm *Hymenolepis diminuta* (Fig. 1.14.1), which develops successfully to the adult stage in the rat. In nature it is a parasite of wild rodents, but some isolates have been maintained in laboratory animals for many decades. Adult worms often achieve a length of over 80 cm. The worms can therefore be dissected from the rat intestine and handled with relative ease to show the outer structure of tapeworms, movement and general body organisation.

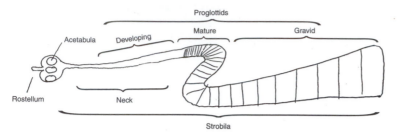

Fig. 1.14.1 Schematic illustration of a tapeworm showing the major parts of the scolex and strobila (not drawn to scale; adult *H. diminuta* recovered from single-worm infections are often about 60 cm in length, occasionally stretching to 80 cm). Note that *H. diminuta* does not have hooks on its rostellum, but the related tapeworm *Rodentolepis* (previously *Hymenolepis*) *microstoma* has a circle of rostellar hooks.

The metacestode stage developing in beetles is called a cysticercoid. The beetles pick up tapeworm eggs whilst feeding on flour, the embryos hatch in the intestine of the insect and penetrate the tissues to reach the haemocoel, where further development to the cysticercoid stage is completed (Fig. 1.14.2). The cysticercoids lie freely in the haemocoel and can be readily observed when the body cavity of the beetle is exposed.

Figure 1.14.3 shows the general shape of the cysticercoid and indicates some of the interesting features. Essentially, it comprises a number of protective layers and a fluid-filled bladder within which the head of the future tapeworm (the scolex) is protected. This is necessary to ensure the survival of the larva in the insect's haemocoel, where otherwise it may be surrounded by host phagocytic cells and eventually destroyed. Protection for the scolex is also necessary to enable it to survive the digestive enzymes in the rodent stomach, when the beetle is consumed.

Laboratory equipment and consumables
(per student or group)

Equipment

Compound microscope
Dissecting microscope
Hand lens

Dissecting dish
50 ml beaker
Dissecting instruments

Consumables

Pasteur pipettes and bulbs
Petri dishes (large and standard)
Slides and coverslips
Pins (for dissection)

Slides and coverslips
Hanks' saline
Paper towel
Lens tissue
Disposable gloves

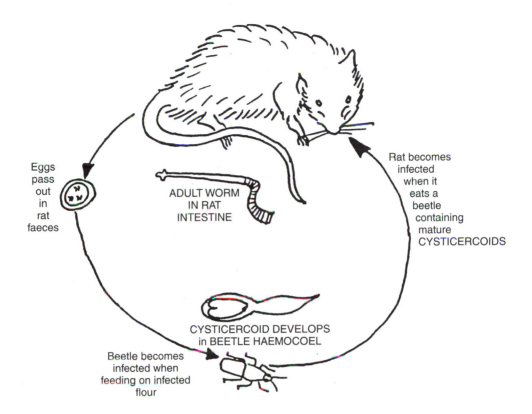

Eggs
pass
out
in
rat
faeces

ADULT WORM
IN RAT
INTESTINE

Rat becomes
infected
when it
eats a
beetle
containing
mature
CYSTICERCOIDS

CYSTICERCOID DEVELOPS
in BEETLE HAEMOCOEL

Beetle becomes
infected when
feeding on infected
flour

Fig. 1.14.2 Schematic illustration of the life cycle of *H. diminuta*. Not drawn to scale.

Sources of parasite material

Adult parasites are obtained from infections of at least 17 days' duration in rats (usually convenient to kill rats 3 weeks after infection). The cysticercoids can be obtained from infected beetles that have fed on gravid proglottids at least 2 weeks earlier and been maintained throughout at 20–25 °C. Beetles can be infected by starving them for 2–5 days in an empty Petri dish, and then placing a small filter paper with a few gravid proglottids into the dish for about a day. After this period, the filter paper is removed and flour added.

Safety

Freshly obtained worms do not have infective stages directly infectious to man, but there are records of *H. diminuta* infections in human beings. The infective stage is the cysticercoid, so care should therefore be taken to ensure that infected beetles and larvae extracted from them are not swallowed during the class.

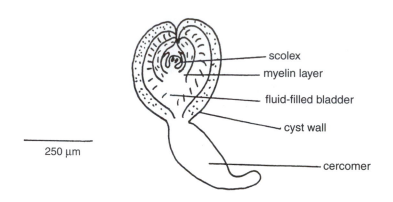

scolex

myelin layer

fluid-filled bladder

cyst wall

cercomer

250 µm

Fig. 1.14.3 Illustration of the metacestode (cysticercoid) of *H. diminuta*.

Appropriate care should also be taken when handling rats and beetles. Some subjects may be allergic to both species.

Instructions for staff

This exercise is best carried out with freshly obtained material. If it is not appropriate for the rats to be killed during the class, they should be killed immediately before the students gather in the laboratory. The small intestine should be removed in its entirety and the worms then extracted. The adult worms may be removed by careful dissection, opening the small intestine from the back end towards the duodenum. This way the gravid proglottids are encountered first and the fine neck region may then be carefully followed to the solex. Alternatively, if the intestine has not been punctured a large syringe (50 ml) may be used to flush the worms out with cold Hanks' saline. The worms usually emerge intact together with gut contents and may then be separated individually and transferred to warm Hanks' in Petri dishes.

Instructions for students

You are provided with fresh adult *H. diminuta*, or rats infected with *H. diminuta* and beetles containing the cysticercoid stages of the parasite.

Removal of adult worms from the intestine of the rat host

1. Place the rat in a dissecting dish, open the abdominal cavity and examine the small intestine *in situ*. Is there any evidence indicating that the animal is infected?

2. Sever the intestine at the pyloric sphincter and at the rectum and remove it from the animal by cutting away the mesenteries. Take care not to damage/puncture the intestine as you cut it away from the body wall.
3. Carefully unravel the intestine. Cut away the caecum and colon and place the small intestine in a large Petri dish containing warm Hanks' saline.
4. Open the intestine from the ileo-caecal junction taking care not to cut the worms. Work your way anteriorly until you have opened the entire length of the small intestine. How many worms have you found? What is their mean length? In which section of the intestine were they found?

External features of the worm

1. Observe the external features of the worm and measure its length. Note that the body is flat and long with a tapering anterior end.
2. Examine the anterior end of the worm. Cut off a section about 5 mm in length from the anterior end and transfer to a glass slide together with a few drops of saline, cover with a glass cover slip and study carefully under the micrsocope. Observe the suckers. How many are there? Is anything else visible? What are the long ducts running down the length of the neck? How many are there? Does the scolex have any hooks? Why?

Removal of cysticeroids from the haemocoel of the beetle host
You are provided with a Petri dish containing beetles which have been infected with *H. diminuta*.

1. Place 2–3 ml (a Pasteur pipette full approximates to 1ml) of Hanks' saline into the base of a solid watch-glass (NOT a Petri dish).
2. Pick up a living beetle, using fine forceps, and transfer it to the watch-glass.
3. Under the dissecting microscope remove the beetle's head and open the haemocoel. The cysticercoids should become apparent immediately and may be inspected under high power. If you are uncertain ask for help.
4. Count the total number of cysticercoids in the beetle.
5. Using a Pasteur pipette, transfer some cysticercoids to a glass slide together with a few drops of saline, cover with a glass

cover slip and study carefully under the microscope. What can you see? Can you see any features of the adult tapeworm in the larval stage?

Ideas for further exploration

- How do tapeworms move?
- What do tapeworms feed on? How do they feed?
- How does fertilisation of the eggs take place?
- What is the function of the scolex?
- How quickly do tapeworms grow?
- How are eggs released from the worm?
- How do the eggs leave the host's body?
- Worm counts and position in gut

Additional information

Another rodent tapeworm that can be used for a similar class is *Rodentolepis microstoma*, which matures in house mice. In this case, care has to be taken not to damage the worms when removing them because the scolex and much of the neck region are deep in the bile duct, often well into the liver.

If sufficient numbers of worms have been examined, quantitative analysis could include frequency distribution plots of the length of worms, and if several rats have been used with different numbers of worms in each, the length of worms could be related to worm burden. *H. diminuta* is very sensitive to density-dependent constraints on growth (crowding effects) (see Exercise 2.4).

Note that armed rodent hymenolepid cestodes have now been consigned to a new genus, *Rodentolepis*. The parasite *Hymenolepis microstoma* is, therefore, currently known as *Rodentolepis microstoma*.

In the USA, cysticercoids of *Hymenolepis diminuta* can be purchased from Carolina Biological Supply Co., Burlington, NC 27215, or from laboratories where the parasite is maintained for teaching and research purposes.

REFERENCES

Arai, H. P. (1980). Biology of the tapeworm *Hymenolepis diminuta*. New York, Academic Press.

Boddington, M. J. & Mettrick, D. F. (1981). Production and reproduction of *Hymenolepis diminuta* (Platyhelminths: Cestoda). *Canadian Journal of Zoology* **59**, 1962–1972.

Goodchild, C. G. & Harrison, D. L. (1961). The growth of the rat tapeworm *Hymenolepis diminuta*, during the first five days in the final host. *Journal of Parasitology* **47**, 819–827.

Hesselberg, C. A. & Andreassen, J. (1975). Some influences of population density on *Hymenolepis diminuta* in rats. *Parasitology* **71**, 517–523.

Read, C. P. (1951). The 'crowding' effect in tapeworm infections. *Journal of Parasitology* **37**, 174–178.

Schmidt, G. D. (1970). *How to Know the Tapeworms*. Iowa, W. M. C. Brown Company.

Smyth, J. D. (1994). *Introduction to Animal Parasitology*, 3rd edn. Cambridge, Cambridge University Press.

1.15 *Heligmosomoides polygyrus* (Nematoda)

J. M. BEHNKE, L. H. CHAPPELL & A. W. PIKE

Aims and objectives

This exercise is designed to demonstrate:

1. The external and internal morphology of the adult stage of the nematode parasite *Heligmosomoides polygyrus*.
2. The mating behaviour of the adult worms.
3. The distribution of worms in the small intestine and the selection of a preferred site.

Introduction

Intestinal nematode parasites are very common in mammalian hosts and are responsible for human disease as well as for losses to the agricultural industries through their effects on domestic animals. *H. polygyrus* (Fig. 1.15.1) is an intestinal nematode parasite of mice that is very easy to maintain in the laboratory and provides convenient material to demonstrate some of the adaptations that have evolved in nematodes for survival in their hosts.

Unfortunately, this parasite has been the subject of a long-standing taxonomic debate as to the most appropriate name for the species. These problems are discussed by Behnke *et al.* (1991) and the reader is referred to this publication for further details. The approach used here is that the alternative name for the species, *Nematospiroides dubius,* is no longer used. *H. polygyrus bakeri* is the strain maintained in domestic/laboratory mice, *Mus musculus*. In Europe, however, wild field or wood mice, *Apodemus sylvaticus,* carry the subspecies called *H.p. polygyrus* and voles, *Clethrionomys glareolus, H. glareoli*. All of these, if available, can provide useful teaching material. This protocol is based on the laboratory passaged subspecies *H. polygyrus bakeri*.

The life cycle of *H. polygyrus* is similar to that of many other trichostrongyloid parasites of mammals, particularly in

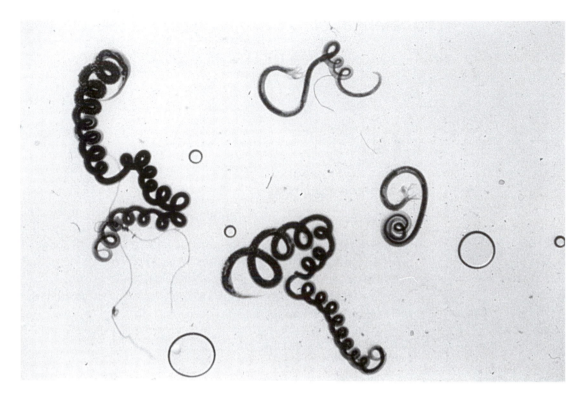

Fig. 1.15.1 Photograph of two adult male (right) and female (left) worms.

relation to the free-living phase of the cycle. The eggs are shed in host faeces and at room temperature take about 36 h to develop into the first larval stage, the L1, which hatch and feed on bacteria. About 28 h later these larvae moult to the second stage, L2, and develop further. The final pre-parasitic moult to the L3 occurs about 20 h later, that is, about 3–4 days after deposition in host faeces. The rate of development is dependent on temperature. The third stage larva is infective to mice and survival in the environment is enhanced by the retention of the second stage cuticle, which forms a sheath around the larva. When swallowed by mice, larvae ex-sheath in the stomach/small intestine and penetrate the intestinal mucosa, usually in the anterior to mid small intestine. They migrate through the muscles, to the outermost layers, where they coil up just under the serosa and begin further development. The moult to the L4 stage occurs 3–4 days after infection and the moult to the pre-adult stage on days 7–8. The pre-adults migrate back into the intestinal lumen and generally accumulate in the anterior of the small intestine, in the duodenal region near the pyloric sphincter. The adult worms begin to copulate on day 9 and eggs generally appear in host faeces by day 10, increasing in numbers until

2–3 weeks after infection. The adult worms are very long-lived and in most commonly employed mouse strains can survive for up to 10 months. Some mouse strains, however, can reject worms as early as 7 weeks after infection.

The present exercise is concerned with the morphology of adult male and female worms and allows an opportunity to explore aspects of the biology of this parasite.

Laboratory equipment and consumables
(per student or group)

Equipment

Compound microscope   Dissecting dish
Dissecting microscope   50 ml beaker
Hand lens   Dissecting instruments

Consumables

Pasteur pipettes and bulbs   Pins (for dissection)
Petri dishes (standard and   Hanks' saline (GIBCO)
   10 small for each student)   Paper towel
Slides and coverslips   Lens tissue
Disposable gloves

Sources of parasite material

The L3 stages can be grown easily in the laboratory on Petri dishes containing a piece of filter paper with a diameter about 1 cm smaller than that of the Petri dish. The faecal slurry is spread evenly but very thinly (overloading the filter paper will reduce the number of larvae developing) in the middle of the filter paper and the Petri dishes kept in a humidity chamber at 20–25 °C. For the chamber, a plastic sandwich-box or cake-box with some wet tissue paper will suffice. The Petri dishes can be stacked on top of the wet paper and the box is then sealed for 1 week. As the L3s develop they migrate from the filter paper to the edges of the Petri dish and they can be collected easily by introducing a few drops of water with a Pasteur pipette, washing these along the edge so as to maintain a distance from the filter paper in order to avoid faecal contamination, and then transferring the fluid to another container. The volume can then be adjusted to give the required number of larvae/ml. Infective larvae can be stored in water at 4 °C for many months without loss of infectivity. The larval culture can be used immediately on removal from the refrigerator for 6 months or so.

Mice are given 100–200 L3 larvae orally, usually by using a blunted syringe needle (21G) and a 1 ml syringe. The volume of fluid containing the larvae should not exceed 0.2 ml otherwise the mice may regurgitate some of the material. Adult parasites are obtained from infections of at least 10 days duration in mice, but preferably 14–21 days. If it is intended to demonstrate 4th stage larvae *in situ*, the mice should be killed 6 days after infection. The parasitic L3 stages are too small to demonstrate easily during a practical and it is hardly worth while attempting to do so. In order to maintain the parasite in the laboratory, fresh batches of mice should be infected every 3 months or so depending on the strain of mouse employed.

Safety

As far as is known, *H. polygyrus* is not directly infectious to humans, although appropriate care should be taken when handling mice, adult worms and faecal culture plates. Some subjects may be allergic to mice and the parasites. Environmental pathogens associated with animals and animal faeces need to be considered as potential contaminants with some associated risk factor to human health. If wild animals are used as a source of parasite material, suitable precautions will need to be taken to ensure that students are not exposed to other rodent-associated pathogens.

Instructions for staff

This exercise is best carried out with freshly obtained material. If it is not appropriate for the mice to be killed during the class, they should be killed immediately before the students gather in the laboratory. The small intestine should be removed in its entirety, opened and incubated in Hanks' saline (or some other suitable physiological saline) for a few minutes. The worms are located entwined among the villi and will soon detach from the mucosa and begin to move around in the saline.

Instructions for students

You are provided with fresh adult *H. polygyrus* or mice infected with *H. polygyrus*.

Removal of adult worms from the intestine of the mouse host

1. Place the mouse in a dissecting dish, open the abdominal cavity and examine the small intestine *in situ*. Is there any evidence indicating that the animal is infected?
2. Sever the intestine at the pyloric sphincter and at the rectum and remove it from the animal by cutting away the mesenteries. Take care not to damage/puncture the intestine as you cut it away from the body wall.
3. Carefully unravel the intestine.
4. Cut away the caecum and colon and place the small intestine in a large Petri dish containing warm Hanks' saline.
5. Cut the intestine into 10 sections of equal length.
6. Open each section in turn in a separate Petri dish in warm Hanks' saline, taking care not to cut the worms; label each dish so that you can recall which section of the small intestine it represents. Work your way through all the 10 sections until each has been opened up.
7. Record the number of worms in each section and also note the number *in copulo*. How many worms have you found? What is their mean length? In which section of the intestine were the majority found?
8. Construct a bar chart showing the number of male and female worms in each of the ten sections of the small intestine.

External features of the worm

1. Observe the worms in the Petri dishes. Are they active? Note that the body is cylindrical and that the sexes appear quite different. The small thin worms are males, the larger coiled specimens females (Fig. 1.15.1). Why are the worms red in colour?
2. Place a male and a female worm onto a glass slide in a drop of saline, cover with a coverslip and examine under the microscope.
3. Draw a male and a female worm, identifying all the features, pointing out those which distinguish the sexes. Note in particular the following: the mouth and pharynx, intestine, rectum, anus, uterus, ovary, oviduct, testis, copulatory bursa, spicules (Fig. 1.15.2).
4. Select a pair of worms *in copulo*. Observe them under the microscope, noting how the two are held together.
   How many eggs does the female worm have in her uterus?

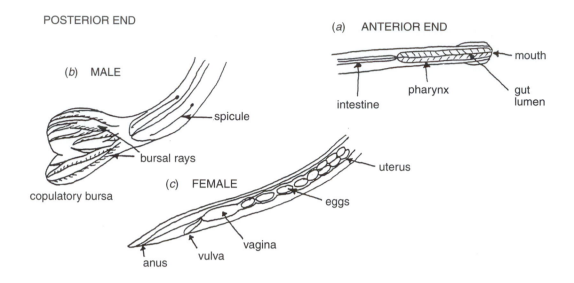

What stage of development have the eggs reached in different regions of the uterus? Use the diagram above to understand all the different features of the male's copulatory bursa Fig. 1.15.3).

Fig. 1.15.2 Schematic illustration of (a) anterior end of *H. polygyrus*; (b) posterior end of a male worm; (c) posterior end of a female worm. Drawn only approximately to scale. Compare with Fig. 1.15.1.

Ideas for further exploration

- How do nematodes move?
- What does *H. polygyrus* feed on? Do they have an intestine?
- Why are the worms red in colour?
- Why are the worms coiled?
- What was the ratio of male to female worms? If not 1:1, why?
- In which sections of the intestine were the worms located and why?
- How do the worms locate their optimal/preferred site?
- What are the spicules of the male used for?
- How many developmental stages can be found in eggs within the uterus?
- Into how many cells has the ovum divided when the eggs are released by the worm?

Additional information

As the adult worms emerge from their sites, coiled around the intestinal villi, they begin to coil around each other in the saline and quite large knots of worms can soon build up. These may be difficult to unravel but can be pulled apart more easily if

Fig. 1.15.3 Schematic diagram of the arrangement of the rays in the copulatory bursa of a male *Heligosomoides* sp. nematode, with associated nomenclature. Drawn only approximately to scale. Compare with Fig. 1.15.1.

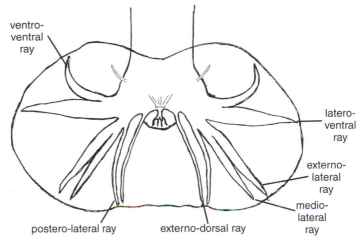

ventro-ventral ray

latero-ventral ray

externo-lateral ray

medio-lateral ray

postero-lateral ray   externo-dorsal ray

Ventral view of male bursa

the worms are fixed first by the addition of formalin or 70% alcohol.

The class can be expanded to demonstrate all of the key stages in the life cycle of this parasite. Students could be provided with Petri dishes incubated for varying lengths of time to coincide with the L1, L2 and L3 stages. The first two stages can be found by taking a small amount of faecal slurry from the plate and transferring this onto a microscope slide in a drop of water. The larvae could be measured and their internal features described. The L3s are best examined initially around the edges of the filter paper where they often stand upright attached by their posterior ends. The larvae move from side to side when stimulated by air movement or warmth, a behaviour termed 'nictation'. After these initial observations some larvae could be placed onto a slide and examined under the microscope to show the sheath surrounding the L3 cuticle. Exsheathment is also easily induced by adding a few drops of dilute bleach to a test-tube containing L3s in water. The parasitic L4 stages are best observed on small sections of the gut, taken from mice carrying a 6 day infection, squashed between two microscope slides

Another rodent species that can be used for a similar class is *Nippostrongylus brasilienis*. However, this species is more expensive to maintain because it is best grown in rats and it has to be passaged more regularly because the adult worms are expelled 2–3 weeks after infection by an intestinal immune response.

The practical with *H. polygyrus* could be expanded in a number of different directions. Infection with this species causes major changes in blood leukocytes with time post-infection. There are effects on the spleen and mesenteric lymph nodes, which could be quantified (See Section 4).

Both *H. polygyrus* and *N. brasiliensis* in their rodent hosts are popular laboratory model systems and the parasites are maintained routinely for research and teaching purposes in many different laboratories in the USA, in Europe and throughout the world. Several pharmaceutical companies also maintain them for use in anthelmintic screens. If you do not have ready access to these parasites, the best course of action would be to conduct a literature search and write to laboratories that have published recent research on them. The larvae of both species travel well, and can be dispatched in sealed Universal tubes (plastic or glass, but pack well!) in water (*H. polygyrus*) or in Petri dishes attached to filter paper (*N. brasiliensis*). Great care should be taken with the latter species because the L3 are skin-penetrating stages and may cause irritation to handlers if the containers break open in transit and the larvae escape. In the USA, the infective larvae of *N. brasiliensis* can be purchased from Wards of Rochester, New York.

REFERENCES

Bansemir, A. D. & Sukhdeo, M. V. K. (1994). The food resource of adult *Heligmosomoides polygyrus* in the small intestine. *Journal of Parasitology* **80**, 24–28.

Bansemir, A. D. & Sukhdeo, M. V. K. (1996). Villus length influences habitat selection by *Heligmosomoides polygyrus*. *Parasitology* **13**, 311–316.

Becket, R. & Pike, A. W. (1980). Mating activity of *Nematospiroides dubius* in laboratory mice. *Journal of Helminthology* **54**, 87–91.

Behnke, J. M., Keymer, A. E. & Lewis, J. W. (1991). *Heligmosomoides polygyrus* or *Nematospiroides dubius*? *Parasitology Today* **7**, 177–179.

Bryant, V. (1973). The life cycle of *Nematospiroides dubius*, Baylis, 1926 (Nematoda: Heligmosomidae). *Journal of Helminthology* **47**, 263–268.

Lewis, J. & Bryant, V. (1976). The distribution of *Nematospiroides dubius* within the small intestine of laboratory mice. *Journal of Helminthology* **50**, 163–171.

Sukhdeo, M. V. K., O'Grady, R. T. & Hsu, S. C. (1984). The site selected by the larvae of *Heligmosomoides polygyrus*. *Journal of Helminthology* **58**, 19–23.

Section 2
Ecology

2.1 Pinworms (Nematoda, Oxyuroidea) in the American cockroach, *Periplaneta americana*

W. M. HOMINICK & J. M. BEHNKE

Aims and objectives

This exercise is designed to demonstrate:

1. The general morphology of oxyuroid nematodes of the family Thelastomatidae.
2. Site specificity and niche segregation within the host.
3. Relationships between parasite species and parasite numbers (intensity of infection), and the stage of development and sex of the host.
4. The use of appropriate statistical analysis to test hypotheses arising from your observations.

Introduction

Like other organisms, parasites are faced with both inter- and intra-specific competition for resources, and both food and space within the host are potentially limiting. During the course of evolution, competition may lead to niche segregation and specialisation in the exploitation of particular resources, in order to minimise competitive interactions. The distribution of two related nematode parasites of the cockroach, namely *Leidynema appendiculata* and *Hammerschmidthiella diesingi,* both in respect of their location within the host and among juvenile and adult hosts of both sexes, suggests that niche segregation has occurred.

Laboratory equipment and consumables
(per student or group)

Equipment

Dissecting microscope
Compound microscope
Objective lenses $\times 40$, $\times 100$ (oil)

High intensity lamp for microscope
Dissecting tray with a base for pins
Dissecting instruments

Consumables

Insect Ringer*	Immersion oil
Ordinary fine thread	Glass sides and coverslips
Several glass Petri dishes	Tea strainer**
Solid watch-glasses	Bucket of diluted bleach for
Pasteur pipettes and bulbs	glassware
A 250 ml glass or plastic beaker	Animal waste bin

*For example, 6.5 g NaCl, 0.14 g KCl, 0.12 g $CaCl_2$, 0.1 g $NaHCO_3$, 0.01 g Na_2HPO_4, per litre distilled water.

**To strain animal waste from fluid for separate disposal.

Sources of parasite material

A breeding colony of cockroaches is required since most cockroach colonies will be naturally infected with both pinworms. Transmission is direct via embryonated eggs, which are then eaten by cockroach nymphs and adults when feeding. See also Additional information in Exercise 1.8.

Safety

There are few health hazards associated with this exercise providing normal laboratory practice is followed. As far as we are aware neither the parasites in question nor any others carried by laboratory maintained cockroaches are likely to cause medical/health problems in humans. The only aspect worth pointing out is that some students may be allergic to insects and appropriate precautions should be taken for such persons. Staff should also take care when handling chloroform.

Instructions for staff

A male, female and nymph cockroach should be available for each participating student and these can be killed a few minutes before the start of class by putting them into a chloroform jar or one containing ethyl acetate. The jar should contain tissues or cotton wool soaked in chloroform or ethyl acetate in the base and a grid separating the anaesthetic from the cockroaches so that no liquid makes contact with the insects.

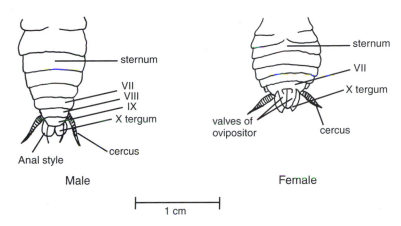

Fig. 2.1.1 Drawings of the ventral view of the abdomen of a male and female cockroach to show features used in distinguishing between the sexes. Note the anal style on males.

Instructions for students

You are provided with freshly killed cockroaches. These will be found in three jars, containing nymphs, adult females or adult males.

1. Working individually, obtain one specimen from each jar. Confirm sex of adults and nymphs. Fig. 2.1.1 illustrates the ventral view of adult male and female cockroaches.
2. Cut off the legs and pin the cockroach ventral side downwards in a dissecting dish full of saline, using pins through the outstretched wings and through the thorax.
3. Gently push the point of a pair of fine scissors beneath the posterior edge of the penultimate abdominal tergum near the side of the body, not in the mid-dorsal line, and cut the tergum forwards, being careful to avoid damaging anything beneath. Continue to cut forward to the posterior margin of the thorax and then proceed in the same way up the opposite side.
4. Using fine forceps, lift the terga and remove them.
5. Remove muscle and fat body to expose the alimentary canal. Before moving it, note the crop and proventriculus in the anterior of the abdomen. At the posterior end, in the region of the 7th abdominal segment and behind a constriction in the gut, note the rectum (see Fig. 2.1.2).
6. Observe carefully any peristaltic movements displayed by the various parts of the digestive tract.
7. Having identified the various parts of the gut, free the whole intermediate part from its attachments – it is largely held in position by tracheae – and spread it out to one side

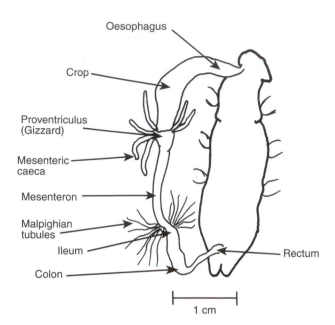

Oesophagus

Crop

Proventriculus
(Gizzard)

Mesenteric
caeca

Mesenteron

Malpighian
tubules

Ileum

Colon

Rectum

1 cm

Fig. 2.1.2 Drawing of the unravelled intestinal tract, to show the different sections.

without causing damage. Note the mesenteron or mid-gut, a comparatively narrow, whitish tube, bearing on its anterior end 7 or 8 mesenteric caeca. At the posterior end of the mid-gut there is a thickening, the pyloric ring, from which numerous fine tubes, the Malpighian tubules, arise. This ring is the boundary line between the mid-gut and the proctodeum or hind gut. The hind gut consists of a very short ileum, a long, usually dark-coloured colon, and the rectum.

8. Cut off the head and gently pull on the crop. The anterior part of the alimentary canal should now be free of the carcass.

9. Cut the posterior part of the rectum and remove the entire digestive tract intact. Stretch it out and pin it through the proventriculus and the rectum.

10. Using two pieces of thread, tie off the colon into thirds.

11. Carefully dissect each third of the colon and note the numbers, species, stage and sex of any nematodes present.

12. Examine the rectum and the ileum and posterior mid-gut.

13. Draw the adults of each species of nematode that you find. Pay particular attention to their stomas and buccal cavities, using oil immersion.

14. At the end of the exercise you will have dissected at least an adult male, adult female and one of the nymphs. When you

dissect the nymphs, measure the length of the tibia of the hind-most leg (i.e. the metatibia). This will provide a measure of the instar of the animal (see Fig. 2.1.3c). Also note its sex (the posterior sternites of the two sexes differ) (see Fig. 2.1.3a).

15. Add your results to those of the rest of the class, analyse the entire data-set and prepare a report of your work.

Ideas for further exploration

- Calculate the mean and standard deviation for the total numbers (i.e. disregard species) of adult female nematodes present in each stage of the host (male, female and nymph). Are there any significant differences in total parasite burden between the stages and sexes of the host? Plot total worm burden against length of the tibia (as an indicator of age). Are parasites more abundant in nymphs or adults?

- Plot separately a frequency distribution of the total worm burden in nymphs, male and female cockroaches. Which stage harbours the heaviest infections?

- Now do the same for each species in turn. What features distinguish the species? Does any species predominate in any particular stage or sex? Test your hypothesis by appropriate statistical analysis.

- Calculate the percentage of cockroaches of each stage harbouring one species of nematode, then the other species, then both species (i.e. single species and concurrent infections). Do any of these data suggest that one or other of the species of nematodes predominates in any particular stage of the host? Test your hypothesis by appropriate statistical analysis.

- Analyse the location of the parasites in the host, using the class results. Does the distribution of the two species differ? Is there any evidence for habitat specificity or is the colon uniformly occupied by the pinworms?

- Do you think that the nematodes are competing or co-existing? Discuss the results, referring to the references provided.

Brief details of expected results

Preparations of the two species, on glass slides in saline and covered by a cover slip, reveal the differences in internal morphology (Fig. 2.1.3a,b) . The shape of the oesophagus is quite different and *L. appendiculata* has a gut diverticulum, which is

(a)

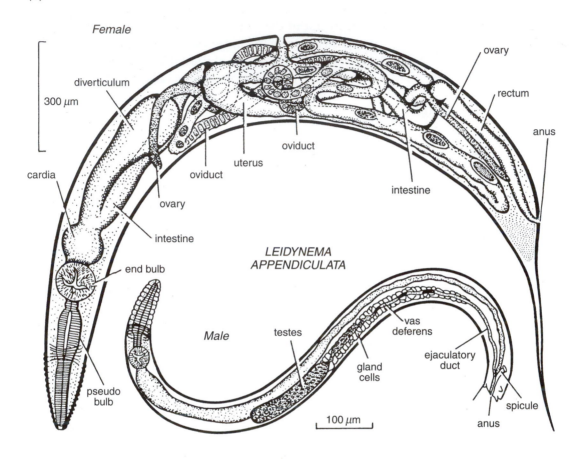

Female

300 μm

diverticulum

cardia

end bulb

pseudo
bulb

ovary

intestine

oviduct

ovary

oviduct

uterus

*LEIDYNEMA
APPENDICULATA*

Male

testes

gland
cells

vas
deferens

ovary

rectum

anus

intestine

ejaculatory
duct

spicule

anus

100 μm

(b)

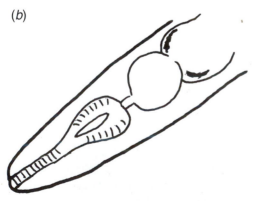

Anterior of *Hammerschmidthiella diesingi.*
Note oval, vase-shaped pseudo bulb

(c)

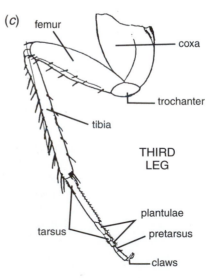

femur

coxa

trochanter

tibia

THIRD
LEG

plantulae

pretarsus

tarsus

claws

Fig. 2.1.3 (opposite) (a) *Leidynema appendiculata* male and female. (b). Anterior end of *Hammerschmidthiella diesingi* to show the different shape of the pseudobulb. Drawn to approximately the same scale. (c). Drawing of the third leg of a cockroach to indicate the different parts and in particular the tibia, which can be used to estimate the instar of the animal.

rare in nematodes. The shape of the eggs is also different. Whilst pointing these differences out, the instructor should also interest the students in the general organisation of the nematodes as a taxon of animals. The internal morphology of these nematodes is particularly clear and usually the excretory pore and the H-shaped excretory system are visible. Adult female *L. appendiculata* are rare in nymphs whereas *H. diesingi* adults are very common (ratio of about 1/18). *H. diesingi* worm burdens increase from the 7th instar, with maximum worm burdens in the 9th and a reduction in adults. Both are common in adults, although worm burdens of *L. appendiculata* are often heavier, particularly in female hosts. Thus adult female cockroaches generally carry heavier worm burdens of *L. appendiculata* compared with males. In contrast, male cockroaches may have heavier infections with *H. diesingi*, particularly in the 8th instar nymphs. The prevalence of parasites and the intensity of infection are higher in adult female hosts compared with adult males. *L. appendiculata* favours a more anterior position and is more restricted in its distribution in the colon. *L. appendiculata* feeds on large food particles and is a lumen dweller. *H. diesingi* selects fine particulate material and its mouth parts are more intimately associated with the gut mucosa.

Information on similar exercises

A similar class exercise could be carried out on laboratory or house mice infected with pinworms. The mice would have to be conventionally maintained animals from suppliers who do not take precautions to eliminate naturally transmitted helminth parasites, or wild caught animals. House mice are parasitised by two species of pinworms, *Syphacia obvelata* and *Aspiculuris tetraptera*. *S. obvelata* lives mainly in the caecum although the adult females migrate through the colon to lay their eggs externally; *A. tetraptera* lives in the anterior third of the colon although larvae up to 7 days old live in the mid colon.

REFERENCES

Adamson, M. L. & Noble, S. (1992). Structure of the pinworm (Oxyurida: Nematoda) guild in the hindgut of the American cockroach, *Periplaneta americana. Parasitology* **104**, 497–507.

Behnke, J. M. (1974). The distribution of larval *Aspiculuris tetraptera* Schulz during primary infection in *Mus musculus, Rattus norvegicus* and *Apodemus sylvaticus. Parasitology* **69**, 391–402.

Hominick, W. M. & Davey, K. G. (1972). The influence of host stage and sex upon the size and composition of the population of two species of thelastomatids parasitic in the hindgut of *Periplaneta americana*. *Canadian Journal of Zoology* **50**, 947–954.

Hominick, W. M. & Davey, K. G. (1973). Food and the spatial distribution of adult female pinworms parasitic in the hindgut of *Periplaneta americana*. *International Journal for Parasitology* **3**, 759–771.

2.2 Distribution and microhabitat of a monogenean on the gills of mackerel

C. R. KENNEDY

Aims and objectives

This exercise is designed to demonstrate:

1. Site selection by a parasite.
2. Niche segregation by a parasite in the absence of competitors.
3. Relationships between parasite numbers, site selection and host size.
4. The manner in which parasites are dispersed throughout a host population.

Introduction

Although we often talk loosely about parasites living in or on a host, this should not be taken to imply that they can live in all tissues or organs of a host. Indeed, all parasites show some degree of site specificity in that they are only found in a particular organ or tissue and are able to select that site. Many of the adaptations they exhibit, e.g. an attachment mechanism, are actually adaptations to this particular site rather than to the host as such.

Interpretations of site selection within an ecological framework are not generally agreed. There is a large measure of agreement that the site of a parasite can be used as an indication or measure of its niche, but far less agreement about the ecological significance of, especially, very specific site selection (i.e., narrow niche breadth) and its relationship to resource partitioning and intra- and inter-specific competition between parasites. Many species appear to increase their niche breadth at higher densities and this is often taken as a manifestation of intra-specific competition. Similarly, species packing within an organ and a narrower niche breadth is considered by many workers to be a likely indication of inter-specific competition. Attempts to explain narrow niche segregation in the absence of any other species is problematic: this has been explained as (1) independent adaptation by a species; (2) a consequence of inter-specific competition in evolutionary time (i.e., the ghost of competition past) and (3) an adaptation to increase the probability of mating or cross-fertilisation.

Monogeneans living on the gills of fish often show very specific microhabitat preferences to the level of a particular gill arch, hemibranch or even group of filaments and position on a filament, whether tip or base. Their site preference may relate to the number of parasites on the individual fish and/or to the size of the fish. In the practical described here, you will look at the distribution and site selection of *Kuhnia scombri* on the gills of the common mackerel *Scomber scombrus*, and then later relate your results to competition.

Laboratory equipment and consumables
(per student or group)

Equipment

Dissecting microscope with zoom or changeable objectives.
Dissection board and instruments.

Consumables

8 Petri dishes and a supply of seawater
(If it is desired to examine the parasites for any other reason, then a compound microscope, slides and coverslips, etc. will be required).

Sources of material

All that is required is a supply of mackerel. These may be fresh, but are preferable if frozen as soon as possible after capture. Fish obtained from a fishmonger/supermarket can be used as long as they are 'fresh' and not a couple of days old and thawed out on the slab. Look for good, non-sunken eyes and other signs of freshness. Fish may be of the same or different sizes. See also Additional information in Exercise 1.3.

Safety

There are no health or safety hazards associated with this exercise provided good laboratory practice is followed. The only dangers are of students cutting themselves on mackerel teeth or with scalpels/scissors.

Instructions for staff

One mackerel per student should suffice. Fish should be thawed out just before the class starts. Depending on your students, you can either give them a whole fish each or just a head (remove just behind the operculum). If you give them only a head, make sure the length of the fish is determined first and stapled to it. If the class is very large, or the number of mackerel low or funds very low, the practical can be adjusted such that each student is only given a pair of gill arches, e.g. Arch I left and right sides: in this case put the arches in a plastic bag labelled I and with the length of the fish. The full procedure will be described below, although parts can be omitted where appropriate.

Instructions for students

You are provided with a whole mackerel.

1. Measure its length from the tip of the snout to the fork in the tail (= fork length).
2. Cut off the head immediately behind the operculum.
3. Remove the operculum from each side of the fish to expose the opercular cavity.
4. Cut forward from the rear of the head towards the mouth on the ventral side in the central line. If done cleanly, this will separate the base of the gill arches ventrally on each side. The arches are conventionally numbered I–IV starting from the mouth (ignore the reduced posterior arch V).
5. Remove each arch separately and transfer it to a Petri dish; cover with water.
6. Label each dish with the number of the arch and the side of the fish (left or right).
7. Examine each arch in turn under a dissecting microscope. Remember parasites may be hidden between the filaments, which can be 'turned over' with a seeker.
8. Count the number of parasites on each arch, making sure you also record zeros.
9. If so instructed, transfer the parasites to watch-glasses of water or fixative for further processing and/or examination.

Data recording and handling

Record your results so that you have a note of the number of parasites per arch (left and right being treated separately); the number of parasites per arch (left and right arches combined); the total number of parasites per fish (left and right sides separately and then combined); and of course the size of the fish

In most samples of mackerel you will only find *Kuhnia scomberi*; in some you may also find *Grubea cochlea*, easily identified as being much larger than *Kuhnia*. When present, record exactly the same data for this species.

Results are best collated on a class basis, with all individual results passed to a demonstrator. It is then possible to:

1. Determine the mean (and variance) number of parasites per fish on the left and right sides, and test (ANOVA or '*t*') if these differ significantly.
2. If they differ, proceed with the following steps keeping left and right data separate. If they do not differ, combine all data for both sides of the fish.
3. Determine the mean and variance (or just total) of number of parasites per arch and display as a histogram. If desired, test differences in mean for significance (ANOVA or as seems appropriate).
4. Divide the data set into arbitrary groups of, for example, light, medium and heavy infections per fish. Determine the mean and variance of number of parasites per arch for each group of fish and display the data as a histogram.
5. Examine the relationship between number of parasites per fish and fish length (only feasible if there is reasonable variation in fish length in the sample). Determine the correlation coefficient for this relationship.
6. Determine the mean number of parasites per fish (abundance, so include the zeros) and its variance, and so the ratio of variance to mean.
7. If *Grubea* is also present, treat the data on this independently but in exactly the same way as above.

Each student can then prepare their own report, comparing their own data-set with that of the combined data-set for the class.

Expected results

1. There should be no difference between left and right side of the fish.

2. If there is a reasonable variation in fish size, abundance should increase with host size.
3. Parasites should show a preference for arches I, II, III and IV in that order.
4. The variance to mean ratio should exceed 1, indicating an aggregated distribution.
5. Use of arches III and IV should increase at very high densities only.
6. *Grubea* may show a different distribution if its density is sufficiently high.

Ideas for further exploration

- Do you think the site preference results from the infection process, i.e. water current patterns, differential mortality on arches I and IV or parasite choice? How would you test each possibility?
- Why should abundance increase in larger fish?
- Why should niche breadth widen at high densities and how might this relate to intra-specific competition?
- Why should a parasite exhibit niche segregation in the absence (or presence) of another species?
- Do your results favour adaptation, inter-specific competition or probability of mating as explanations?
- What does the aggregated distribution imply about the probability of inter-specific competition and of finding a mate?

Similar exercises

This exercise can be extended if desired by studying and mapping parasite distribution on a finer scale, e.g. by recording presence on each hemibranch, on anterior, middle and posterior part of each arch or on basal or distal parts of a filament. It can equally be shortened by ignoring right and left sides or by selecting fish of similar size.

Suggested alternatives if mackerel are unavailable are:

- *Salmo trutta* (brown trout): if you have access to a lake or reservoir population with infections of *Discocotyle sagittata* on the gills.
- *Anguilla anguilla* (eel): if you have access to almost any eel population, you can use eels and *Pseudodactylogyrus* as an alternative. This is more difficult as the dactylogyrids are smaller and

harder to detect. However, there are normally two species present, *P.anguilla* and *P.bini*, and they show differences in site location. They need to be distinguished under a compound microscope and even then are hard to separate: so they can be lumped together as one species if desired.

- Sprats: a simpler version of the exercise can be conducted on sprats and larval anisakines. Sprats can be bought from any local supermarket (do *not* reveal its identity to students!), and the encysted nematodes found in different organs of the body cavity. In data handling, treat each organ as equivalent to a gill arch in the present exercise. If you transfer the nematodes to warm Ringer it can usually be seen that they are still alive, but care must be taken as they can infect humans (suitable discussion points).

REFERENCES

Buchmann, K. (1988). Spatial distribution of *Pseudodactylogyrus anguillae* and *P. bini* (Monogenea) on the gills of the European eel, *Anguilla anguilla*. *Journal of Fish Biology* **32**, 801–802.

Llewellyn, J. (1956). The host-specificity, micro-ecology, adhesive attitudes and comparative morphology of some trematode gill parasites. *Journal of the Marine Biological Association of the United Kingdom* **35**, 113–127.

Rohde, K. (1991). Intra- and inter-specific interactions in low density populations in resource-rich habitats. *Oikos* **60**, 91–104.

Rohde, K. & Watson, N. (1985). Morphology, microhabitats and geographical variation of *Kuhnia* spp. (Monogenea: Polyopisthocotylea). *International Journal of Parasitology* **15**, 569–586.

2.3 Population dynamics of *Gyrodactylus* on stickleback

R. C. TINSLEY

Aims and objectives

This exercise is designed to demonstrate:

1. The dynamic interactions occurring between host and parasite under the influence of external environmental conditions.
2. The balance that exists in the host–parasite relationship between the potential for parasite-induced pathogenic damage, including host death, and the capacity of the host response to regulate infection levels.
3. The value of numerical data with which to develop an understanding of the distribution of parasite populations within host populations (and the significance for transmission, pathology, etc).
4. Adaptation to parasitism, the skills and patience required for manipulation, and the rewards of microscopic examination.

Introduction

Assessment of infection levels in the host (and host population) provides important information on the success of the parasite life cycle, the severity of pathogenic effects on the host, and the effectiveness of host immunity. Combined with information on parasite age and developmental rate, infection levels may also enable reconstruction of the recent history of recruitment of parasites into the individual host and population: current infection levels reflect successful parasite invasion moderated by the host defensive response and other influences such as competition. Measurement of infection levels, however, represents only a 'snapshot' of the dynamic processes that may affect the host–parasite interaction. The illustration of dramatic fluctuations in parasite numbers within individual hosts can be observed in microparasite infections multiplying and being **147**

moderated *in situ*, as in the cycles of parasitaemia of trypanosomes. However, for most macroparasites, changes in infection
levels normally take place over much longer time scales. These
are influenced by variations in external invasion, which are in
turn regulated by host behaviour (changes in diet, population
density, intraspecific interactions including sex and breeding),
external environmental factors (including seasonal changes in
temperature and water availability), etc. Invasion levels may subsequently be moderated by immune defences, which are in turn
influenced by host physiological state including reproduction
and nutrition, host age and experience of infection, and environmental temperature, especially in the case of ectothermic
hosts. It is clear from this complexity that the 'snapshot'
showing numbers of parasites per host is a product of many
interacting variables.

The monogenean genus *Gyrodactylus* provides a system that is
highly amenable for demonstrating the dynamic interactions
that determine infection levels at any point in time. The genus
comprises about 400 species, which are primarily ectoparasites
of fish. Although a platyhelminth, *Gyrodactylus* behaves like a
microparasite in reproducing *in situ* on the host, with the potential for rapid, massive population growth. *Gyrodactylus* provides
a good system for student practical work because the worms are
relatively easily observed and counted on their hosts, which are
readily available. Hosts include sticklebacks, which can be collected in local ponds and streams, and guppies, which can be
purchased in aquarium pet shops. Dramatic changes in infection levels can be observed over the course of a few weeks. Parts
of a succession of practical classes can be used for this purpose,
or a staggered sequence of experimental infections can be made
in the weeks preceding a single practical session, during which
students reconstruct the time series of changes.

It is well established in textbooks of general parasitology that
the reproduction of *Gyrodactylus* is unique within the Animal
Kingdom (Kearn, 1998 provides an excellent account). The parasites lack a swimming infective stage typical of the Monogenea
and instead give birth to fully grown offspring. Remarkably,
each already contains a further developing embryo, this in turn
will develop another, and so on. The series of 'Russian dolls' provides the capacity for exponential population increase with
each offspring released alongside the parent attached on the
skin or gills of the fish host (see review by Harris, 1993). Transfer
to new hosts relies on contact. The parasites feed on epithelial

cells and in very heavy infections cause irritation and damage that has the potential to kill the host. This pathogenic potential is illustrated by an epidemic of *G. salaris* that has devastated wild Norwegian salmon stocks over the past 15 years (reviewed by Mo, 1994). Infected fish can mount a defensive reaction to eliminate infrapopulations of *Gyrodactylus*. As a consequence of these features, the population biology of *Gyrodactylus* is more akin to that of microparasites such as bacteria and protozoans than to other helminths, and this has led to its extensive use as a model in experimental epidemiology (see, for instance, Scott, 1985).

As an ectoparasite of fish, *Gyrodactylus* is directly exposed to changes in the external environment. The rate of production of offspring is temperature dependent and therefore faster population growth would be predicted at higher temperatures. However, the host is also affected by the same factors; in particular, the development of an immune response in fish is temperature dependent. The interaction between these two processes provides the main theme for this exercise. Circumstances under which the parasites multiply fastest can facilitate their most effective rejection by the host.

In vitro studies indicate that complement (Buchmann, 1998; Harris *et al.*, 1998), together with interleukin and reactive oxygen metabolites (Buchmann & Bresciani, 1999), are involved in the host immune response. Following infection, the fish has a period of temporary resistance to re-infection. Parasites detached from the skin by the host response may not be killed; simple laboratory experiments show they can survive attached to the substrate for at least 24 h (depending on temperature) and if they contact a susceptible host during this time they may re-establish.

Laboratory equipment and consumables
(per student or small group)

Equipment

Dissecting microscope with zoom or changeable objectives
High intensity light source (fibre optic or other above-stage
    oblique lighting)
Compound microscope
Basic dissecting kit (including small scissors, fine forceps,
    mounted needles)
Aquarium tanks (approx. 12 l capacity for each group of 15 fish)
Facilities to maintain fish at constant temperatures (in this
    exercise: 5, 10, 15, 20 °C)

Consumables

Slides, coverslips
Petri dishes
Supply of dechlorinated water (tap water left to stand for 24 h)
Pasteur pipette (for transferring drops of water), filter or
   blotting paper

Sources of material

This exercise will refer to three-spined sticklebacks, *Gasterosteus aculeatus,* freshly caught from local populations, but the study could also be carried out on guppies, *Poecilia reticulata*, obtained from aquarium suppliers or other small fish if these are infected with *Gyrodactylus* spp. Fish should be separated into small groups (in this exercise 15 fish per aquarium tank) as soon as possible after collection to restrict parasite transfer to within each subset of the overall host population sample. Three-spined sticklebacks are found throughout northern USA, and can be caught locally in shallow waters with a standard hand-held net. It may be possible to trace potential suppliers through recent publications in one of the biological publications databases or through the Internet.

Safety

If good laboratory practice is observed, there are no special health or safety hazards.

Instructions for staff

Under the UK Animals Act 1986, some of the procedures that follow, including manipulation of infections on wild caught fish and infection of fish by transfer of parasites, can only be carried out if an appropriate Home Office project licence is held by the staff/institute. Those carrying out the procedure will require a personal licence. Sticklebacks can be maintained in good condition with 2–3 feeds/week of bloodworm, with accompanying water changes, and a photoperiod equivalent to that occurring naturally at the time of collection.
Experimental infections can be established in two ways:

1. To obtain a uniform starting point, all *Gyrodactylus* (and other ectoparasites) can be removed from a wild-caught sample of

sticklebacks (or from infected 'pet shop' guppies) by immersion of the fish for 3 h in very dilute formalin (1 : 8000), repeated 2 or 3 times. The fish should then be left for about a week to allow recovery of the epidermis. To establish experimental infections, parasites can be removed from a freshly killed host and transfered to a recipient fish anaesthetised briefly with 0.2% tricaine methane sulphonate (MS222). Individual worms can be induced to attach to the tip of a mounted needle and the transfer observed under a binocular microscope. It is best to establish a two-worm infection initially in case one worm should fail to establish or reproduce. The infected host can be introduced to an aquarium tank containing uninfected fish (15 fish per 12 l tank provides conditions for good population growth over a 1–3 week experimental period).

2. Random samples of sticklebacks can be taken from a large freshly caught field collection and made up into 'populations' of 15 fish per aquarium. These are maintained without further manipulation under defined environmental conditions.

The first approach involves more complicated preparation, including disinfestation and particularly the manipulation of parasites to create experimental infections. However, starting with a known 'inoculum' of two parasites provides more exact data on the course of parasite population growth. A Home Office licence (and designated premises) is necessary to carry out the procedures. The second approach can be set up without advance preparation, but the course of population growth must be interpreted by comparison of infection levels against a 'baseline' of the natural infection levels established by dissection of an initial freshly caught host sample. This lacks the precision of the experimental infections but avoids the need for a Home Office licence. (Investigators are advised to consult their institution's Home Office Inspector for interpretation of the need for a licence for subsequent laboratory maintenance. Strictly, the wild-caught fish are simply maintained in the laboratory, at different temperatures, without any regulated procedure or other interference.) The fish are killed by an approved method before the exercise, and this means that students are not involved with work on living animals.

Up to four species of *Gyrodactylus* occur on three-spined sticklebacks, and it is possible that these have different population growth characteristics. Identification requires examination of each individual parasite with a compound microscope; for most practical class work, all *Gyrodactylus* recorded may be combined.

Sticklebacks should be killed immediately before the exercise by a skilled assistant, cutting through the spinal cord at the level of the operculum and destroying the nervous tissue with a needle. The fish should be placed in individual Petri dishes and immersed in dechlorinated water; dishes must be numbered to enable reference to the relevant aquarium 'population'. Care should be taken to avoid issuing students with any fish that show residual nervous twitching.

Given the care and patience required with delicate parasite material, class results are likely to include some errors and guesses, even in a committed undergraduate group. Trends are sufficiently distinct, however, that minor miscalculations are swamped in the very large data set. 'Quality control' can be conducted by instructing the class to leave all Petri dishes, labelled with host number, etc., for inspection after the practical: this usually indicates a high degree of accuracy in the first of a series of practicals and, after practice, complete reliability by the second.

Instructions for students

Fish must be immersed in water at all times: the ectoparasites are delicate and will be damaged if they dry. Carry out all observations under the binocular dissecting microscope.

1. Examine the stickleback intact, looking carefully over the external surface.
2. Cut off the fins in turn and transfer to water in a separate Petri dish. Bright illumination is essential, shining obliquely across the field of view, but take care that the light source does not overheat the material; arrange the preparation so that there is a dark background (place a black surface beneath the Petri dish) and the parasites will be illuminated in silhouette on the host skin, initially distinguishable by their leech-like movement.
3. Make initial observations on the parasites *in situ*, noting methods of attachment and locomotion and the reaching movements with which worms test the substrate before changing position.
4. Make temporary preparations of live worms for study with

the compound microscope. The transfer of *Gyrodactylus* to a microscope slide requires patience! Induce a worm to attach to the end of a mounted needle and quickly transfer this to a drop of water on a slide. With luck (and skill) the worms detach in the water drop and a coverslip can be carefully lowered onto the preparation.

5. If these manipulations fail, cut off a piece of host fin with a parasite attached and transfer this to a drop of water on a slide.

6. Take care not to burst the parasite with the coverslip pressure: this can be regulated by adding water to the side of the coverslip with a Pasteur pipette or withdrawing excess water with filter (or blotting) paper. With the correct pressure, the worm can be restrained for observation at low and then progressively higher magnification.

The posterior haptor of the parasite is armed with 16 marginal hooks and two hamuli with connecting bars. If the worm has an offspring in the uterus, this may be recognised initially by a set of haptoral hooks in the midbody of the parent. If this embryo has, in turn, an offspring in its uterus then this may also have a set of partly developed hooks. Different microscope preparations will show different stages in this 'Russian doll' sequence: a class demonstration can be set up with selected specimens showing this developmental series.

Data recording and handling

- Record fish length, make accurate counts of *Gyrodactylus,* and add data to class records for each 'population' of sticklebacks maintained for the specified times at different temperatures.

- After compilation of class data by the demonstrators, calculate for each population: prevalence (percentage of fish infected), mean intensity of infection (average burden amongst infected hosts), abundance (mean number of parasites per fish, including those uninfected); display as histograms the frequency distribution of infection levels at each time and temperature; determine the ratio of variance to mean and compare the degree of aggregation of worms in each population; use appropriate statistical tests to test differences between populations for significance; compare total numbers of parasites in each population as a measure of net reproductive increase.

- Write a report which attempts to interpret the changes of the *Gyrodactylus* populations in relation to temperature and time.

Results and interpretation

Under the conditions of temperature variation, which are more or less the maximum seasonal range experienced by the host in nature, the changes in infection levels are highly dynamic. Some of the key events may be observed only by chance, determined by the sampling interval (weekly in this exercise), attempting to capture a continuous process by successive 'snapshots'. It is useful therefore to keep data separately for the series of aquarium populations that may be maintained at each time and temperature. The events may not necessarily be synchronised and some 'populations' may carry evidence of change, such as the appearance of maximum burdens that have already passed or not yet occurred in other populations. After inspection for these features, data for each time and temperature can be pooled.

The following account is based on practical class data derived from several years in which a relatively simple manipulation was undertaken. Sticklebacks were collected in early/mid January and one sample was examined by the class as soon as possible after capture to record natural infection levels. Groups of 15 fish were maintained in aquaria at 5, 10, 15 and 20 °C for 1, 2 and 3 weeks; some exercises were extended to 4 weeks. There was no further manipulation of the material, but the original infections on these groups of fish would be expected to multiply in the intervals before examination.

Natural infection levels

Mid-winter samples commonly show a prevalence of around 60% and mean burdens of around 2.5 worms/ fish overall (3.5 worms/ infected host). Infection levels within the population approximate to an overdispersed distribution, with most fish uninfected or with low worm burdens (1–3 parasites/ host). A small proportion carry relatively higher infestations (10s, 20s, sometimes around 30 worms/ host). Plotted as a frequency distribution, the majority of the fish population appear on the left of the distribution (uninfected or lightly infected) and a few host individuals extend a tail of the distribution to the right. It can be calculated that these few heavy infestations may contain total parasite numbers more or less comparable with the total harboured by all the other fish in the sample.

Gyrodactylus infection levels would not be expected to conform to a negative binomial distribution since the parasites multiply *in situ* (typical of microparasites): there should be an 'all or nothing' effect with hosts that have acquired infection subject to a progressively increasing number of parasites (and therefore few light infestations). Despite this prediction, the pattern described above is routinely encountered, so low worm burdens must remain stable as a result of losses that counteract the inherent capacity for *in situ* multiplication. The factors responsible must be a regular feature of population dynamics in nature, at least at low water temperatures.

Infection levels after one week in controlled environmental conditions
(1) 5 °C
Total parasite numbers increase by a factor of at least 2, sometimes 3. Prevalence increases, confirming host-to-host transmission within the confines of the aquarium and indicating that lack of infection in a significant proportion of individuals in natural populations is likely to be attributable largely to chance rather than to resistance. A frequency distribution of infection levels shows a marked shift towards higher worm burdens with very few hosts carrying the 1–3 worms that are typical of freshly caught hosts: around half of the hosts have burdens exceeding 10 with a maximum commonly twice that recorded one week earlier and a mean intensity of about 7 worms/fish.

(2) 10 °C
Infection levels reflect an increase in comparison with both the original sample and the 5 °C data. A frequency distribution demonstrates a further shift towards higher worm burdens in an increasing proportion of the population. Maximum levels may exceed 100 worms/host and mean intensity around 11 worms/fish).

(3) 15 °C
Data for this temperature and interval tend to be variable, indicating a 'crisis' in population growth. Typically, prevalence is 100%, demonstrating that all fish are still susceptible, and confirming that lack of infection in the wild population may be attributable largely to stochastic factors rather than host resistance. The 'populations' of 15 fish commonly contain a few very heavily infected individuals (with maxima exceeding 500 worms/ host) and a majority with very low burdens (only 1–3

worms). Indeed, if the very heavy infestations are excluded, the distribution of infection levels approaches that seen in the wild-caught sample. In some practical class exercises, the 1 week/15 °C populations show a marked decline in prevalence (in comparison with 'week 0') suggesting that the low infection levels reflect a progressive trend towards zero.

(4) 20 °C

These data also tend to be variable between different replicates of the practical exercise suggesting that the sampling interval may be critical in capturing a similar 'crisis' in the parasite populations. In some exercises, all fish are uninfected after one week at 20 °C. In others, there may be some very heavy burdens (several hundred worms/host) alongside uninfected fish and hosts with small burdens (presumably declining). Within the confines of the aquaria, where the active phase of population growth leads to transmission to all fish within the 'population', it must be assumed that the appearance of uninfected fish reflects a lack of susceptibility to infection in the face of challenge from worms shed from heavily-infected fish.

The possibility that prevalence falls because 15 and 20 °C are lethal to *Gyrodactylus* can be excluded by the fact that some fish accumulate enormous burdens at these temperatures.

Comparison of infection levels (both abundance and total numbers of parasites) indicates that the rate of population growth at 15 °C is about twice that at 5 °C, and, in exercises where infections persist, the growth rate at 20 °C is about twice that at 10 °C. This, of course, accords with the Q_{10} of 2 for most physiological processes. However, at the higher temperatures, the parasite numbers achieved by the growth process are highly aggregated (commonly 50–80% of the parasite population on just one or two hosts). This cannot result from *in situ* reproduction but must reflect active transfer from fish with declining burdens (or from the substrate) onto fish that remain susceptible.

Infection levels after two weeks
(1) 5 °C and (2) 10 °C
At these temperatures, trends are generally consistent. There is a progressive increase in infection levels with all, or almost all, fish infected and an increase in the numbers of fish with higher burdens (mean intensity around 10 at 5 °C and 15 at 10 °C). The differential between these temperatures (around 50%) is the

same after 2 weeks as after 1 week, suggesting that the differences can be attributed largely to the effect of temperature on metabolic rate and reproductive processes.

(3) 15 °C
Prevalence falls by 60% in some exercises, even lower in others, and heavy worm burdens are atypical; indeed, the maximum infestations are sometimes only 5–10 worms/host. Direct comparison with infection levels in the original wild population would suggest that there has been no net population growth in 2 weeks but, of course, the intermediate sample shows that parasite numbers peaked at around 1 week (represented by the chance 'capture' of records of several hundred worms per host) and subsequently crashed.

(4) 20 °C
Strikingly, infection levels are zero. Given the very heavy burdens sometimes evident after 1 week, it must be concluded that worms shed from these hosts were unable to re-infect any of the other members of the population. After initial, rapid population growth, total parasite mortality must be very considerable within the 2 week period, although it is not clear from the 'snapshots' how many fish accumulate burdens of up to 500 worms/ host before the population crashes.

Infection levels after three and four weeks
In experiments continued for 4 weeks at 5 °C, infection levels continue to grow at more or less the same rate, and prevalence remains at or near 100%, suggesting that hosts do not develop resistance. At 10 °C, there may be a decline in parasite numbers after 3 weeks but these may rise again after 4 weeks, suggesting the possibility of oscillating susceptibility/resistance to infection in the host population. Parasite populations have not been found to become extinct within the 4 week maximum observation period at these temperatures.

In the 15 and 20 °C aquarium populations, parasites that are thrown off successive sticklebacks as immunity develops eventually have 'nowhere to go' and the parasite population becomes extinct. This is likely to happen in small ponds in summer, but flooded environments in winter will re-introduce infected hosts into parasite-free populations and the cycle will begin again. This scenario indicates that the overall parasite population survives through time by being passed successively from fish to fish, with host individuals acquiring temporary respite from infection whilst the parasites are maintained on others.

Postscript
It is part of the challenge of attempting to capture highly dynamic natural events, that unexpected outcomes may occur. This exercise has been repeated with variations over a period of 25 years and, in general, has produced highly consistent results (although this can never be taken for granted!). Results from a recent exercise, however, have proved more difficult to interpret and, in retrospect, may have been influenced by seasonal factors. Parasite population growth and its termination by the host response may be strongest in material collected in winter when, at low water temperatures, sticklebacks may have minimal levels of acquired immunity before the start of the laboratory procedures. Fish collected in early autumn, when water temperatures are still relatively high, may have variable levels of immunity gained during exposure to infection during the summer. This would affect their ability to 'support' the parasite population growth that is so dramatic in cold-adapted fish. These and many other aspects of the host–parasite interaction require further investigation.

Ideas for further exploration

- The simple procedures employed in this exercise are amenable to inventive manipulation. A range of other parameters influencing transmission could be investigated including the effect of varying host density.
- Parasite populations become extinct when small host populations gain resistance simultaneously; it would be interesting therefore to test the prediction that periodic addition of naive fish to an aquarium population would serve to maintain the infection.
- The mortality of *Gyrodactylus* during a population crash must be dramatic (over 500 worms dying over the course of a few days in the exercises described): a series of Petri dishes placed in the bottom of the aquarium and removed periodically for microscope examination could provide a means of sampling the extent of this mortality.

Similar exercises

As well as a wide range of modifications that provide a more detailed view of this host–parasite interaction, highly simplified procedures also produce interesting results. If a wild-caught sample of sticklebacks is divided in two, with half examined in one practical class and half examined in a second class a week later, students will obtain data to show that infection levels double at 10 °C (and they will have twice as many parasites for microscope examination!).

A range of alternative host–gyrodactylid systems could be investigated. There is an extensive literature on the *Gyrodactylus* species infecting guppies, *G. turnbulli* and *G. bullatarudis*; the former has a generation time of 2.4 days at 25 °C (see Scott, 1985), and population cycles are correspondingly faster than in the stickleback parasites. Experiments with this tropical host–parasite system would require higher temperatures and shorter intervals between observations. Other aquatic vertebrates also carry gyrodactylid species, including the African clawed toad, *Xenopus laevis*, which is infected by a related form *Gyrdicotylus gallieni*. This worm lives in the oral cavity of its amphibian host, protected from desiccation when the toad emerges occasionally on land, and it invades this site via the nostrils (Harris & Tinsley, 1987). *Gyrdicotylus* exhibits corresponding population cycles to those of fish gyrodactylids but reproduction is very slow (at 20 °C) and it takes up to 4 months before peak numbers develop; at this point there is an exactly equivalent population crash leading to a temporary state of resistance to re-infection (Jackson & Tinsley, 1994). Laboratory colonies of *X. laevis* are sometimes infected with this parasite (imported with their host from South Africa), and these make interesting material for experimental study, but if small groups of toads are kept together in an aquarium they eventually develop immunity simultaneously and *G. gallieni* becomes extinct (exactly as in the stickleback *Gyrodactylus* populations described above).

REFERENCES

Buchmann, K. (1998). Binding and lethal effect of complement from *Oncorhynchus mykiss* on *Gyrodactylus derjavini* (Platyhelminthes: Monogenea). *Diseases of Aquatic Organisms* **32**, 195–200.

Buchmann, K. & Bresciani, J. (1999). Rainbow trout leucocyte activity: influence on the ectoparasitic monogenean *Gyrodactylus derjavini*. *Diseases of Aquatic Organisms* **35**, 13–22.

Cable, J., Harris, P. D. & Tinsley, R. C. (1998). Life history specializations of monogenean flatworms: a review of experimental and microscopical studies. *Microscopy Research Techniques* **42**, 186–199.

Harris, P. D. (1993). Interactions between reproduction and population biology in gyrodactylid monogeneans – a review. *Bulletin Français de la Pêche et de la Pisciculture* **328**, 47–65.

Harris, P. D. & Tinsley, R. C. (1987). The biology of *Gyrdicotylus gallieni*, an unusual viviparous monogenean from the African clawed toad, *Xenopus laevis. Journal of Zoology, London.* **212**, 325–346.

Harris, P. D., Soleng, A. & Bakke, T. A. (1998). Killing of *Gyrodactylus salaris* (Platyhelminthes, Monogenea) mediated by host complement. *Parasitology* **117**, 137–143.

Jackson, J. A. & Tinsley, R. C. (1994). Infrapopulation dynamics of *Gyrdicotylus gallieni* (Monogenea) infecting *Xenopus laevis. Parasitology* **108**, 447–452.

Kearn, G.C. (1998). *Parasitism and the Platyhelminths.* London, Chapman & Hall.

Mo, T.A. (1994) Status of *Gyrodactylus salaris* problems and research in Norway. In: *Parasitic Diseases of Fish* (eds. A. W. Pike & J. W. Lewis). Dyfed, Samara Publishing.

Scott, M. E. (1985). Experimental epidemiology of *Gyrodactylus bullatarudis* (Monogenea) on guppies (*Poecilia reticulata*): short- and long-term studies. In: *Ecology and Genetics of Host–Parasite Interactions* (eds. D. Rollinson & R. M. Anderson), pp. 21–38. London, Academic Press.

2.4 Intraspecific competition in the cestode, *Hymenolepis diminuta*, in rats

C. R. KENNEDY & J. M. BEHNKE

Aims and objectives

This exercise is designed to demonstrate:

1. The effect of parasite population density on adult cestode size.
2. The effect of parasite population density on parasite fecundity.
3. The relationship between parasite size and fecundity.

Introduction

It has been known for a long time that there is a relationship between the number of parasites of a species in a host (infra-population density) and parasite size and fecundity. The relationship originally referred to as a crowding effect is most easily observed in parasites such as cestodes, in which growth is relatively indeterminate and adult size may be large. This effect can be detected in other groups of parasites such as nematodes and acanthocephalans, but these are often smaller in size and growth is more determinate.

The crowding effect has been studied most intensively in cestodes, and whilst there is a measure of agreement on its manifestation, and its population consequences, there is less agreement on its causation. One school of thought believes it is an example of intra-specific competition, probably for carbohydrate resources, which become increasingly scarce as infra-population density increases. Another school believes that it is a result of a host immune response, which increases in intensity and effectiveness as infra-population density increases. A third school believes that it is caused by secretions from the parasites that suppress growth. Whatever the cause, there is no doubt that it is potentially a very effective regulator and stabiliser of parasite population density, since the effect on fecundity is density-dependent.

In this exercise, you will look at the relationships between individual parasite size and fecundity, and between infra-population density and mean parasite size and fecundity, in adult *Hymenolepis diminuta* in laboratory-bred and -reared rats.

Laboratory equipment and consumables (per student or group)

Equipment

Compound microscope

Dissecting microscope

Dissecting board, pins and instruments

Access to suitable micro or top-pan balance

500 ml flask

Consumables

Mammal Ringer solution*

Pasteur pipettes and bulbs

Slides and coverslips

Large Petri dish

Stirring rods

Watch-glasses

Filter paper

*Mammalian Ringer; 9.0g NaCl, 0.42g KCl, 0.24g $CaCl_2$, 0.5g $NaHCO_3$, per litre distilled water.

Sources of material

A population of laboratory bred rats of the same age, strain and sex, preferably young adults, is required. In addition, *Tribolium* sp. infected with cysticercoids of *H. diminuta*, preferably of the Texas strain (the normal laboratory strain in the UK). Do not use wild rats or parasites recovered from wild rats. See also Additional information in Exercise 1.14.

Safety

There should be no serious hazards associated with this exercsie provided good laboratory practice is followed. Freshly obtained worms do not have infective stages directly infectious to man, but there are records of *H. diminuta* infections in human beings. The infective stage is the cysticercoid, so care should therefore be taken to ensure that infected beetles and larvae extracted from them are not swallowed during the class. Appropriate care should also be taken when handling rats and beetles. Some subjects may be allergic to both species.

Instructions for staff and preparation

Prior to the practical, infect groups of rats with 1, 2, 5, 10 and 20 cysticercoids each: numbers per group will depend on the number of students, but there should ideally be a minimum of five rats per group of students. Infections are carried out by stomach tube, a procedure under the Animals (Scientific Procedures) Act 1986, which requires that the handler has an appropriate licence in the UK. Rats are kept individually, in labelled cages, for at least four weeks before the exercise to ensure the cestodes have grown to maturity.

Approximately 24 h before the practical, clean out all faecal material from each cage and leave the rats for 24 h before sacrifice (the fresher the faeces the better). After this time, remove the faeces produced in each cage and place in a grip-seal bag labelled with the number of the rat. Sacrifice the rats in the approved manner as close to the time of the class as possible.

Students can be given one rat each and asked to dissect out the intestine, or they can be presented with the intestine in one bag and the appropriate faecal collection in another.

Instructions for students

1. Take the faeces produced over 24 h and macerate in a little tap water in the 500 ml flask with the stirring rod. Leave on one side to soak for at least 1 h.
2. Pin out the intestine on a dissecting board and open up carefully from the posterior to the anterior, using sharp scissors. **Keep the points up to avoid cutting any parasites, which are often visible through the wall of the intestine.** If you use a dissecting dish, do this under saline.
3. Carefully transfer the parasites to a Petri dish filled with Ringer solution. Agitate gently with a seeker to free the parasites from any faeces or mucus, changing the solution if necessary.
4. Count the number of parasites by counting the number of scolices. Determine the wet weight of each parasite (damp dry on filter paper) as rapidly as possible, then return each to a separate watch-glass or Petri dish labelled with the weight.
5. Remove the terminal proglottid(s) from each cestode and transfer to a microscope slide with a drop of saline and coverslip. Apply gentle pressure to the coverslip, and count the number of eggs in a proglottid. Repeat on at least two more proglottids to obtain an estimate of the number of eggs per proglottid. Carry out the same procedure for each cestode.

6. Return to the flask with faeces. Break up any debris further and dilute to 500 ml or 100 ml as instructed. Stopper the flask, thoroughly mix by shaking and **immediately** remove a 0.1 ml random sample to a microscope slide. Cover and count the number of eggs. Repeat on two further random samples and calculate the number of eggs produced per 24 h.

Data handling and calculations

You should have the following data:

- Number of parasites administered per rat.
- Number of parasites recovered per rat.
- Total number of eggs shed per rat/24h.
- Wet weight of each cestode.
- Number of eggs per proglottid for each cestode.

Results can be best collated on a class basis as follows. For each infra-population density group, determine:

- Prevalence of infection (% of rats infected).
- Mean (and variance) wet weight per cestode.
- Mean (and variance) number of eggs per proglottid.
- Mean (and variance) total number of eggs shed per rat per 24 h.
- Mean (and variance) number of eggs shed per cestode per 24 h.
- Mean (and variance) number of proglottids shed per rat per 24 h.

These results can be tabulated or plotted as desired.

For all individual tapeworms, plot number of eggs per proglottid against wet weight of cestode. Calculate the correlation coefficient. Note particularly changes in variance of these parameters as parasite density increases. Each student can then prepare their own reports.

Expected results

1. Prevalence should remain steady or decline in animals that received the higher doses of cysticercoids.
2. As density increases, mean weight per worm, mean number of eggs shed per worm per 24 h, mean number of eggs per proglottid and mean number of proglottids shed per worm per 24 h should decline.
3. As worm burden increases, the total number of eggs shed per rat per 24 h should rise to a peak and then decline.

4. There should be a significant positive correlation between fecundity and size of cestode.

Ideas for further exploration

- Can size of parasite be used as an index of its fecundity?
- Is the decrease in fecundity at higher densities a result of decreased egg production as such, or does it merely reflect the decrease in cestode size?
- Do the results provide any evidence in favour or against each of the three explanations, i.e. competition, immunity or secretions? Can you devise any experiments that would distinguish between the three?
- At high densities are all cestodes of equal weight, or are some large and some stunted? How would you explain the latter situation?
- Why do you think prevalence does/does not change?
- Is there an optimum infra-population size per rat?

Similar exercises

It is difficult to find a similar exercise as suitable as this one, but *Rodentolepis microstoma* and *R. nana* in mice can be used. A laboratory bred population of hosts of similar strain and size is necessary. Cestodes are better than nematodes, but the latter may be used, taking into account that higher densities are then needed to induce any crowding effect. Wild hosts cannot be used, as differences in parasite size could merely reflect differences in age/time of infection. It is possible to use wild hosts and turn the aims around, i.e. make the point of the exercise the fact that you cannot demonstrate competition.

The exercise can be extended by using higher levels of infection, but many parasites then destrobilate and it takes great skill to detect the minute scolices. It can also be extended by simultaneously determining parasite position along the intestine and establishing whether this changes with density.

REFERENCES

Hesselberg, C. A. & Andreassen, J. (1975). Some influences of population density on *Hymenolepis diminuta* in rats. *Parasitology* **71**, 517–523.

Read, C. P. (1951). The crowding effect in tapeworm infections. *Journal of Parasitology* **37**,174–178.

2.5 Transmission dynamics and the pattern of dispersion of the cestode *Hymenolepis diminuta*, in the intermediate host population

C. R. KENNEDY & J. M. BEHNKE

Aims and objectives

This exercise is designed to demonstrate:

1. The actual transmission rates of cestode eggs to cysticercoids in a beetle host.
2. The manner in which parasites are dispersed throughout a host population.
3. The influence of egg density on transmission rates and patterns of dispersion.

Introduction

Transmission of parasites from a free-living stage to a host or from one host to another is always hazardous and associated with high parasite mortality. The probability of a single egg giving rise to a cysticercoid larva in a beetle is very low, in the region of $P = 10^{-2}$ or less. In nature, much of the mortality is due to beetles failing to encounter eggs, or, having encountered one, failing to become infected. Additional factors reducing transmission rates include shortage of time in which to infect and/or encounters between parasite and resistant host. Even under ideal laboratory conditions, however, the probability of successful infection is low. It can be predicted on the basis of intuition and mathematical models that the probability of a successful infection will relate to the density of parasite eggs relative to that of beetles, such that the greater the egg density, the higher the probability of transmission.

What is not always appreciated is that transmission (a birth/recruitment process) rate can also influence the dispersion pattern of a parasite in a subsequent host population. An understanding of parasite dispersion patterns within their host populations is vital to an understanding of parasite population dynamics, epidemiology and control. There are a number of ways of describing the frequency distribution of parasites within a host population, their pattern of dispersion and quantification, e.g. by fitting the observed distributions to mathematical models of known properties and testing for goodness of fit. A widely used and simpler method makes use of the ratio between parasite abundance (mean number of parasites per host, infected and uninfected) and its variance. If s^2(variance) $= x$ (mean), the parasite population is randomly dispersed throughout its hosts: if $s^2 < x$, the parasite is underdispersed (even, uniform), and if $s^2 > x$, it is aggregated (over-dispersed).

The most important causes of aggregated distributions are heterogeneities in transmission processes. Some hosts may be more susceptible to infection because their immune system is less efficient, or they are physiologically more suitable, or they may carry heavier parasite burdens because they have been exposed to more parasite eggs for behavioural or dietary reasons. Similarly, some parasites and hosts may or may not make contact; if contact is established, parasites may or may not have sufficient energy to infect successfully. Whatever the reason, heterogeneity in transmission is widespread, the distribution of parasites throughout a host population is variable, and typically some hosts are uninfected, many harbour a few parasites and a few harbour heavy infections, i.e. the pattern of parasite dispersion within a host population is normally aggregated.

The exercise described here investigates the dispersion pattern of cysticercoids of the rat tapeworm, *Hymenolepis diminuta*, in the intermediate host, *Tribolium confusum*, in a laboratory environment. The Index of Dispersion (s^2/x) will be determined and related to egg density and transmission rates.

Laboratory equipment and consumables
(per student or group)

Equipment

Dissecting microscope (zoom   250 ml beaker
   or changeable objectives)

Consumables

Solid watch glasses   Pasteur pipettes and bulbs
Insect Ringer*   Disposable gloves
   *Insect Ringer: 6.5 g NaCl, 0.14 g KCl, 0.12 g $CaCl_2$, 0.1 g $NaHCO_3$, 0.01 g Na_2HPO_4, per litre distilled water.

Sources of material

A breeding colony of *Tribolium* sp. (*confusum* or *castaneum*) and a supply of fresh, gravid proglottids of *H. diminuta*, obtained from infections of one tapeworm per rat are required. If the laboratory does not keep infected rats, it may be possible to obtain proglottids from a neighbouring University or College. See also Additional information in Exercise 1.14.

Safety

There are only minor hazards associated with this exercise if good laboratory practice is followed. *H. diminuta* has been reported in humans, so care should be taken when handling cysticercoids; they should be handled with a pipette, and students should wear gloves and wash hands before leaving the laboratory.

Instructions for staff and for preparation

Recover proglottids of *H. diminuta* either by sacrificing a rat with a single parasite infection at least 30 days post-infection, or by extracting proglottids from faeces from such a rat. Beetles should be young adults of similar age: it is preferable to starve them for 2–5 days before infecting them.

Infections are carried out in artificial arenas, e.g. honey jars or specimen pots of no more than 5 cm diameter (base). The bases of the jars are lined with damp filter paper and known amounts of proglottid, teased apart, are introduced into each arena to provide egg densities of 700, 1500, 3000, 6000, 12 000, 24 000 and 48 000 eggs per respective arena. Thirty beetles are then introduced immediately into each arena and kept there for 24 h *in the dark*. They can then be returned to normal cultures and fed (keep in separate labelled jars, each with a gauze top) at 20–25 °C and 70% RH (relative humidity). They will be ready for a class 2–3 weeks post infection.

The numbers of replicates of each arena can be adjusted to the size of the class such that each student examines one jar, i.e. all 30 beetles at a given egg density, or more as required.

Instructions for students

Jars of beetles infected in the manner described above, each labelled with an egg density, will be available.

1. Select one or more jars as instructed, preferably one at a time and note carefully the density of eggs to which the beetles were exposed.
2. Work individually, and place a solid watch-glass containing insect Ringer on the stage of the dissecting microscope.
3. Transfer a single beetle to the watch-glass, remove the head and dissect open the abdomen so that any cysticercoids are released into the fluid.
4. Count and record the number of cysticercoids.
5. Repeat until you have examined all the beetles in a jar.
6. Record cysticercoid densities for each beetle separately, including the zero counts.
7. After examining each beetle, clean out the watch-glass and pipette and dispose of the parasites in the manner instructed.

Data handling and calculations

For each jar (egg density):

1. Construct a frequency distribution of the number of cysticercoids per beetle (either as a table or histogram, according to instructions).
2. Determine the mean number of parasites per beetle (abundance) and the variance of this mean, and so the Index of Dispersion (s^2/x).
3. Calculate the transmission rate of the parasite (egg density/total number of cysticercoids recovered/30).

Results can then be collated on a class basis to produce a table/figure showing parasite egg density, parasite transmission rate and Index of Dispersion. Ideally, plot separately transmission rate against egg density and abundance and Index of Dispersion against egg density, and calculate the correlation coefficients for the relationships (you will probably have to use the non-parametric Spearman's coefficient).

Each student can then prepare their own report, using their own frequency distribution and the class results.

Expected results

1. At each egg density, the parasites should show an aggregated distribution with $s^2 > x$, i.e. parasites are overdispersed.
2. The Index of Dispersion should initially increase with egg density, but the rate of increase should slow down or tail off at higher densities, i.e. infective stage density affects the degree of overdispersion.

3. Transmission rates will be very low, but they and abundance should increase with egg density in a curvilinear manner and approach an asymptote, i.e. even under 'ideal' circumstances, transmission rates are low but do relate to egg density.

Ideas for further exploration

- Why were parasites overdispersed even when beetles were of the same age and condition and transmission conditions constant?
- Why does the degree of aggregation increase with egg density?
- Why are the relationships between egg density and (a) transmission rates, (b) Index of Dispersion and, (c) parasite abundance all curvilinear and why do they approach an asymptote? Why are they not linear?
- Why are transmission rates (probabilities of infection) so low even under 'ideal' conditions for transmission?
- What can these results tell you about transmission rates, parasite mortalities and patterns of parasite dispersion in natural situations?

Similar exercises

If desired, or if experimental infections are impossible, the exercise can be focused on dispersion patterns only, by using naturally infected host populations. It is then clearly impossible to determine transmission rates, but the aim of the exercise can be altered to illustrate other and different aspects of parasite dispersion patterns.

Suggested alternatives

- *Mytilus edulis* (mussels); if you have easy access to a natural population of mussels you can collect a sample spanning a range of ages/sizes. Mussels harbour metacercariae of *Himasthla elongata* in the foot and of *Renicola roscovita* in the palps and stomach, and adults of the copepod *Mytilicola intestinalis* in the intestine. It is thus possible to look at the changes in Index of Dispersion with age/size of host, and to demonstrate that different species in the same host show different patterns of dispersion. This could be extended to establishing whether there are interactions between the species of parasite. Other bivalves such as cockles could be used in a similar manner.

- *Gammarus pulex*: if you have access to a *G. pulex* population infected with larval acanthocephalans e.g. *Polymorphus* spp. or *Pomphorhynchus laevis*, a similar exercise can be undertaken, using size as a measure of host age. Here, however, the degree of aggregation is likely to decline with host size since (a) the parasite may retard host growth, and (b) the parasite does improve the probability of an infected host being eaten, i.e. there is parasite-induced host mortality. It is thus possible to look at age abundance curves in association with changes in Index of Overdispersion.

Any of the exercises can be extended to attempt to fit Poisson and negative binomial models (and calculating k, for degree of overdispersion – a low k indicates a high level of clumping) to the observed data.

REFERENCES

Anderson, R. M. & Gordon, D. M. (1982). Processes influencing the distribution of parasite numbers within host populations with special emphasis on parasite-induced host mortalities. *Parasitology* **85**, 373–398.

Keymer, A. & Anderson, R. M. (1979). The dynamics of infection of *Tribolium confusum* by *Hymenolepis diminuta*: the influence of infective stage density and spatial distribution. *Parasitology* **79**, 195–207.

Section 3
Physiology and Biochemistry

3.1 Hatching *in vitro* of oncosphere/hexacanth larvae of *Hymenolepis diminuta*

C. E. BENNETT

Aims

This exercise is designed to demonstrate:

1. The importance of various physicochemical factors in the activation and hatching of the oncosphere/hexacanth via dissolution of the embryophore.
2. The behaviour of freshly hatched larvae.

Introduction

Hymenolepis diminuta is a tapeworm of rodents (for life cycle, see Exercise 1.14). *Hymenolepis* eggs (Fig. 3.1.1) contain a single oncosphere/hexacanth larva (six hooks). These eggs are infective to the beetles *Tenebrio* and *Tribolium*.

Sources of parasite material

A small number of infected rats held in one or two institutions of higher education can provide enough tapeworm eggs to supply large practical classes run at many other institutions. Eggs remain infective in faeces for several weeks and can be supplied by post from those universities and institutions that maintain the infections. Faeces should be refrigerated on arrival. See also Additional information in Exercise 1.14.

Safety

This exercise is entirely safe for humans since the eggs are only infectious to beetles! Tapeworm eggs will have been thoroughly washed and students will normally only touch the sides of clean microscope slides. It is good laboratory practice to ensure that hand to mouth movements do not occur and that students wash their hands and use a nail-brush before leaving the laboratory.

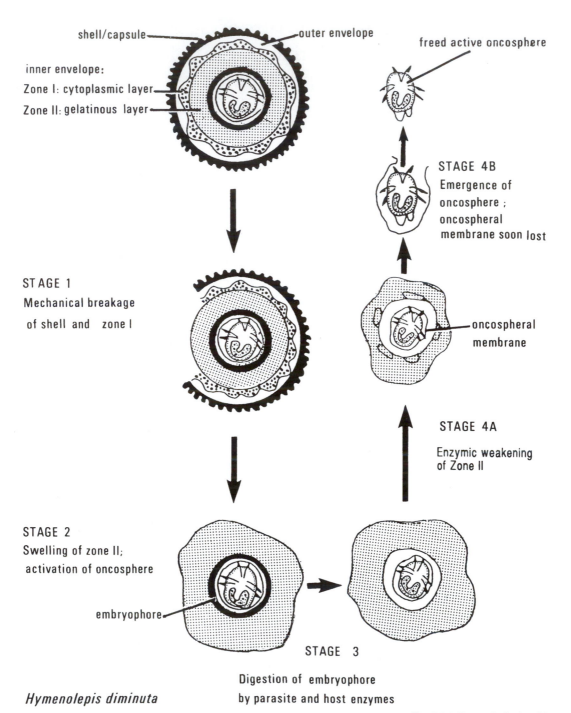

shell/capsule — outer envelope

freed active oncosphere

inner envelope:
Zone I: cytoplasmic layer
Zone II: gelatinous layer

STAGE 4B
Emergence of
oncosphere ;
oncospheral
membrane soon lost

STAGE 1
Mechanical breakage
of shell and zone I

oncospheral
membrane

STAGE 4A

Enzymic weakening
of Zone II

STAGE 2
Swelling of zone II;
activation of oncosphere

embryophore

STAGE 3

Hymenolepis diminuta

Digestion of embryophore
by parasite and host enzymes

Fig. 3.1.1 Stages in the hatching
of the egg of *Hymenolepis
diminuta* (after Smyth, 1994).

Laboratory equipment and consumables
(per student or group)

Equipment

Compound microscope, preferably with phase contrast

Consumables

Tyrode's saline*
Bacterial alpha amylase (e.g.
   Sigma A6380) 6.5 mg/ml
   = 13 000 units/ml
Porcine crystallised Trypsin
   (e.g. Sigma T7418) 1 mg/ml
   = 13 000 units/ml

Neutral red
Saturated solution of sodium
   chloride

*Tyrode's saline: 8.0 g NaCl, 0.20 g KCl, 0.20 g CaCl$_2$, 0.10 g MgCl$_2$, 0.05 g NaH$_2$PO$_4$, 1.0 g glucose, 1.0 g NaHCO$_3$ (added immediately before use as indicated below), per litre de-ionised water.

Instructions for staff

This is a relatively inexpensive practical to run. The significant expense is the enzymes, which must be bought fresh each year, prepared immediately before the practical and kept on ice. The exercise is designed to minimise the amounts of enzymes used, but these should be of good quality and recently purchased.

Hymenolepis diminuta eggs used in this practical are recovered and cleaned from the faeces of infected rats. Faeces are soaked for 24 h in 750 ml tap water, preferably in a fume hood. They are then stirred vigorously into a homogenous suspension and poured through two layers of muslin gauze stretched across a 1 litre beaker. The gauze and debris are discarded (to a bucket containing diluted bleach) and the contents of the beaker left to settle for 3 h. The supernatant layer is then drawn off and discarded and the sediment is re-suspended in 500 ml of tap water. This process is repeated three times on the following day, the final wash being with distilled water. Next the sediment is mixed with saturated NaCl solution in a large translucent centrifuge tube (20 ml) and spun at 200*g* for 20 min. On removing the tubes carefully from the centrifuge, a dark surface ring should be visible at the top of the centrifuge tube. This will contain the eggs. Draw off the eggs with a Pasteur pipette and wash with Tyrode's saline × 3 (10 ml) by 2 min spins at 150*g*. Resuspend the pellet of eggs in Tyrode's saline, which should have a pH of about 7.0. Store for up to 48 h at 5 °C.

Before use, spin again and re-suspend in 3 ml Tyrode's saline. When dispensing the eggs in the various solutions, make sure that the liquid is continuously agitated to ensure that all dispensed drops carry approximately equal numbers of eggs (they will settle to the bottom of tubes in a few seconds if this is not done).

Divide the agitated concentrate into three and dilute one of the aliquots as necessary for the numbers in the class to receive one drop on a microscope slide for Part 1 of the practical.

Take one of the remaining 1 ml aliquots of concentrated eggs and spin again at 150 g. Draw off the supernatant and add sufficient volume of hatching solution (Soln H, also used in Part 3 below): alpha amylase and trypsin in Tyrode's with sodium bicarbonate (see consumables) to the sedimented eggs such that (with constant agitation) each student can receive one drop of solution with eggs for Part 2.

For Part 3, take the remaining 1 ml and aliquot 100 µl of egg suspension into each of the enzyme solutions to make up the final required concentrations.

A procedure of distributing the material on pre-marked slides ensures good hygiene and that there is no contamination of the enzyme solutions by students cross-pipetting.

Instructions to students

Two distinct processes are involved in emergence of parasite larvae from their eggs in the insect intermediate hosts. This exercise seeks to replicate these processes and to determine which enzymes and conditions are optimal for hatching. You will be provided with the eggs and concentrations of the enzymes used.

The two processes are:

(a) 'Cracking' of the egg capsule. This is a mechanical process normally exerted by the jaws of the host beetle and is an essential prerequisite for 'hatching' under the control of enzymes.

(b) 'Hatching' of the oncosphere/hexacanth. This is accomplished with enzymes that break down the embryophore layer and allow the oncosphere/hexacanth to paddle free of the embryophore and vitelline cellular debris.

Part 1

1. Collect a drop of Tyrode's solution containing some eggs on a microscope slide from the laboratory supervisor.
2. Cover with a coverslip and examine, preferably by phase-contrast microscopy.
3. Draw and label fully. You should be able to identify the outer capsule of the egg, which is made of tanned protein. Inside, is a thick vitelline layer of gelatinous material with a limiting outer membrane and the inner embryophore layer. The embryophore is a protective layer made of sulphur-rich protein blocks. The oncosphere/hexacanth lies in the centre of the egg, surrounded by a thin oncospheral membrane (which will not be visible when the oncosphere/hexacanth is constrained inside the embryophore).
4. Draw and measure the thickness of the visible layers.
5. Tap the coverslip 4 or 5 times, quite gently at first, with the blunt end of a pencil (this procedure mimics the action of the insect mouthparts and should crack the capsules). Too violent tapping will damage the enclosed oncospheres/hexacanths.
6. Examine the slide after tapping and, if necessary, tap more firmly until a high percentage (>90%) of the egg capsules are cracked open.
7. Observe the oncospheres/hexacanths for 10–15 min.
8. Draw and describe any movements, remembering that this preparation is only in Tyrode's saline and has not been exposed to any enzyme.

Part 2

1. Collect some eggs, which have been suspended in a hatching solution of Tyrodes + amylase + trypsin + $NaHCO_3$ solution, on a microscope slide.
2. Cover with a coverslip and repeat the procedure of cracking the capsule (as in part 1).
3. Make careful observations of the sequence of events: i.e. what happens to the respective layers during hatching. Your observations should continue for several minutes until you have seen some hexacanths emerge fully.
4. Count the percentage of cracked eggs that have hatched at 5 min intervals up to 30 min. 'Hatching' in this case refers to the dissolution of the embryophore and the free movement

of the hexacanth. With time the larvae will move clear of the surrounding membranes.

5. Make notes on the dissolution of the outer gelatinous layers.
6. Draw the hatched oncosphere/hexacanths and describe their movement with additional sketches.
7. Introduce a drop of 1% neutral red solution under the cover-slip as a means of staining the penetration gland cells of the free oncosphere/ hexacanth larvae. This will kill the hexa-canths but should give a strong protein staining of the secre-tory enzyme granules in the mid-posterior region of the body of the larva.

Part 3
For this section a group of 8 students should work together. The following are laid out on labelled microscope slides, for you to collect, and are already covered with coverslips. All are made up in Tyrode's solution without $NaHCO_3$, which is added immedi-ately before use as designated:

A. Tyrode's lacking $NaHCO_3$ (CONTROL A)
B. Tyrode's with $NaHCO_3$ (0.1%) (CONTROL B)
C. α-amylase in Tyrode's lacking $NaHCO_3$
D. α-amylase in Tyrode's with $NaHCO_3$ (0.1%)
E. Trypsin in Tyrode's lacking $NaHCO_3$
F. Trypsin in Tyrode's with $NaHCO_3$ (0.1%)
G. α-amylase and trypsin in Tyrode's lacking $NaHCO_3$
H. α-amylase and trypsin in Tyrode's with $NaHCO_3$ (0.1%)

1. Nominate two students to start the experiments by tapping on the coverslips of the set of slides to crack the capsules, as in part 1 and part 2.
2. Cover the slides with appropriately sized lids to avoid drying out and include a piece of damp tissue if necessary. Leave for 40 min.
3. After the 40 min, each student in the group will be respon-sible for obtaining the results from one of the above hatching solutions A–H.
4. Estimate the percentage of cracked eggs that have hatched from the embryophore. Ignore uncracked eggs in your count-ing. As mentioned above, 'hatching' refers to the dissolution of the embryophore and the free movement of the hexa-canth, particularly the free extension of the hooks, which will no longer be constrained to clawing at the inner surface of the embryophore.

5. Collect and copy down the data from the entire class. The data from the experiment must be analysed by appropriate statistical tests and should be summarised in tables, graphs or histograms as appropriate.

Expected results

There will be movement of hexacanths even in uncracked eggs. Hatching is an active process but is aided by externally applied enzymes. Trypsin is more effective than amylase in helping to break down the embryophore. Following breakdown of the embryophore, the hexacanths will move freely in the surrounding cytoplasm of the egg and finally trail the cytoplasm and oncospheral membrane. Consider all of these individuals as hatched for the purpose of the experiment. When you analyse the class data there will be similarity between some of the treatments, but you will be able to test for significant differences. Consider the source and purity of the enzymes used and discuss them in relation to the naturally occurring enzymes in the guts of *Tribolium* and *Tenebrio.*

Ideas for further exploration

- Design new experiments to consider the physical conditions of hatching.
- Compare different dilutions of the enzymes used.
- Use beetle extracts rather than commercially prepared enzymes.

REFERENCES

Berntzen, A. K. & Voge, M. (1965). In vitro hatching of four hymenolepid cestodes. *Journal of Parasitology* **51**, 235–242.

Holmes, S. D. & Fairweather, I. (1982). *Hymenolepis diminuta*: the mechanism of egg hatching. *Parasitology* **85**, 237–250.

Lethbridge, R. C. (1971). The hatching of *Hymenolepis diminuta* eggs and penetration of the hexacanth in *Tenebrio molitor* beetles. *Parasitology* **62**, 445–456.

Lethbridge, R. C. (1972). *In vitro* hatching of *Hymenolepis diminuta* eggs in *Tenebrio molitor* extracts and in defined enzyme preparations. *Parasitology* **64**, 389–400.

Smyth, J. D. (1994). *Introduction to Animal Parasitology*, 3rd edn. Cambridge, Cambridge University Press.

Voge, M. (1970). Experiment 10. Hatching *in-vitro* of oncospheres. In: *Experiments and Techniques in Parasitology* (eds. A. J. MacInnis & M. Voge). San Francisco, W. H. Freeman and Co. (Out of print)

Voge, M. & Berntzen, A. K. (1961). *In vitro* hatching of oncospheres of *Hymenolepis diminuta* (Cestoda: Cyclophyllidea). *Journal of Parasitology* **47**, 813–818.

3.2 Activation of the cysticercoids of *Hymenolepis* species *in vitro*

J. M. BEHNKE

Aims and objectives

This exercise is designed to demonstrate:

1. The importance of various factors in the activation and excystation of the cysticercoids.
2. Changes in the appearance of the cysticercoids during excystation
3. The behaviour of freshly excysted larvae.

Introduction

Many parasites use more than a single host species during the course of their life cycle. Often the larval stages develop in one host and the adults in another. Thus, the parasite must be transmitted from one host to the next in order to complete development. The most common relationship between hosts exploited by parasites for transmission, particularly those living in the intestine as adults, is that involved in the predator–prey food chain. For example, some tapeworms develop as larvae in small flour beetles, *Tribolium confusum*, but parasitise rodents such as mice and rats as adults. Rodents are essentially grain feeders, but will consume insects also and therefore by depending on two host species, both of which have habitats closely associated with cereals and grain, the tapeworms ensure that their transmission cycle is completed successfully.

This practical is based on two species of tapeworm, *Hymenolepis diminuta* and *Rodentolepis microstoma* (formerly known as *H. microstoma*; see Exercise 1.14). The former develops successfully to the adult stage in the rat only (Fig. 3.2.1), the latter matures in mice. Both utilise the flour beetle, *Tribolium confusum*, as the intermediate host.

The larval (metacestode) stage developing in beetles is called a cysticercoid. The beetles acquire tapeworm eggs whilst feeding on grain, the larvae hatch in the insect's intestine and penetrate **183**

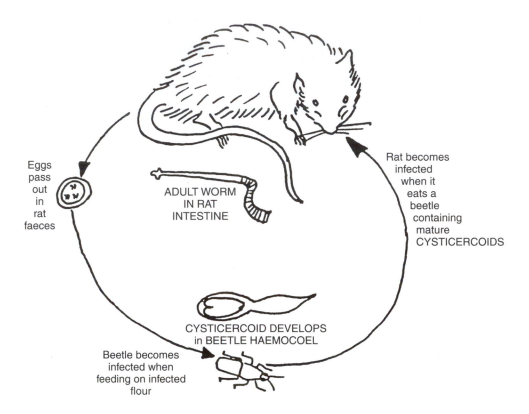

Eggs
pass
out
in
rat
faeces

ADULT WORM
IN RAT
INTESTINE

Rat becomes
infected
when it
eats a
beetle
containing
mature
CYSTICERCOIDS

CYSTICERCOID DEVELOPS
in BEETLE HAEMOCOEL

Beetle becomes
infected when
feeding on infected
flour

Fig. 3.2.1 Schematic illustration of the life cycle of *H. diminuta*. Not drawn to scale.

the tissues to reach the haemocoel, where further development to the cysticercoid stage is completed. The cysticercoids lie freely in the haemocoel and can be readily observed when the body cavity of the beetle is exposed. Figure 3.2.2 shows the general shape of the cysticercoids of the two species and indicates some of the interesting features.

Essentially the cysticercoid comprises a number of protective layers and a fluid-filled bladder within which the head of the future tapeworm (the scolex) is protected. This is necessary to ensure survival of the larva in the insect haemocoel, where otherwise it may be surrounded by host phagocytic cells and eventually destroyed. Protection for the scolex is also necessary to enable it to survive the digestive enzymes in the rodent stomach, when the beetle is consumed.

The cysticercoids are resting stages. Once development is complete, the larvae become dormant, saving energy reserves for the period following transmission, when development will resume and rapid growth will be necessary in a new host. In nature, the signals required to initiate further development of the resting larvae are normally provided by factors in the internal environment of the appropriate host. When the proper

Fig. 3.2.2 Illustration of the cysticercoids (metacestodes) of *H. diminuta* (a) and *Rodentolepis microstoma* (b). The scale is approximate.

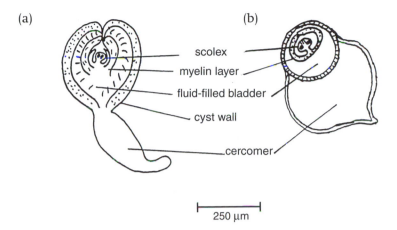

(a)

(b)

scolex

myelin layer

fluid-filled bladder

cyst wall

cercomer

250 μm

signal or combination of signals impinge on the larvae, they respond by activation, which may involve movement, or in other ways that may assist in establishing the parasites in the new host. In the case of cysticercoids the scolex, which is hidden within the protective layers, emerges and begins to search for the intestinal villi to which it would normally attach. The response to the correct sequence of triggers ensures that the parasite establishes only in a suitable host.

Laboratory equipment and consumables
(per student or group)

Equipment

Dissecting microscope

High intensity lamp for use with microscope

Pasteur pipettes and bulbs

Fine forceps such as watch-makers' forceps

Compound light microscope with at least × 40

Several glass Petri dishes and solid watch-glasses

A 250 ml glass or plastic beaker

Consumables

Insect Ringer* or Hanks' saline (GIBCO)

Paper towelling

Soap and towels

Glass slides and coverslips

Tea strainer**

Bucket containing diluted bleach***

Animal waste bin

Disposable gloves

*Insect Ringer: 6.5 g NaCl, 0.14 g KCl, 0.12 g $CaCl_2$, 0.1 g $NaHCO_3$, 0.01 g Na_2HPO_4 per litre distilled water.

**to strain animal waste from fluid for separate disposal

***for used glassware

Enzyme solutions and salines labelled as follows:

A. 2 g of pepsin in 100 ml of 0.85% NaCl
B. 1.0 ml conc. HCl in 100 ml of 0.85% NaCl
C. 1 g of trypsin in 100 ml of Tyrode's saline****
D. 1 g of sodium tauroglycocholate in 100 ml of Tyrode's saline
E. 0.85% NaCl
F. Tyrode's saline

Calibrated measuring pipettes, one for each solution above labelled appropriately.

****Tyrode's saline: 8.0 g NaCl, 0.20 g KCl, 0.20 g $CaCl_2$, 0.10 g $MgCl_2$, 0.05 g NaH_2PO_4, 1.0 g glucose, 1.0 g $NaHCO_3$ (added immediately before use as indicated below), per litre de-ionised water.

Sources of parasite material

Tribolium beetles can be obtained from laboratories where these are used for the maintenance of tapeworms. Likewise, both species of tapeworms can be obtained from institutions where they are currently maintained. See also Additional information in Exercise 1.14.

Safety

There are only minor health hazards associated with this practical providing normal good laboratory practice is followed. As far as is known, laboratory-maintained beetles are unlikely to cause medical/health problems in human beings, other than in those already showing allergies to insects. Appropriate precautions should be taken for such persons. *Hymenolepis diminuta* has been reported from humans so care should be taken when handling these cysticercoids. The only route for infection is through consumption of the infected beetles or the cysticercoids themselves. Students should wear gloves and wash hands before leaving the laboratory.

Instructions for staff

It is best to prepare the enzyme solutions fresh on the day. Older solutions do not work as well. Students should be provided with live infected beetles and warned not to select any dead ones as their cysticeroids will also be dead. Beetles should be infected with tapeworm eggs at least 3 weeks before the class, and maintained throughout at room temperature, to allow all the cysticercoids to complete development. Infection involves isolation of a batch of beetles from the colony, in a Petri dish with a filter-paper base, followed by starvation for 2–5 days. Then place a few fresh gravid proglottids onto the filter paper and when these have been totally consumed add flour for food.

Instructions for students

You should work in teams of 4–6 individuals. Students on each side of a laboratory bench should organise themselves into a single team. You are provided with a Petri dish containing beetles which have been infected with either *H. diminuta* or *R. microstoma*. Each team should work on one species only and each of the two teams on a laboratory bench on a different species.

Preparation of cysticercoids

1. Place 2–3 ml (a Pasteur pipette full approximates to 1 ml) of Hanks' saline into the base of a solid watch-glass (NOT PETRI DISH!).
2. Pick up a living beetle, using fine forceps, and transfer it to the watch-glass.
3. Under the dissecting microscope remove the beetle's head and open the haemocoel. The cysticercoids should become apparent immediately and may be inspected under high power. If you are uncertain ask for help!
4. Transfer the cysticercoids with a Pasteur pipette to another solid watch-glass with minimal contamination and wash out the first with tap water. Repeat the process until you have collected about 100 cysticercoids in total.

Hatching procedure

You are provided with the following solutions:

A. 2 g of pepsin in 100 ml of 0.85% NaCl
B. 1.0 ml conc. HCl in 100 ml of 0.85% NaCl
C. 1 g of trypsin in 100 ml of Tyrode's saline
D. 1 g of sodium tauroglycocholate in 100 ml of Tyrode's saline
E. 0.85% NaCl
F. Tyrode's saline

NB: EACH OF THESE SOLUTIONS HAS A SEPARATE MARKED PIPETTE. DO NOT MIX THE PIPETTES, OTHERWISE CONTAMINATION WILL OCCUR AND SOME OF THE SOLUTIONS WILL PRECIPITATE

1. Prepare 5 ml of 1% pepsin in acid 0.85% NaCl by adding 2.5 ml of solution A to 2.5 ml of solution B in a small Petri dish. **This will be the acid pepsin solution.**

2. Prepare 5 ml of 0.5% sodium tauroglycocholate by mixing 2.5 ml of solution C with 2.5 ml of solution D in a separate Petri dish. **This becomes the Tyrode's tauroglycocholate-trypsin solution.**

3. Preheat these solutions on a bench incubator adjusted to 37 °C.

4. Select 20–30 cysticercoids (leaving some for a second experiment) from your stock and transfer these to a clean solid watch-glass. Remove excess fluid with a Pasteur pipette, until the cysticercoids are left in minimal Hanks' saline in the centre of the depression in your watch-glass. Do not allow to dry.

5. Now add about two Pasteur pipettes full (i.e. about 2 ml) of the preheated acid pepsin solution (Step 1 above).

6. Place on the bench-top incubator, cover to prevent evaporation and wait for 5 min. Examine the cysticercoids under the binocular microscope for any changes to their appearance and for signs of movement within the cysticercoid walls.

7. Place back onto the bench-top incubator and wait for a further 5 min.

8. Examine again and when you have noted any changes (do not spend more than 1 min on this), draw up as much of the fluid as possible and discard it. Add a few drops of Tyrode's saline (Solution F). Aspirate the fluid and discard. Repeat this washing procedure 3 times in total, discarding the saline after each wash but ensuring that your cysticercoids remain in the centre of the watch-glass after each wash.

NB: THIS IS A VERY IMPORTANT PROCEDURE BECAUSE NO TRACE OF ACIDITY MUST REMAIN. IF YOU DO NOT WASH THOROUGHLY AT THIS STAGE, THE SOLUTIONS IN THE SECOND STEP WILL PRECIPITATE OUT AND YOUR EXPERIMENT WILL BE RUINED.

9. After you have removed the Tyrode's from the last wash, replace with Tyrode's tauroglycocholate–trypsin (again add only two Pasteur pipettes full).
10. Place back onto the bench-top incubator and leave for 1 min.
11. After 1 min, examine under the dissecting microscope for signs of movement or any other changes which may have occurred. This should be as brief as possible – not more than 1 min.
12. Place back onto the incubator and leave for a further 1 min and repeat observations until all the cysticercoids have hatched or for a maximum of 30 min if no signs of movement or hatching are evident. Record at each time interval any change you can detect and the number of hatched cysticercoids.
13. Calculate the time taken for 50% of the larvae to excyst.
14. What happens to the outside membranes of the cysticercoids in the various combinations of solutions?
15. Make detailed labelled drawings of all stages of the excystation process.
16. Having familiarised yourselves with all stages in the hatching procedure design experiments to answer some of the points listed below. You should work in the same groups as before and decide among yourselves the task of each participant. Your experiments should focus on the species of tapeworm you used earlier, but throughout you should compare notes with your colleagues on the opposite side of the bench who should be working on the other species.

Further questions

- What is the role of the preincubation in acid pepsin? Is it necessary for activation and hatching? Is the acid a vital component of the preincubation step?
- What role does sodium tauroglycocholate play?
- How are activation and hatching affected by trypsin? Is the effect of trypsin dose-dependent?
- What role does temperature play in activation and hatching?

The data from your experiment must be analysed by appropriate statistical tests and should be summarised in tables, graphs or histograms as appropriate.

Information on similar exercises

Various parasites have dormant transmission stages, which can be activated and hatched with appropriate stimuli. The eggs of various species including cestodes, nematodes and trematodes respond to a variety of signals, which differ between the species. The cystacanth of acanthocephalans is another possibility. Nematodes have ensheathed third-stage larvae, which can be stimulated to exsheath with appropriate stimuli. Information on some of these can be found in the references listed below.

Additional information

In the USA, viable eggs and cysticercoids of *Hymenolepis diminuta* can be purchased from Carolina Biological Supply Company (see Appendix 3) or from laboratories where the parasite is maintained for teaching and research purposes.

REFERENCES

Anya, A. O. (1966). Experimental studies on the physiology of hatching of eggs of *Aspiculuris tetraptera* Schulz (Oxyuridea: Nematoda). *Parasitology* **56**, 733–744.

Berntzen, A. K. & Voge, M. (1965). *In vitro* hatching of four hymenolepid cestodes. *Journal of Parasitology* **51**, 235–242.

Caley, J. (1975). The functional significance of scolex retraction and subsequent cyst formation in the cysticercoid of *Hymenolepis microstoma*. *Parasitology* **68**, 207–277.

Goodchild, C. G. & Davies, B. D. (1972). *Hymenolepis microstoma* cysticercoid activation and excystation *in vitro* (Cestoda). *Journal of Parasitology* **58**, 735–741.

Lackie, A. M. (1975). The activation of infective stages of endoparasites of vertebrates. *Biological Reviews* **50**, 285–323.

Mueller, J. F. (1959). The laboratory propagation of *Spirometra mansonoides* as an experimental tool. 1. Collecting, incubation and hatching of the eggs. *Journal of Parasitology* **45**, 353–361.

Perry, R. N. (1989). Dormancy and hatching of nematode eggs. *Parasitology Today* **5**, 377–383.

Ratcliffe, L. H. (1968). Hatching of *Dicrocoelium lanceolatum* eggs. *Experimental Parasitology* **23**, 67–78.

Rothman, A. H. (1959). Studies on the excystment of tapeworms. *Experimental Parasitology* **8**, 336–364.

Wilson, R. A. (1968). The hatching mechanisms of the egg of *Fasciola hepatica* L. *Parasitology* **55**, 124–133.

3.3 Membrane transport in the cestode *Hymenolepis diminuta, in vitro*

C. ARME

Aims and objectives

This exercise is designed to:

1. Demonstrate nutrient acquisition using a model tapeworm maintained outside the host (*in vitro*).
2. Provide experience of dissection to remove tapeworms from a rat.
3. Introduce and provide instruction in the handling of radio-isotopes.
4. Develop skills in data analysis and interpretation.
5. Encourage development in the process of hypothesis testing.
6. Provide sufficient data for the production of a detailed laboratory report.

Introduction

The tegument plays an important role in the physiology of the Cestoda. Since tapeworms lack a mouth, gut and anus, all nutrients must enter across the external body wall. Before commencing this exercise you should list the possible mechanisms that could be involved in nutrient absorption. In the following work the uptake of amino acid nutrients by the rat tapeworm, *Hymenolepis diminuta*, is examined.

Laboratory equipment and consumables
(per student or group)

Equipment

Dissecting microscope
Scintillation counter
Scintillation vials
Shaking water bath
Beakers
Access to a drying oven
   (80–100 °C)

Test-tubes
Erlenmeyer flasks
Petri dishes
Fine scissors and forceps
Worm-hook (seeker with
   recurved tip)

Consumables

Krebs Ringer-Tris (KRT)*	Alanine (unlabelled)
Ethanol (70%)	Scintillation cocktail
Radiolabelled [^{14}C]-cycloleucine	Plastic-backed paper sheets
(spec. act. 0.5 µCi/µmole; 2 mM	Aluminium foil
stock)	Disposable syringes (25 ml)
Coarse filter paper	Disposable gloves

*KRT: 120 mM NaCl, 4.8 mM KCl, 2.6 mM CaCl$_2$, 1.2 mM MgSO$_4$, 25 mM tris (hydroxymethyl) aminomethane-maleate buffer pH 7.2.

Safety

Radiochemicals will be used in these experiments. The amount of isotope used will not constitute a radiation hazard if handled correctly. **For your safety the following procedures must be followed**:

1. Protective clothing must be worn.
2. No hand-to-mouth movements, including smoking, eating.
3. Use disposable paper towels and paper tissues.
4. Isotopes to be handled only over plastic-backed sheets.
5. All pipetting to be with safety devices; do not use your mouth.
6. Report any accidental spillage, contamination of hands, clothing, etc. immediately. Remain at your place until you have received instructions from a demonstrator.
7. Separate containers will be provided for contaminated waste and glassware.
8. Wash hands thoroughly before leaving the laboratory.
9. All persons must be monitored for radioactivity before leaving the class.

Note to staff

This practical is composed of two parts. Experiment 1 will take approximately 4–5 h while experiment 2 should take no more than 2–2.5 h.

Laboratory maintenance of *H. diminuta*

In an attempt to standardise experimental material, all parasites are cultured in male rats, 80–100 g in weight. Obtain cysticercoids from *Tenebrio molitor* or *Tribolium confusum* and feed to the final host at a dose rate of 30 per rat. Worms are used for experiments 10 days following infection. See also Additional information in Exercise 1.14.

Instructions to students

Before the start of each experiment in this series ensure that you have:

1. Sufficient (at least 20 for Expt 1, 40 for Expt 2) numbered 25 ml beakers for preincubation and a similar number for incubation. A supply of test-tubes, containing 2 ml ethanol and bearing numbers corresponding to those on the incubation beakers.
2. Three 250 ml beakers containing KRT for washing the worms; filter paper for blotting parasites.
3. Several Petri dishes, dissecting instruments and a 20 ml syringe, for collecting tapeworms from the rat.
4. A worm-hook for handling the parasites.

Basic outline of experimental procedures

1. The saline used throughout is KRT (Krebs Ringer buffered with Tris (hydroxymethyl) aminomethane maleate).
2. Recover worms from the infected rats and preincubate them in KRT alone for 30 min.
3. Following preincubation, transfer the parasites to an incubation medium consisting of the radiolabelled nutrient to be studied, dissolved in KRT.
4. After incubation, wash the worms in at least 3 changes of KRT, blot and place in test-tubes containing 2 ml 70% ethanol. The ethanol serves to kill the worm and also to extract the chemical under investigation. An aliquot of the ethanolic extract is then assayed for radioactivity to determine the amount of nutrient absorbed by the worm.
5. Remove the parasites from the ethanol, dry to a constant weight.

Experiment 1: Does *H. diminuta* absorb amino acids *in vitro*?

In this experiment, *H. diminuta* will be incubated for varying time periods in a labelled amino acid, cycloleucine. It has been shown that this compound is not metabolised by *H. diminuta* – are there any advantages in the use of metabolically-inert compounds in transport studies?

Recovery of worms

In this experiment, you will incubate the parasites in 0.1 mM [^{14}C]-cycloleucine (specific activity 0.5 µCi/µmole) for periods of 2, 5, 10, 20 and 40 min. For each time interval studied, 4 replicate observations will be taken. Each replicate sample will comprise 5 tapeworms. Thus, the total number of tapeworms required for the complete experiment is $5 \times 4 \times 5 = 100$. At an infection rate of 30 worms per rat, 4 animals should provide a sufficient number of parasites.

Remove the small intestine from a freshly killed rat, between the pylorus and caecum. Insert the tip of a 20 ml disposable syringe into the cut anterior end of the gut and inject about 15 ml warm KRT into the intestinal lumen. Gently work the saline down the intestine using a 'finger over finger' technique and collect the expelled worms in a beaker. Using a worm-hook, wash the parasites in at least 3 changes of KRT to remove faecal material and place the worms in a Petri dish. If less than 26 worms have been recovered from any particular rat, reject that sample. Continue to collect worms from infected rats until sufficient have been obtained for the experiment. Keep the parasites from each rat in separate dishes.

Preincubation

Arrange 20 numbered 25 ml beakers on the bench and place 10 ml of KRT into each one. These are the preincubation beakers. Place five tapeworms into each beaker in a random manner, so that each beaker contains worms originating from different rats. When all the preincubation beakers have been prepared, begin placing them in the water bath according to the timing schedule given in Table 3.3.1. While some members of the group are preparing the parasites for preincubation, others should prepare the incubation media.

Incubation medium

For each time interval you will need 4×25 ml beakers each containing 10 ml of 0.1 mM [^{14}C]-cycloleucine in KRT = a total of 40 ml. In order to allow a safety margin prepare 50 ml incubation medium for each time interval = $50 \times 5 = 250$ ml. The cycloleucine has been prepared as a 2 mM aqueous stock solution. Therefore the KRT must be made up at a greater than normal strength in order to allow for its dilution by the aqueous cycloleucine. It is convenient to prepare the KRT at twice normal strength.

1. To prepare 200 ml of 2× strength KRT, take 40 ml of 10× strength Krebs Ringer and 40 ml of 10× strength Tris buffer and add 120 ml distilled water.

2. To prepare 250 ml of 0.1 mM cycloleucine solution in KRT you need: 12.5 ml 2 mM stock cycloleucine-1-[^{14}C] + 125 ml 2× KRT + 112.5 ml water. Add the 2× KRT and water to a 500 ml flask, and, using a bulb or automatic pipette, add the labelled amino acid. Perform this and subsequent operations involving radiochemicals over the plastic backed sheet provided. After use, place the contaminated pipette into the receptacle provided for radioactive glassware.

Arrange 20 numbered beakers on the bench and pipette 10 ml incubation medium into each one. These are the incubation beakers. Numbers 1–4 will be incubated for 2 min, 5–8 for 5 min, 9–12 for 10 min, 13–16 for 20 min and 17–20 for 40 min. For details of the timing sequence consult Table 3.3.1. Note that all incubation beakers must be placed in the water bath at least 5 min before they are due to receive worms to allow the incubation medium to reach 37 °C.

Experimental procedure

Consult the timing schedule; the first beaker of worms to start preincubation was beaker number 17. This was placed into the water bath at time 0 min. At time 30 min, the worms are due to be transferred from the preincubation to the incubation medium. Therefore at about time 25 min place the incubation beaker number 17 in the bath to preheat. About 15 sec before time 30 min, lift the preincubation beaker number 17 from the bath and, using a swirling action, collect the parasites on the worm hook. At about 30 min minus 5 sec, lift the worms from the preincubation medium, touch them on the side of the beaker to remove excess saline and at time 30 min exactly, place the worms in incubation beaker number 17. Repeat this procedure with the other parasites at the times indicated on the timing schedule.

You can see in the schedule that worms in incubation number 5 are due to be removed at time 39 min, i.e. after 5 min incubation. A few seconds prior to time 39 min aggregate the worms on the hook and, at time 39 min exactly, remove the worms. Keeping them on the hook wash them quickly in three successive saline washes. After washing, blot the parasites on

Table 3.3.1 *Timing schedule for Experiment 1*

Sample number	Begin pre incubn. (min)	Begin incubn. (min)	Stop incubn. (min)	Incubn. time (min)
17	0	30*	70	40
18	1	31	71	40
19	2	32	72	40
20	3	33	73	40
5	4	34	39	5
6	5	35	40	5
7	6	36	41	5
8	7	37	42	5
13	8	38	58	20
14	13	43	63	20
15	14	44	64	20
16	15	45	65	20
9	16	46	56	10
10	17	47	57	10
1	18	48	50	2
11	19	49	59	10
12	21	51	61	10
2	22	52	54	2
3	23	53	55	2
4	30*	60	62	2

Note:
* Note that these two times coincide; if you are performing this experiment alone you will require assistance at time 30 min.

hard filter paper and place them into ethanol in the appropriately numbered test-tube – in this case test-tube number 5. Stopper the tube and leave in the test-tube rack.

Extraction of [C^{14}]-cycloleucine and determination of radioactivity
Almost all the cycloleucine will be extracted into 70% ethanol within 1 h but parasites may be left for longer periods if necessary. After extraction, count a 0.5 ml extract from each tube using the available scintillation counter and cocktail provided. In addition to the above, counts per minute (CPM) from medium samples and standards must be obtained.

Initial medium sample
Although you should check the final cycloleucine concentration in every incubation beaker, you may limit the samples to one

beaker from each time interval studied. Repeat the dilution steps listed above for the initial incubation medium for final medium samples from incubation beakers 1, 5, 9, 13 and 17.

Standard
Place 0.5 ml of the original 10 mM aqueous stock cycloleucine solution in a 100 ml volumetric flask and make up to the mark with water. Assay radioactivity in 0.5 ml of this solution.

Dry weights
After removal of an aliquot for determination of radioactivity, remove the worms from the test tubes and transfer to numbered aluminium tares. After drying at 80–100 °C overnight, weigh the parasites.

Calculations

1. Correct all counts for background radiation (CPM_{sample} – CPM_{bkgd})
2. Determine the CPM in 1 mole of cycloleucine using the results obtained from the aliquot of the 10 mM stock solution.
3. Determine CPM/mole cycloleucine using the data obtained from the counts from the initial medium sample. The result should be the same as that obtained in 2 above.
4. To determine the amount of cycloleucine absorbed by the worms apply the following formula to each tube:

$$\frac{(CPM/aliquot) \times 4}{(CPM/mole\ cycloleucine) \times dry\ wt\ (g)} = moles\ cycloleucine\ absorbed/g$$

5. Calculate final cycloleucine concentrations for the media.
6. Calculate cycloleucine concentrations in worm water, given that the ethanol extracted dry weight/fresh weight ratio is 19.2% and that the water content of a live worm is 75% of fresh weight.

Question
Does *H. diminuta* absorb cycloleucine by a carrier-mediated transport system? Is it possible fully to answer this question using the data obtained in experiment 1?

Experiment 2: Concentration effects and inhibition

In this exercise an experimental design will be employed that will permit the examination of two features of amino acid absorption by *H. diminuta*. First, what is the relationship between cycloleucine uptake and concentration? Second, what is the effect of a potential inhibitor on the uptake systems? If cycloleucine is absorbed via a carrier-mediated transport system, then it is likely that uptake would be inhibited by other neutral amino acids that compete with cycloleucine for binding to the carrier. This can be investigated over, for example, a concentration range of 0.1–0.9 mM in the presence of another neutral amino acid, alanine. **Recovery of worms and preincubation procedures are identical to those used in experiment 1.** Incubation periods for this experiment will be reduced to 2 min and the volume of incubation medium to 4 ml. A total of 40 beakers will be required for preincubation, 40 for incubation and, of course, 40 test-tubes for extraction.

Preparation of incubation medium

Table 3.3.2 is a guide for the preparation of incubation media. There are 10 lots of medium to be prepared and these should be stored in separate stoppered 50 ml Erlenmeyer flasks. Label the flask with the numbers of the incubation beakers to be filled from that flask, e.g. 1–4 for the first lot of medium in the Table below. Make sure that you understand how the various values have been calculated and attempt the following calculations:

1. What is the final alanine concentration in beakers 21–40?
2. What is the total number of moles cycloleucine in flasks 9–12?
3. How many mg of alanine are required to make 113 ml of a 6 mM solution and how many moles alanine would there be in 4.7 ml of such a solution?

Incubation

After you have placed 5 worms in each of the 40 preincubation beakers arrange them on the bench as follows:

1	2	3	4	5	6	7	8
9	10	11	12	13	14	15	16
17	18	19	20	21	22	23	24
25	26	27	28	29	30	31	32
33	34	35	36	37	38	39	40

Table 3.3.2 *Incubation media for Experiment 2*

Sample numbers	2 mM cycloleucine ^{14}C (ml)	2 × KRT (ml)	Water (ml)	40 mM alanine (ml)	Final cycloleucine concn. (mM)
1–4	1	10	9	–	0.1
5–8	2	10	8	–	0.2
9–12	4	10	6	–	0.4
13–16	6	10	4	–	0.6
17–20	9	10	1	–	0.9
21–24	1	10	8	1	0.1
25–28	2	10	7	1	0.2
29–32	4	10	5	1	0.4
33–36	6	10	3	1	0.6
37–40	9	10	0	1	0.9

and arrange the incubation beakers in a similar manner by their side. To achieve a good randomisation do not place the beakers in the bath in numerical sequence but follow vertical columns of five: 1, 9, 17, 25, 33, then 5, 13, 21, 29, 37, then 2, 10, 18, 26, 34, then 6, 14, etc. Since in this experiment we are incubating for 2 min in all cases, the time sequence in Table 3.3.2 may be used which, since certain times overlap, requires at least two people.

Calculations

1. Using appropriate standards, calculate the total moles cycloleucine absorbed by each group of worms. Calculate the mean of the data for each group of 4 replicates and plot against concentration. Is uptake linear with concentration over the range 0.1–0.9 mM? If not, suggest a possible explanation.

2. Plot your data using the method of Lineweaver & Burk. What is the nature of the alanine inhibition of cycloleucine uptake? Calculate V_{max} and K_t for cycloleucine and the K_i for alanine as an inhibitor of cycloleucine using the following formula:

$$\text{Slope of inhibited line} = \frac{K_t}{V_{max}}\left(\frac{1+I}{K_i}\right)$$

where:

K_t = transport constant for cycloleucine determined in the absence of inhibitors

Table 3.3.3 *Timing schedule for Experiment 2*

Sample number	Begin preincubn. (min)	Begin incubn. (min)	Stop incubn. (min)
1	0	30	32
9	1	31	33
17	2	32	34
25	3	33	35
33	4	34	36
5	5	35	37
etc.	etc.	etc.	etc.

V_{max} = maximum velocity for cycloleucine
I = inhibitor concentration
K_i = inhibition constant

Further experiments – hypothesis testing

You have already determined that *H. diminuta* absorbs a neutral amino acid via a mediated transport system and that other neutral amino acids (e.g. alanine) inhibit the uptake of cycloleucine. In the light of experience gained in experiments 1 and 2, devise a method of investigating the site specificity of the amino transport systems in *H. diminuta*. What other information would you consider it desirable to acquire in an investigation of transport processes in this worm?

REFERENCES

Chappell, L. H. (1993). Physiology and nutrition. In: *Modern Parasitology – a Textbook of Parasitology*, 2nd edn. (ed. F. E. G. Cox), pp. 157–192. Oxford, Blackwell.

Pappas, P. W. (1983). Host–parasite interface. In: *Biology of the Eucestoda*, vol. 2 (eds. C. Arme & P. W. Pappas), pp. 297–334. London, Academic Press.

Pappas, P. W. & Read, C. P. (1975). Membrane transport in helminth parasites: a review. *Experimental Parasitology* **37**, 469–530.

3.4 Glycogen utilisation and deposition in flatworm parasites

D. W. HALTON

Aims and objectives

This exercise is designed to demonstrate:

1. Utilisation of ethanol-precipitable carbohydrate (glycogen) as an energy store in an endoparasitic flatworm (trematode or cestode), using simple spectrophotometry.
2. The distribution of glycogen in the various organ systems of the parasite, using histochemical procedures.

Introduction

The bioenergetic pathways of endoparasitic flatworms generally function anaerobically, with emphasis on synthetic metabolic capacities rather than complete substrate breakdown. The worms are generally facultative anaerobes, deriving their energy largely from the catabolism of glucose and glycogen and excreting highly reduced end products. In contrast to free-living organisms, terminal oxidative pathways are either absent or largely abbreviated in parasitic flatworms, precluding the use of proteins or lipids as energy sources. Thus, cestodes and most digenean trematodes have a pronounced carbohydrate metabolism. They contain and turnover large quantities of endogenous carbohydrate, have a high rate of transport of exogenous sugars into the tissues, and are known to produce substantial amounts of fatty acids *in vitro*. For example, the cyclophyllidean tapeworm, *Hymenolepis diminuta,* may contain as much as 50% of its dry weight as glycogen. When the host (rat) is deprived of carbohydrate in its diet, the glycogen content of the worm falls rapidly, and when such starved worms are given glucose *in vitro*, glycogen synthesis (glycogenesis) also proceeds rapidly. Similarly, the liver fluke, *Fasciola hepatica,* generally contains 15–20% glycogen per dry weight and under conditions of starvation uses up some 20% of this reserve within 5 h. Again, glycogen stores can be rapidly replenished when exogenous sugars become available.

The principal site of glycogen deposition in these worms is the parenchyma which serves as a carbohydrate reserve analogous in some ways to the liver of vertebrates. Glycogen also occurs in muscles, particularly those of the suckers, pharynx (in trematodes) and vitellaria, and in the developing embryos within the eggs. This exercise is in two parts:

1. The biochemical determination of ethanolic precipitable carbohydrate (glycogen).
2. The histochemical distribution of glycogen.

The exercises are based on findings with *F. hepatica*, but similar results can be obtained with *H. diminuta*.

Investigation 1: Biochemical determination of the glycogen content in *Fasciola*

Laboratory equipment and consumables
(per student or group)

Equipment

Spectrophotometer (preferably double-beam carrying two cuvettes)
Centrifuge
Four test-tubes and rack per student
Accurate 1 ml and 5 ml pipettes (preferably automatic)
Safety glasses

Consumables

Mammalian saline (0.9% NaCl w/v)	95% ethanol
Standard glucose solution (0.15 mg/ml)	30% potassium hydroxide
	Fresh liver fluke extract
	Starved liver fluke extract
GOD/Perid reagent (Boehringer)	Starved/refed liver fluke extract
Disposable gloves	

Source of parasite material

Live specimens of *F. hepatica* can be recovered from infected sheep and cattle livers following slaughter at local abattoirs and transported to the laboratory in a large volume (*c.* 100 worms per 3 litres) of warm (37 °C) mammalian saline), using a Dewar flask. Livers infected with *F. hepatica* are routinely condemned by meat inspectors on the production line and are easily identifiable. See also Additional information in Exercises 1.6 and 1.14.

Safety

There are few health hazards or risks of infection associated with adult liver fluke. Nevertheless, in recovering the parasite it is essential to practice good basic hygiene, making use of disposable gloves, laboratory coat and a face mask. In the UK it is advisable to telephone the abattoir and ask permission from the head veterinary officer or head meat inspector on site prior to arrival. They will insist that you are aware of local safety regulations and have the appropriate clothing, such that, on site, you wear a hard hat and hair-net, Wellington boots and a clean laboratory coat; it may be possible to borrow a hard hat at the abattoir, but check in advance. Admission will probably become easier once a satisfactory collection routine has been established. Great care should be taken in the laboratory when digesting worms in hot KOH (see below).

Instructions for staff

Of the live worms collected, one-third should be quickly transferred to 95% ethanol to provide material for a 'fresh fluke' extract. The remaining two thirds are maintained overnight (10 worms per litre) in a glucose-free mammalian saline at 37 °C. Of these, half are then placed in 95% ethanol to provide material for a 'starved fluke' extract; the remaining worms are transferred to mammalian saline containing 10 mM glucose (5 worms per litre) for a further 24 h at 37 °C and then placed in 95% ethanol as material for a 'refed fluke' extract. Two or three worms from each of the three conditions (fresh, starved, refed) should be selected for fixation and processing for histochemistry (see later).

Following ethanolic treatment, all of the flukes for the biochemical investigation are dried to constant weight and the weights noted. Worms are then digested individually in 2 ml 30% KOH solution in a Pyrex boiling tube, using a boiling water bath in a fume hood, after which the digest is cooled and the extracted glycogen precipitated by the addition of excess (3 ml) 95% ethanol; a gentle boiling of the ethanolic mixture will

facilitate precipitation of the glycogen. **Ensure these procedures with KOH and alcohol are carried out with great care
in a fume hood and that safety glasses are worn throughout.**
When cool, each sample digest is centrifuged (300 g) for 15 min,
the supernatant decanted, and the residue (glycogen) drained-
dry on filter paper. Each glycogen sample is then dissolved and
reduced to glucose by acid hydrolysis followed by neutralisation of the mixture with alkali. The final volume for each
sample after hydrolysis and neutralisation is adjusted to 20 ml.
Thus, the glucose derived from hydrolysis of the total glycogen
in each sample is contained in 20 ml of neutral solution.

Instructions for students

Each student is provided with a 1 ml sub-sample from each of
the three 20 ml test samples, together with 1 ml of a standard
glucose solution.

Procedure

Estimate the glucose content of each sample as follows:

1. Place 0.1 ml of each test sample and of the standard into an
   individual clean dry test-tube.
2. Add 5 ml of GOD/Perid reagent to each tube and leave the
   tubes for 30 min at room temperature while the colour
   develops.

During this period, glucose is oxidised to gluconate in the presence of molecular oxygen and the enzyme glucose oxidase
(GOD). Hydrogen peroxide (H_2O_2) is produced in the reaction,
and this in turn oxidises the dye, di-ammonium 2,2′-azino-bis
(3-ethylbenzothiazoline-6-sulphonate) (ABTS), which is also
present in the test reagent, to a coloured product. This stage of
the reaction proceeds in the presence of periodate (Perid). The
reactions may be summarised as follows:

$$\text{glucose} + O_2 + H_2O \xrightarrow{\text{GOD}} \text{gluconate} + H_2O_2$$
$$H_2O_2 + \text{ABTS} \xrightarrow{\text{Perid}} \text{coloured complex} + H_2O$$

3. Read the optical density for each sample at 436 nm using the
   spectrophotometer provided. Blanks for adjustment of the
   instrument are provided.
4. Tabulate your results and calculate:
   (i) The total amount of glycogen (mg) present in each of the
       three original samples of fluke material:

$$\text{mg glycogen in worm } (c_1) = \frac{OD_1}{OD_2} \times c_2 \times \frac{DF}{1.11}$$

OD_1	=	optical density of test
OD_2	=	optical density of standard
c_2	=	concentration of standard
DF	=	dilution factor (i.e. volume in which the total glucose derived from the sample is contained)
1.11	=	conversion factor for glucose to glycogen

(ii) the percentage dry weight of each fluke sample which represents glycogen.

Investigation 2: Histochemical demonstration of glycogen in *Fasciola*

Laboratory equipment and consumables
(per student or group)

Equipment

Staining jars or troughs
Source of running water
Timer

Consumables

Histoclear or similar wax solvent
Alcohol series (100%, 90%, 70%)
Distilled water
Mountant (e.g. DPX)
Source of amylase (filtered human saliva will do)
0.5% periodic acid

0.5% light green or methyl green
Glass slides and coverslips
Paper towelling
Alcoholic Bouin's fluid
Schiff reagent (Sigma)
Mammalian saline (0.9% w/v NaCl)

Instructions for staff

Prepare separate paraffin wax blocks of the three samples of *Fasciola* (fresh, starved, starved–refed), following routine fixation for 4h in alcoholic Bouin's fluid, dehydration, clearing and embedment in paraffin wax (see standard histological handbook for procedure). Orientate so as to cut sagittal serial sections (5–7 μm) of the worms and mount on standard glass slides (etch-labelled with diamond pencil), using adhesive (e.g. Mayer's egg albumin). Allow to dry overnight. Provide each student with two slides (for test and control) of sections (one or two sections per slide) from each of the three test conditions.

Instructions for students

The method used is the periodic acid–Schiff (PAS) reaction with suitable controls. The reaction is based on the fact that aqueous periodic acid will oxidize 1,2 glycol groups in tissues, largely in materials that consist of or contain carbohydrates, to produce aldehydes that are coloured by Schiff's reagent. Glycogen is hydrolysed by the enzyme amylase; thus, if an amylase-treated slide is stained in parallel with the test slide, comparison between these slides will show specifically the presence of glycogen. Human saliva is a ready source of amylase.

Procedure

1. Bring test slide and control slide for each worm sample through Histoclear (removes wax), then progressively through 100%, 90%, 70% alcohol (2 min each)
2. Wash the control slide in water; cover the sections with filtered saliva (or other source of amylase) and leave at room temperature for 20–30 min. Leave test slides in 70% during this time.
3. After 30 min, wash both test and control slides in running water (2 min).
4. Place both test and control slides in 0.5% periodic acid (5 min).
5. Rinse slides in tap water and then in distilled water.
6. Place in Schiff's reagent for 15–20 min.
7. Wash in running tap water (15–30 min).
8. Counterstain with light green or methyl green (1–2 min).
9. Wash in water, dehydrate in graded alcohols (2 min each), clear in Histoclear and mount in DPX mountant. Label.

Expected results
Carbohydrates in test sections will be coloured magenta or purple; they are PAS-positive. Any staining (PAS) in control slide is not due to the presence of glycogen (this has been removed by amylase). **For distributon of glycogen, therefore, subtract any staining in control slide from that given in test slide.**

   Enter your results in the Table (Fig. 3.4.1) so as to compare the distribution and relative amounts of glycogen in the given

Tissue	Fresh worm	Starved worm	Refed worm
Tegument			
Parenchyma – surface			
Parenchyma – deep			
Body musculature			
Oral sucker			
Acetabulum			
Pharynx			
Gut caeca			
Ovary (oocytes)			
Vitellaria			
Testis (sperm)			
Eggs			
Excretory system			
Key: +++, intensely stained; ++, moderately stained; +, slightly stained; o/+, occasionally stained; o, not stained; –, not known.			

Fig. 3.4.1 Histochemical distribution of glycogen in *Fasciola hepatica*.

samples. Pool results to cover all of the organ systems in the worms. Use a points system or O, +, ++, +++, as a measure of the relative amounts of glycogen in the various tissues (see the key). Discuss your results and relate them to the findings of the previous experiment.

Ideas for further exploration

- What are the main sites of glycogen deposition in the worms examined?
- Are the glycogen deposits diffuse and/or granular in appearance?
- Are they largely intracellular or intercellular as far as you can determine?
- What is the significance of any glycogen in the sperm and/or eggs?

- From which sites is glycogen depleted following starvation?
- Are these depleted reserves made good with refeeding over 24 h?

REFERENCES

Halton, D. W. (1967). Glycogen deposition in Trematoda. *Comparative Biochemistry and Physiology* **23**, 113–120.

Mansour, R. E. (1959). Studies on the carbohydrate metabolism of the liver fluke, *Fasciola hepatica. Biochimica Biophysica Acta* **34**, 456–464.

Smyth, J. D. & Halton, D. W. (1983). *The Physiology of Trematodes*. Cambridge, Cambridge University Press.

Smyth, J.D. & McManus, D.P. (1989). *The Physiology and Biochemistry of Cestodes*. Cambridge, Cambridge University Press.

von Brand, R. & Mercado, T. (1961). Histochemical glycogen studies on *Fasciola hepatica. Journal of Parasitology* **47**, 459–464.

3.5 Effects of classical transmitters on the motility of parasitic roundworms and flatworms

A. G. MAULE, N. J. MARKS & J. W. BOWMAN

Aims and objectives

This exercise is designed to demonstrate:

1. The physiological effects of classical transmitter substances on the motor activity of roundworm (nematode) parasite somatic musculature.
2. The physiological effects of classical transmitter substances on the motor activity of flatworm (platyhelminth) parasite somatic musculature.

Introduction

The common liver fluke, *Fasciola hepatica*, is of great economic importance, occurring worldwide, with the exception of Africa and South Asia, where it is replaced by *F. gigantica*. It parasitises all domestic ruminants and causes a wide range of clinical symptoms. Although not normally regarded as an important parasite of humans, there are exceptions, for example, in parts of Bolivia. The most common type of infection is chronic fascioliasis, which occurs mainly in cattle and sheep and causes anaemia, oedema (bottle jaw), digestive disturbances (constipation and diarrhoea) and general weight loss. Most of the damage is caused by juvenile *F. hepatica* as they migrate through liver tissue en route to the bile ducts, where they develop to maturity as sexually reproducing adult worms.

Ascariasis is the most prevalent human and livestock helminth-parasite infection and is caused by the sibling species *Ascaris lumbricoides* and *A. suum*. The latter is the large gastrointestinal parasitic roundworm of pigs and is responsible for a large economic burden on farming communities worldwide. Adult female *A. suum* are 20–35 cm in length, the males are somewhat smaller, 15–30 cm. The life cycle of *A. suum* is direct. Infective larvae develop within the egg and upon digestion by the host animal hatch and migrate through the gut wall and liver, ultimately reaching the lungs. Once the larvae have reached the lungs they penetrate the alveoli and proceed to the trachea, where they are swallowed, finally reaching the small intestine, where they mature and reproduce. The mature female worm is capable of producing up to 200 000 eggs a day.

Cytochemical studies have shown both of these parasites to have well developed nervous systems. As an ever-increasing number of reports of resistance to front-line anthelmintics become apparent there is need to find novel drug targets. The neuromuscular systems of these parasites are obvious potential targets, in view of their importance in the worms' movement, attachment, feeding and reproductive activities. However, in identifying likely targets, it is important to understand the pharmacological effects of classical neurotransmitters on the motility of flatworm and roundworm parasites.

Laboratory equipment and consumables
(per student or group)

Equipment

Stereomicroscope with external light source (magnification $\times 200$)

Sylgard-lined dissecting dish

Water bath, 37 °C

Isometric force transducer (e.g. # 797159-1 or 369500-8640)

Accurate automatic pipettes (e.g. Gilsons or equivalent, range 10–500 μl, depending on the volume of the recording chamber)

Disposable gloves

Face mask

Fine pair of dissecting scissors and forceps

Chart recorder

Fine wire hooks

Tissue chamber (jacketed or placed within a waterbath)

Anaerobic gas supply (5% CO_2, 10% H_2 and 85% N_2)

Consumables
For Ascaris experiments:

Ascaris Ringers Solution (ARS) comprising: NaCl, 4 mM; $CaCl_2$, 5.9 mM; $MgCl_2$, 4.9 mM; $C_4H_{11}NO_3$ [Tris], 5 mM; $NaC_2H_3O_2$, 125 mM; KCl, 24.5 mM, pH 7.4 (with HCl).

Classical transmitters: acetylcholine (ACh), γ-aminobutyric acid (GABA), prepared as 10 mM stock solutions.

For Fasciola experiments:

Hedon Fleig solution (all values in mM unless otherwise indicated) mix 100 ml stock solution A, 100 ml stock solution B, 800 ml double-distilled water and glucose, 15; pH 7.6.
Stock solution A: NaCl, 120.6; KCl, 4; $MgSO_4.2H_2O$, 1.9; $CaCl_2.2H_2O$, 0.9; 4 litres double-distilled water.
Stock solution B: $NaHCO_3$, 18.4; 4 litres of double-distilled water.
Classical transmitters: 5-HT, ACh, prepared as 10 mM stock solutions.
High K^+ solution (90 mM KCl)
Mammalian saline (0.9%)

Sources of parasite material

Ascaris suum specimens can be obtained from abattoirs, especially those which process pig gut for use as sausage skins. Guts are removed and passed through mangles, which force out the intestinal contents, including *Ascaris*. The worms can be collected from the exudate and are placed in a flask containing ARS (37 °C) and transported to the laboratory.

Fasciola hepatica specimens are generally available from the infected livers of sheep and cattle slaughtered at local abattoirs located mainly in the wetter North Western parts of the UK. Worms can be dissected easily from the liver and bile ducts and placed in a large volume (100 worms/3 litres) mammalian saline (0.9% NaCl (w/v), 37 °C) in a Dewar flask for transport back to the laboratory. Livers infected with *F. hepatica* are routinely condemned by meat inspectors on the production line and are easily identifiable. See also Additional information in Exercises 1.5 and 1.6.

Safety

There are few health hazards or risks of infection associated with adult liver fluke. Nevertheless, in recovering the parasite it is essential to practice good basic hygiene, using disposable gloves, laboratory coat and a face mask. Collecting or working with specimens of *Ascaris* does pose a health risk, principally because of sensitisation to the pseudocoelomic fluid and excretory products, which are very allergenic; accidental infection is extremely rare.

1. **Infection of humans** with embryonated larvae. This may occur if hygiene is not adequate during handling. Most eggs produced by female *Ascaris* accumulate in the bottom of the vessels in which they are contained because they settle under gravity. Eggs may survive in a moist environment for at least 12 months. They do not survive desiccation, so a hot drying oven may be used to clean vessels that have contained *Ascaris*. The eggs are also very sticky and may be transferred from hands to other objects such as door handles. Use gloves during handling but remove them before using a microscope or other equipment. Always wash your hands before leaving the laboratory.

2. The **body fluid** of *Ascaris* **is very allergenic**. Fresh *Ascaris* are turgid because they have a positive pressure in their body cavity. The first cut during dissection may produce a fine spray of body fluid that could be inhaled. If this happens, it may initiate an asthma-like condition and there are reports describing individuals so sensitized to *Ascaris* antigens as having collapsed while carrying the Thermos flask from the abattoir to the laboratory. Initial dissection should be carried out behind the hood of an effective fume cupboard and under saline and wearing a face mask.

Instructions for staff

It is possible to maintain *Ascaris* for up to 4 days *in vitro* as long as the ARS is changed twice daily. Where possible for safety reasons, the *Ascaris* should be maintained in a water bath at 37 °C in a fume hood (see above). *F. hepatica* remain viable for physiological study for a few hours only. Again, regular changing of the Hedon–Fleig saline is important (alternative and more expensive media, including RPMI-1640, have been employed for the maintenance and physiological analyses of flukes).

It is advisable to telephone the appropriate abattoir and ask permission from the veterinary officer or senior meat inspector on site prior to arrival. They will ensure that you are aware of local safety regulations and have the appropriate clothing. Most abattoirs will insist that, on site, you wear a hard hat and hair-net, Wellington boots and a clean laboratory coat. It may be possible to borrow a hard hat at the abattoir, but check in advance. Admission will become easier once a satisfactory collection routine has been established.

If it is not possible to obtain a jacketed tissue chamber, a tissue chamber placed upright in a Perspex tank or water bath maintained at 37 °C will suffice.

Female *A. suum* are larger than male worms and more amenable to dissection and physiological analysis; males may be identified also by their hooked posterior ends. It is advisable to provide students with female worms.

Note that drug solutions should be prepared on the day of the practical and the stock solutions should be prepared in double-distilled water and stored in the refrigerator prior to use. If longer term storage is required, stock solutions should be maintained at −20 °C in the dark.

Instructions to students

A. *Ascaris suum*

1. Choose a worm that is turgid, pink in colour with well-defined reddish lateral lines as these signs indicate that the worm is healthy.
2. Locate the gonopore, which is situated on the ventral surface about one-third of the worm's length from the head, midway between the lateral lines; it appears as a small indentation.
3. Make a transverse incision through the body wall approximately 2 mm below the gonopore and another approximately 2 cm above. A cylindrical segment of worm can then be removed and placed in a Sylgard dissecting dish containing ARS, preferably at 37 °C.
4. The segment is then opened up by cutting longitudinally along the length of one of the lateral lines and pinning down the edges of the segment, outer face downward. At this stage, the intestine should be clearly visible as a soft brown or green tube-like structure situated in the centre of the body-wall segment.
5. Gently tease away the intestine from the muscle field.
6. A second cut is made along the length of the remaining lateral line effectively separating the muscle into a dorsal and ventral (containing the gonopore) muscle-strip preparation.
7. Insert the fine metal hooks midway between the cut lateral lines and 3 mm from the anterior and posterior end of one of the muscle segments. Using the inserted hooks, this can be transferred to the tissue chamber containing ARS at 37 °C and attached to a stationary holdfast (in the tissue chamber) and a force transducer (Fig. 3.5.1).

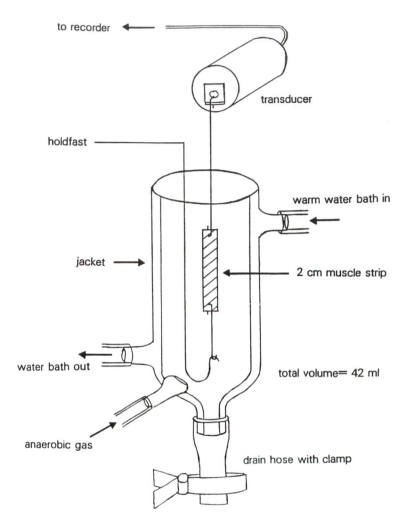

to recorder

transducer

holdfast

warm water bath in

jacket

2 cm muscle strip

water bath out

total volume= 42 ml

anaerobic gas

drain hose with clamp

Fig. 3.5.1 Schematic diagram of muscle tension recording apparatus.

8. Apply gentle tension to the muscle strip, taking care not to rip the hooks out of the tissue. Set the gas supply so that a steady stream of bubbles is dispersed within the chamber (the gas will assist in rapid distribution and mixing of the test compounds). Any movement generated by the muscle is then amplified and recorded on a chart recorder.
9. Leave the muscle strip for 10–15 min to equilibrate and establish regular spontaneous contractions.
10. Add the test compound, a classical transmitter, to the experimental chamber using a pipette.
11. Record the activity of the muscle preparation for a further 20 min, stop the recorder and slowly drain the medium from the recording chamber.
12. Gently refill the chamber with fresh medium at 37 °C, restart the recorder, and continue recording activity for a

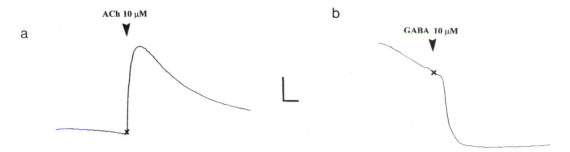

Fig. 3.5.2 Muscle tension recordings showing the effect of (a) 10 μM ACh and (b) 10 μM GABA on *Ascaris suum* muscle preparations. Horizontal scale bar represents 5 min duration and vertical bar 1 g tension.

further 20 min. The activity of the muscle preparation before, during and after drug addition and removal should be recorded for analysis.

Drug addition and results

1. Using the stock solution of ACh (10 mM) add a volume equal to one hundredth the volume of the experimental chamber using a pipette, e.g. if the chamber is 42 ml (see Fig. 3.5.1), add 0.42 ml (= 420 μl).

   ACh is an excitatory neurotransmitter in those nematodes examined and will cause a rapid increase in tension, which will slowly diminish due to desensitisation of the tissue to the drug (Fig. 3.5.2a). ACh opens a nicotinic-like cation channel in *Ascaris* muscle.

2. Using the stock solution of γ-aminobutyric acid (GABA; 10 mM) add a volume equal to one hundredth the volume of the experimental chamber using a pipette, as described above in 1 (see Fig. 3.5.1).

   GABA is an inhibitory neurotransmitter in nematodes and will cause a rapid relaxation of the tissue and cessation of spontaneous contractions (Fig. 3.5.2b). The effect will slowly diminish due to desensitisation of the tissue to the drug. GABA opens a chloride channel in *Ascaris* muscle.

B. *F. hepatica*

1. Select a worm that has regurgitated its intestinal contents (appears uniform in colour, i.e. no red or black pigment visible in gut).
2. Place the worm in a Sylgard dissecting dish containing Hedon–Fleig solution preferably at 37 °C.
3. Using the dissecting scissors, carefully remove the lateral edges (0.5 mm) of the fluke from the widest part of the body; this facilitates penetration of test substances.

4. Insert one of the metal hooks through the ventral sucker and the other approximately 3–5 mm from the posterior end.
5. Transfer the fluke to the tissue bath containing fresh Hedon–Fleig solution at 37 °C and place under slight tension; allow to equilibrate for at least 10 min prior to the addition of a test compound. It may take up to 20 min for regular spontaneous contractility to develop.
6. Once a steady baseline has been established, carefully add the test compound to the chamber using the pipette.
7. Record the muscle activity of the specimen for another 20 min.
8. Stop the recorder and gently drain the fluid from the recording chamber, taking care not to dislodge the tissue, and replace with fresh medium at 37 °C.
9. Restart the recorder and record muscle activity for a further 20 min. The time taken for the muscle preparation to return to pre-treatment levels can then be noted. The activity of the muscle preparation before, during and after drug addition and removal should be recorded for analysis.

Drug addition and results

1. Using the stock solution of ACh (10 mM) add a volume equal to one hundredth that of the experimental chamber using a pipette, as described for *Ascaris* (see Fig. 3.5.1).
   ACh is believed to be an inhibitory neurotransmitter in most trematodes and will cause a rapid decrease in tension and muscle activity.
2. Using the stock solution of 5-HT (10 mM) add a volume equal to one hundredth that of the experimental chamber using a pipette, as described in 1 (see Fig. 3.5.1). 5-HT has excitatory effects on trematode muscle and generally increases the frequency and amplitude (height) of the spontaneous contractions.

Ideas for further exploration

- Experiments could be carried out using serial dilutions of the stock solutions such that the effects of a range of concentrations of the classical transmitters could be examined, e.g. 10 mM, 1 mM and 0.1 mM concentrations of ACh, GABA and 5-HT could be prepared and tested on the muscle preparations. In all cases the lower concentrations of test compound should be tested first.

- The effects of other classical transmitter compounds, including dopamine, adrenaline and noradrenaline on *F. hepatica*, could be examined. Also, cholinergic antagonists (mecamylamine, atropine) and agonists (carbachol, arecoline) could be examined and their effects compared with those of the classical transmitters.

- It is possible to examine the effects of the non-classical transmitters, including neuropeptides, on the motility of both *Ascaris* and *Fasciola* muscle preparations. Examination of the physiological effects of native flatworm (YIRFamide, RYIRFamide) and nematode (KHEYLRFamide, SDPNFLRFamide) FMRFamide-related peptides will provide readily reproducible data (the amino acid sequences of these peptides are presented in single letter format). Unfortunately, neuropeptides may be prohibitively expensive to use in such practicals. It should be noted, however, that they are much more potent on the tissues than the classical transmitters and can generally be used at 100-fold lower concentrations than the classical transmitters.

- The use of denervated muscle strips of *Ascaris* allows the experimenter to determine whether or not the observed physiological response is nerve-cord dependent. To prepare denervated preparations the muscle segments are not trimmed along the lateral lines (as outlined above) but are cut on either side of the ventral and dorsal nerve cords (these appear white at the mid-points between the lateral lines). Note that two very narrow strips that contain the ventral and dorsal nerve cords are produced by this method and are discarded. These dissections are more difficult to perform than that in standard innervated muscle strip preparations.

- Investigation of ionic dependencies of drug responses, i.e. application of the drug in high K^+ media (ARS containing 125 mM KCl), Ca^{++}-free ARS (in which $CaCl_2$ is replaced by $CoCl_2$) and finally in Cl^--free media [$C_3H_5O_2Na$ (4 mM), $C_3H_5O_2 1/2Ca$ (5.9 mM), $MgSO_4$ (4.9 mM), Tris (5 mM), $NaC_2H_3O_2$ (125 mM) and $KC_2H_3O_2$ (24.5 mM)]. In these experiments, the standard ARS is replaced with one of these modified solutions. An example experiment is the examination of the effects of GABA on an *Ascaris* muscle strip maintained in Cl^--free ARS. Upon the addition of GABA, Cl^- channels are opened and, since the external Cl^- has been removed, Cl^- will move out of the muscle cells making them more positively charged and causing a contraction. This is the opposite response to that observed in standard ARS media where the [Cl⁻] outside the cell is greater than the [Cl⁻] inside the muscles such that when GABA opens the channels, Cl⁻ will move into the muscle cells and cause them to hyperpolarise and thus relax.

Information on similar exercises

Other helminth taxa can be used for these practicals, although the bathing medium will need to be altered accordingly. Examples include the cyclophyllidean tapeworm, *Moniezia expansa*, from sheep gut and the monogenean fish-gill parasite, *Diclidophora merlangi*, from whiting.

REFERENCES

Chance, M. R. A & Mansour, T. E. (1953). A contribution to the pharmacology of movement in the liver fluke. *British Journal of Pharmacology* **8**, 134–138.

Holmes, S. D. & Fairweather, I. (1984). *Fasciola hepatica*: the effects of neuropharmacological agents upon *in vitro* motility. *Experimental Parasitology* **58**, 194–208.

Marks, N. J., Johnson, S., Maule, A. G., Halton, D. W., Shaw, C., Geary, T. G., Moore, S. & Thompson, D. P. (1996). Physiological effects of platyhelminth RFamide peptides on muscle-strip preparations of *Fasciola hepatica* (Trematoda: Digenea). *Parasitology* **113**, 393–401.

Marks, N. J., Maule, A. G., Halton, D. W., Geary, T. G., Shaw, C. & Thompson, D. P. (1997). Pharmacological effects of nematode FMRFamide-related peptides (FaRPs) on muscle contractility of the trematode, *Fasciola hepatica*. *Parasitology* **114**, 531–539.

Maule, A. G., Geary, T. G., Bowman, J. W., Marks, N. J., Blair, K. L., Halton, D. W., Shaw, C. & Thompson, D. P. (1995). Inhibitory effects of nematode FMRFamide-related peptides (FaRPs) on muscle strips from *Ascaris suum*. *Invertebrate Neuroscience* **1**, 255–265.

Maule, A. G., Bowman, J. W., Thompson, D. P., Marks, N. J., Friedman, A. R. & Geary, T. G. (1996a). FMRFamide-related peptides (FaRPs) in nematodes: occurrence and neuromuscular physiology. *Parasitology* **113**, S119–S135.

Maule, A. G., Geary, T. G., Marks, N. J., Bowman, J. W., Friedman, A. R. & Thompson, D. P. (1996b). Nematode FMRFamide-related peptide (FaRP)-systems: occurrence, distribution and physiology. *International Journal for Parasitology* **26**, 927–36.

Sukhedo, S. C., Sangster, N. C. & Mettrick, D. F. (1986). Effects of cholinergic drugs on longitudinal muscle contractions of *Fasciola hepatica*. *Journal of Parasitology* **72**, 858–864.

3.6 Electrophysiology of *Ascaris suum* body muscle

R. J. MARTIN

Aims and objectives

This exercise is designed to demonstrate:

1. Some electrophysiological properties of nematode muscle cells.
2. Measurement of intracellular membrane potentials from *A. suum* muscle by recording from the bag (muscle cell body) region.
3. Recording the electrophysiological effects of piperazine, an inhibitory γ-aminobutyric acid (GABA)ergic anthelmintic.
4. Recording the electrophysiological effects of pyrantel, an excitatory nicotinic anthelmintic.

Introduction

Adult *Ascaris suum* are large parasitic nematodes that may be recovered from the intestine of pigs slaughtered at the abattoir. Mature female *A. suum* may be up to 35 cm in length. The number found in each pig varies dramatically from area to area and relates to the husbandry/hygiene on which the pig has been raised. Migrating *A. suum* larvae may cause significant scarring of the liver (milk spot) as they pass through this organ before entering the lungs and reaching the upper intestine where adults are found. After mating, many thousands of *Ascaris* eggs are produced each day by the mature female and are released in faeces.

Human infection with *A. suum* is rare. The species of *Ascaris* that regularly infects humans is *Ascaris lumbricoides*. Over half the world's human population is infected at one time during their lives with *A. lumbricoides*. Treatment of these infections in man or animals involves the use of anthelmintics (i.e. anti-helminth drugs), which have a selective action on the nematode parasite without affecting the host. Some of these effects can be **219**

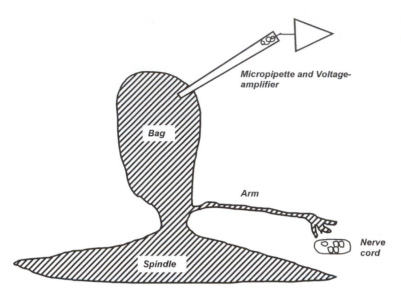

Fig. 3.6.1 Single muscle cell from *Ascaris suum.*

conveniently studied on *A. suum* because it is a large nematode that enables intracellular recordings of body muscle cells to be made, using micropipettes. One commonly used anthelmintic is pyrantel, an acetylcholine agonist that opens nicotinic acetylcholine channels on nematode muscle, producing depolarization and spastic paralysis. The paralysed worm is then eliminated from the host intestine since it cannot maintain its position. Another anthelmintic, piperazine, acts as a GABA agonist and opens GABA-gated Cl^--channels in nematode muscle (Fig. 3.6.1) to produce hyperpolarization, relaxation and flaccid paralysis.

Laboratory equipment and consumables (per group)

Equipment

Dissecting microscope (zoom up to × 80) and fibre optic illumination
Perspex experimental chamber illuminated from below, fitted with a water jacket*
Sylgard base onto which the muscle flap is pinned
Stable (vibration-free) table on which to mount the equipment
Micromanipulator to hold the intracellular micropipettes
Intracellular amplifier to record membrane potentials
Oscilloscope to observe the membrane potential records

Pen recorder to record the membrane potentials

Water bath (30 °C) and large beaker for *Ascaris* after collection from the abattoir

Locke's saline solution** (changed daily) to maintain the *Ascaris****

Micropipette holder

Bath earth electrode

Ag/AgCl wire to place in the intracellular micropipette and connect to the intracellular amplifier to record the membrane potentials

Fine forceps, fine scissors and fine pins

*Warmed water is pumped through jacket to maintain a temperature of ~30 °C.

**Locke's saline: 9.2g NaCl, 0.4g KCl, 0.24g $CaCl_2$, 0.15g $NaHCO_3$, per litre distilled water.

***Normally worms survive 4–5 days after collection.

Consumables

Capillary tubing with a glass fibre (Clark Electromedical) suitable for making intracellular micropipettes

Potassium acetate (2M) solution to fill intracellular micropipettes and filling syringe and fine long needle/catheter

Ascaris saline containing (mM): NaCl, 135; KCl, 5; $CaCl_2$, 3; $MgCl_2$ 7; Tris, 5; adjust to pH 7.2 using maleic acid

Pyrantel tartrate (0.1 mM) and 1 mM piperazine as separate stock solutions

Petri dishes, disposable gloves, face mask

Sources of parasite material

A. suum can be recovered from an abattoir that slaughters pigs and clears the intestines for sausage skins. If the abattoir does not strip the intestine but throws them away then it may be possible to pay the workers in the gut room to collect a few *Ascaris* from infected pigs; they often know by feel which pig gut is infected. The worms should be collected into a stainless steel Thermos flask (glass is too vulnerable under abattoir conditions) containing Locke's solution (warm ~ 35 °C). The flask should be returned to the laboratory and maintained at 30 °C. The Locke's solution should be replaced daily. See also Additional information in Exercise 1.5.

Safety

There are two recognised hazards of working with *A. suum:*

1. **Infection of humans** with embryonated larvae. This may occur if hygiene is not adequate during the handling. Most eggs produced by the female *Ascaris* accumulate in the bottom of the vessels in which they are contained because they settle under gravity. Eggs may survive in a moist environment for at least 12 months. They do not survive desiccation, so a hot drying oven may be used to clean vessels that have contained *Ascaris*. The *Ascaris* eggs are also very sticky and may be transferred from hands to other objects such as door handles. Use gloves during handling but remove them before using a microscope or other equipment. Always wash your hands before leaving the laboratory.
2. The **body fluid** of *Ascaris* **is very allergenic**. Fresh *Ascaris* are turgid because they have a positive pressure in their body cavity. The first cut during dissection may produce a fine spray of body fluid that could be inhaled. If this happens, it may initiate an asthma-like condition. Initial dissection should be carried out behind the hood of an effective fume cupboard and under saline, and wearing a face-mask.

Instructions for staff

As mentioned, *A. suum* may be collected from abattoirs killing pigs and stripping the intestines for sausage skins. It is advised that a meat inspector or veterinary officer at the abattoir is approached and their help in collecting the *Ascaris* sought. It is important to collect worms fresh into warm saline in a thermos flask because cold shock will reduce the viability of the preparation. Fresh specimens can be returned to the laboratory and maintained in Locke's solution at 30 °C for about 5 days if the solution is replaced daily.

Although the electrophysiological experiments described here are not hard to perform, they are most likely to be successful if they are set up and supervised by an electrophysiologist who has some experience of intracellular recording. Not all the details required for electrophysiological experiments to work can be described here. Many small problems, such as pulling and filling a micropipette successfully, finding the beam on the oscilloscope, sorting out 50 Hz noise problems, making a bath earth, preventing salt accumulation on the recording elec-

Fig. 3.6.2 Location of cylindrical 2 cm section of worm for excision.

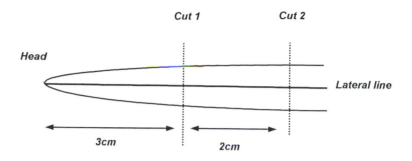

trodes, inserting the micropipette into the muscle cell, and recognising when the experiments are working, are acquired through experience. Students often expect the experiments to work immediately but are capable of finding all sorts of ways of stopping them from doing so. Time, patience and persistence are important ingredients for electrophysiological success.

Instructions for students

Preparation of parasite muscle

1. Collect a mature *Ascaris* using forceps, wearing surgical gloves and face mask, and place it under saline in a Petri dish in a fume cupboard.
2. Pull the fume hood down leaving enough room for your hands to work and turn on the fan. Do not breath in any of the *Ascaris* body fluid that might be produced in a fine spray as you make the first cut.
3. Prepare a 2 cm cylindrical section (Fig. 3.6.2) by making two transverse cuts with scissors across the body: one at a point 3 cm from the head and the other 5 cm from the head. Avoid cutting into the uterus of the female – it will release many eggs!
4. Place the preparation in the experimental chamber and immerse in *Ascaris* saline.
5. View the preparation from above while illuminating with a fibre optic from below.
6. Using fine scissors, carefully cut along one lateral line (the faint brown line) of the *Ascaris* from one end of the cylindrical section to produce a curled flap preparation.
7. This should then be pinned cuticle side down onto the Sylgard base at the bottom of the experimental chamber (Fig. 3.6.3).

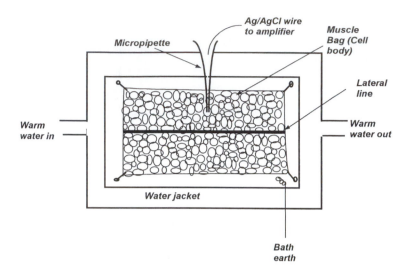

Fig. 3.6.3 Experimental chamber and preparation *in situ*.

8. Remove the lemon-yellow intestine with fine forceps.
9. Stretch the muscle flap to reduce movement and adjust the depth of the *Ascaris* saline to just cover the flap preparation.

Preparation of micropipettes

1. A micropipette is prepared using the micropipette puller and filled with the 2 M potassium acetate. If capillary tubing with an internal glass fibre is used it will fill easily and bubbles usually do not matter. It is sometimes useful to check the micropipette under the microscope for a broken tip before proceeding.
2. Place the micropipette in the pipette holder attached to the micromanipulator with its tip placed in the *Ascaris* saline. There are many types of holder and micromanipulator but because the size of the cell body of the muscle cell is very large, inexpensive manipulators (Prior) are usually adequate.
3. Connect the intracellular amplifier by placing the Ag/AgCl wire into the 2 M potassium acetate in the micropipette.
4. Turn on the amplifier and other devices including the oscilloscope.
5. Centre the beam of the oscilloscope monitoring the potential recording by backing standing potentials so that it reads 0 mV; a gain of 10 mV per 'box' is recommended.
6. Measure the resistance of the micropipette using the intracellular amplifier. A resistance of 15–25 MΩ for the micropipette may be most suitable. A micropipette with too low a resistance indicates a large-tipped pipette that is likely to

Fig. 3.6.4 An intracellular recording trace from *Ascaris* muscle.

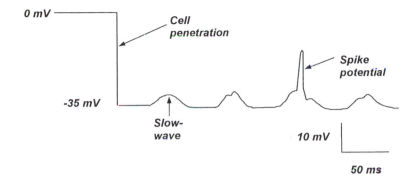

damage the cell during penetration. The larger the size of the *Ascaris,* the lower the resistance of micropipettes can be. Very high resistance pipettes break more easily and are often very noisy. Making a micropipette, placing it in the experimental bath without breaking and then measuring its resistance, are skills that take time to acquire; be patient when you start.

Recording muscle potentials

1. View the preparation and micropipette through the dissecting microscope once the micropipette is in its holder and its resistance is satisfactory. The illumination and magnification of the microscope should be adjusted to maximise visibility.
2. Move the micropipette using the micromanipulator and place directly over a target bag cell (cell body).
3. Lower the micropipette, using the fine adjustment of the micromanipulator, until it just touches the bag membrane; there is usually a small deflection on the oscilloscope trace of 1–5 mV.
4. Penetrate the cell with the micropipette by buzzing (increasing the capacitance feedback of the amplifier to produce oscillation for a very brief period) or tapping the table. Either method may work best depending on the preparation and your set-up. If you are not successful, target another cell. If the microelectrode drops into an undamaged muscle cell bag you will record a membrane potential of about −35 mV. In active fresh preparations at temperatures above 30 °C, the membrane potential shows regular depolarising potentials. There are two types that are most often seen (Fig. 3.6.4):
(i) The fast spike potential.
(ii) The smaller amplitude slow wave.

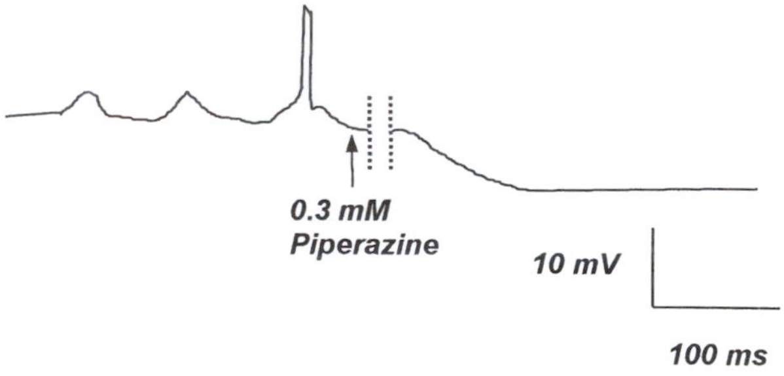

Fig. 3.6.5 An intracellular trace following application of piperazine.

5. Turn on your pen-recorder and adjust speed and gain to make a suitable tracing. By convention, negative potentials are displayed downwards and potentials in a positive direction are displayed upwards.

   There are two types of motor neurone that supply the muscle cell of nematodes: inhibitory GABAergic and excitatory cholinergic. There are inhibitory receptors on *Ascaris* muscle that gate (open) Cl channels when GABA or the anthelmintic piperazine is applied to the preparation. GABA and piperazine produce a hyperpolarisation of the membrane, an increase in membrane potential. There are also excitatory nicotinic receptors that gate non-selective cation channels; stimulation of these receptors with acetylcholine or with a nicotinic anthelmintic, like pyrantel, produces depolarisation, a decrease in the membrane potential.

6. While recording, add an amount of 1 mM piperazine to the edge of the batch, avoiding direct contact with the preparation, such that the final bath concentration of drug is 0.3 mM. You will notice that the effect of piperazine is to increase the membrane potential of the preparation, i.e. hyperpolarise the membrane, and to decrease or abolish the spontaneous slow waves or spike potentials (Fig. 3.6.5). The speed of onset depends on the amount of stirring (circulation in the bath). Sometimes the piperazine produces a sufficient relaxation to displace the intracellular pipette; this may show itself as an artefactual depolarisation.

7. Remove the piperazine from your bath by changing the bath solution at least four times. You are likely to displace your intracellular micropipette during this process so after completion of the washing you will have to target another muscle cell for recording.

8. Record the effects of pyrantel, starting with a stable recording from another cell. It is ideal if your preparation is

Fig. 3.6.6 An intracellular trace following application of pyrantel.

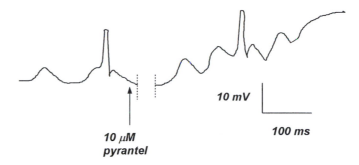

10 mV

100 ms

10 μM
pyrantel

showing spontaneous depolarising potentials after pipera-zine has been washed off. Allow the preparation to warm up again if you have washed your preparation in cold saline.

9. Add an amount of 0.1 mM pyrantel tartrate to the bath, as you did for piperazine, such that the final concentration of drug in the bath is 10 μM. The effect of pyrantel should be to produce a depolarisation associated with an initial increase in the frequency of the depolarising potentials (Fig. 3.6.6). The depolarisation may, after an increase in spike activity, produce a depolarising block of the depolarising waves. Con-traction may sometime displace the microelectrodes. The pyrantel may be removed from the preparation by vigorous washing (four times as before with piperazine).

Ideas for further exploration

*   Adjacent muscle cells are electrically coupled at the arms so that when one muscle cell in the ventral half (one side of the lateral line) depolarises and contracts, its neighbour does so as well. Some evidence for this may be gathered by recording the electrical activity of one cell for 2–3 min and then remov-ing the micropipettes from that cell and then recording from an adjacent cell. It may be possible to notice that the pattern of the depolarising potentials (usually slow waves) is nearly the same. If, however, it is from the dorsal half (the other side of the lateral line) muscle cells that recordings are made, there should be a different pattern of slow waves.
*   The receptors mediating the effects of pyrantel are nicotinic and may be reduced by high concentrations of tubocurarine (100 μM), so it may be interesting to investigate the effects of an application of tubocurarine on the action of pyrantel.
*   Another commonly used nicotinic anthelmintic is levami-sole. You may wish to examine the effects of this compound on the electrophysiology of the preparation.

- The effects of piperazine are mediated by an increase in the Cl^- permeability of the membrane. It is possible to replace all the Cl^- ions in the bathing solution with larger impermeant anions like acetate. If you use an agar bridge for a bath earth instead of the Ag/AgCl wire, which limits changes in standing potentials as the Cl^- ions of the bath solutions are replaced, then effects of Cl^- replacement with acetate anions may be examined. It is often found that immediately after replacing the bath Cl^- ions with acetate, there is a depolarising response to piperazine instead of a hyperpolarising response. This is because Cl^- ions leave the muscle cell instead of entering it under the new concentration gradient when acetate is in the bath. If all the intracellular Cl^- is leached from the cell after prolonged replacement of Cl^- with acetate, then piperazine will have little effect on the membrane potential because there is no Cl^- movement across the cell membrane.
- The organisation of the neuromuscular junction in *Ascaris* has been studied in considerable detail. What details of the anatomy, physiology and pharmacology of this organisation can you determine?

Information on similar exercises

Ascaridia galli or other sufficiently large nematodes may be used for these experiments, providing they can be collected freshly and are large enough for dissection. For *A. suum,* it may be possible to purchase samples from Ridgway Sciences, Rodmore Mill Farm, Alvington, Gloucester, GL15 6AH, Tel: 01594-530204.

REFERENCES

Aubry, M. L., CowelL, P., Davey, M. J. & Shevde, S. (1970). Aspects of the pharmacology of new anthelmintics: pyrantel. *British Journal of Pharmacology* **38**, 332–344.

del Castillo, J., de Mello, W. C. & Morales, T. (1964). Mechanism of the paralysing action of piperazine on *Ascaris* muscle. *British Journal of Pharmacology* **22**, 463–477.

Harrow, I. D. & Gration, K. A. F. (1985). Mode of action of the anthelmintics morantel, pyrantel and levamisole in the muscle cell membrane of the nematode *Ascaris suum. Pesticide Science* **16**, 662–672.

Martin, R. J. (1982). Electrophysiological effects of piperazine and diethylcarbamazine on *Ascaris suum* somatic muscle. *British Journal of Pharmacology* **77**, 255–265.

Martin, R. J. (1985). γ-aminobutyric acid- and piperazine-activated single channel currents from *Ascaris suum* body muscle. *British Journal of Pharmacology* **84**, 445–461.

Martin, R. J. (1993). Neuromuscular-transmission in nematode parasites and antinematodal drug-action. *Pharmacology & Therapeutics* **58**, 13–50.

Martin, R. J., Pennington, A. J., Duittoz, A. H., Robertson, S. & Kusel, J. R. (1991). The physiology and pharmacology of neuromuscular-transmission in the nematode parasite, *Ascaris suum*. *Parasitology* **102**, S41–S58.

Robertson, S. J. & Martin, R. J. (1993). Levamisole-activated single-channel currents from muscle of the nematode parasite *Ascaris suum*. *British Journal of Pharmacology* **108**, 170–178.

Robertson, S. J., Pennington, A. J., Evans, A. M. & Martin, R. J. (1994). The action of pyrantel as an agonist and an open-channel blocker at acetylcholine receptors in isolated *Ascaris suum* muscle vesicles. *European Journal of Pharmacology* **271**, 273–282.

3.7 Immunocytochemical localisation of neuroactive substances in helminth parasites

N. J. MARKS, A. G. MAULE & D. W. HALTON

Aims and objectives

This exercise is designed to investigate:

1. The distribution of the biogenic amine, 5-hydroxytryptamine (5-HT), in selected helminths.
2. The distribution of immunoreactivity to the neuropeptide, FMRFamide, in the nervous system of selected helminths.

Introduction

Immunocytochemistry is a method whereby antibodies, tagged with a visible fluorescent probe or fluorophore, are employed to detect antigens in tissue preparations using a specific antibody–antigen reaction. The most commonly employed immunocytochemical method is the indirect immunofluorescence technique (Coons *et al.*, 1955) in which the primary antibody binds to the antigen in the tissue sample followed by a secondary antibody, labelled with a fluorescent tag, that binds to the primary antibody (Fig. 3.7.1A). The indirect technique allows for more than one secondary antibody to bind to the primary antibody, thereby ensuring amplification of the signal (stronger immunostaining).

The nervous system of helminth parasites is multifunctional and has been shown to be neurochemically complex (Halton & Gustafsson, 1996; Maule *et al.*, 1996). 5-Hydroxytryptamine (5-HT) occurs extensively in the nervous systems of platyhelminth (flatworm) parasites and has been implicated as an excitatory neurotransmitter. FMRFamide is a tetrapeptide amide that was first isolated from the Venus clam, *Macrocallista nimbosa*, by Price & Greenberg (1977). It is now known that peptides with a similar structure to FMRFamide occur widely throughout invertebrate phyla and are commonly referred to as the FMRFamide-related peptides (FaRPs). FaRPs are known to be abundant in both

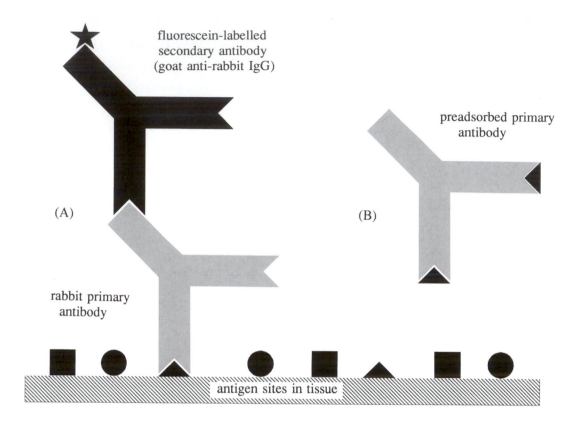

flatworm and roundworm nervous systems and have been shown to influence a range of biological activities, including myoexcitation in flatworms (Day & Maule, 1999).

The indirect immunofluorescence method has been widely used to examine the distribution of classical neurotransmitters and neuropeptides in the nervous systems of helminth parasites, both at light and electron microscope levels. The information obtained has proved important in providing details on the neurochemical complexity of the parasite nervous systems and in revealing the distribution of specific antigens. For example, the localisation of neuropeptide immunoreactivity to a specific part of the nervous system is the first step in directing physiological studies and thus establishing a function.

Fig. 3.7.1A. Schematic showing indirect immunofluorescence technique, whereby the primary antibody (rabbit anti-antigen) binds to the antigen site (▲) in parasite tissue and thus serves as a IgG antigen for the secondary (goat anti-rabbit) antibody, which has been previously labelled with a fluorophore (★), such as FITC or TRITC.
B. Schematic showing liquid-phase preadsorption test of antiserum specificity, whereby the primary antiserum is preincubated with a range of concentrations of antigen. Preadsorbed antibodies have no free antigen binding sites and so do not react (fail to stain) with tissue antigens (after Maule *et al.*, 1997).

Laboratory equipment and consumables (per student group)

Equipment

Fluorescence microscope
Coverslips
Slide trays
Plastic trays
Access to refrigerator/cold
  room (4 °C)

Glass microscope slides
Pasteur pipettes
Cryostat
Slide racks
Fume cupboard

Consumables

Paraformaldehyde
Tinfoil
Glycerol
Antifade (1,4-diazabicyclo-
  [2,2,2] octane)
  (Sigma-Aldrich Co. Ltd)
Rabbit anti-5-HT primary
  antiserum (Sigma-Aldrich
  Co. Ltd)
Cryo-M-Bed (BDH Chemicals,
  Poole, UK)
Bovine serum albumin
Sodium dihydrogen
  orthophosphate
  monohydrate
Sodium hydroxide
Gelatin (Type B)
Sodium chloride

Disposable gloves
Tissue paper
Sucrose
Swine-anti-rabbit IgG (SWAR)
  conjugated to fluorescein
  isothiocyanate (FITC) (Dako
  immunoglobulins, Denmark)
Rabbit anti-FMRFamide
  primary antiserum (Peninsula
Laboratories Europe Ltd)
Triton X100
Anhydrous disodium
  hydrogen orthophosphate
Ammonium hydroxide
Chromium potassium sulphate
  $(CrK(SO_4)_2.12H_2O)$
Mammalian saline (0.9% NaCl
  w/v)

Immunocytochemical media

1. *Four percent (w/v) paraformaldehyde (PFA):* Add 16 g of paraformaldehyde to 200 ml of distilled water, cover with tin foil,
   place in a fume cupboard and heat the mixture while stirring
   continuously with a magnetic stir-bar to 55–60 °C for approximately 1 h. It is important to regulate the temperature carefully. Add 1 M sodium hydroxide (dropwise) until the solution
   clears. Allow the solution to cool and then add 2 × phosphate-
   buffered saline and adjust the pH to 7.4 using sodium hydroxide; filter and store at 4 °C. Use within 2 weeks. **Note: PFA**

should always be prepared in a fume cupboard and handled with care.

2. *Phosphate-buffered saline (PBS):* Take 2 litres of distilled water and add in sequence until dissolved sodium chloride (85 g), sodium dihydrogen orthophosphate monohydrate (134.5 g) and anhydrous disodium hydrogen orthophosphate (107 g); adjust the pH to 7.2 with ammonium hydroxide (1 M). Make up this stock solution to 10 litres by adding distilled water.

3. *Twice strength (2×) PBS:* Dissolve sodium chloride (3.4 g), sodium dihydrogen orthophosphate monohydrate (1.38 g), and anhydrous disodium hydrogen orthophosphate in 200 ml of distilled water.

4. *Antibody diluent:* Add 2.5 g Triton X-100 (or 2.5 ml), 0.5 g bovine serum albumin (BSA) and 0.5 g NaN_3 (sodium azide) to 500 ml of PBS. Make sure all of the ingredients are dissolved; store at 4 °C.

5. *PBS/glycerol antifade (mounting medium):* Take 90 ml PBS and add 10 ml of glycerol and 2.5 g of antifade (1,4-diazobicyclo-[2,2,2] octane). Store antifade in light-proof bottle in refrigerator.

6. *Subbed slides:* Heat 1 litre distilled water to 80 °C and add gelatin (5 g) and chrome alum (0.5 g) until completely dissolved. Filter and store at 4 °C. Place clean microscope slides in slide racks and dip in solution, making sure all of the slides are coated. Remove and place slides in oven at 60 °C to dry. Repeat procedure at least twice.

7. *Five percent sucrose:* To 500 ml of phosphate-buffered saline (PBS), add 0.5 g NaN_3 (sodium azide) and 25 g of sucrose. Stir until dissolved. Store at 4 °C.

8. *Thirty percent sucrose:* To 500 ml of PBS, add 0.5 g NaN_3 and 150 g of sucrose. Stir until dissolved. Store at 4 °C.

Sources of parasite material

While most helminth parasites are suitable for use in immunocytochemistry, the liver fluke, *Fasciola hepatica*, and the cyclophyllidean tapeworm, *Moniezia expansa*, are generally readily obtainable from local sheep abattoirs in sufficient numbers to meet the needs of a class. Specimens of *F. hepatica* are easily recovered from infected livers that have been condemned, while *M. expansa* can be expelled readily from sheep intestines (a procedure normally done mechanically at abattoirs). Both parasites can be transported to the laboratory in Thermos flasks containing mammalian saline at 37 °C. If fresh fish are locally available, then monogeneans such as *Diclidophora merlangi* from whiting or *Discocotyle sagittata* from trout may be collected from the gills and processed for comparison. These worms can be kept alive in aerated water at 4–8 °C for up to 1 week.

Safety

There are few health hazards or risks of infection associated with the collection of adult liver fluke or with the tapeworm, *M. expansa*. Nevertheless, in recovering the parasites it is essential to practice good hygiene, with the use of disposable gloves, laboratory coat and face mask. It is advisable to telephone the appropriate abattoir and ask permission from the veterinary officer or senior meat inspector on site prior to arrival. They will ensure that you are aware of local safety regulations and that, on site, you are wearing appropriate clothing: hard hat, hair-net, Wellington boots and clean laboratory coat. It may be possible to borrow a hard hat at the abattoir, but check in advance. Admission will become easier once a satisfactory collection routine has been established. Recovery and handling of *D. merlangi* or *D. sagittata* do not pose a health risk.

Immunocytochemical procedures

Whole-mount preparations

1. Wash freshly collected worms in either warm mammalian saline (37 °C) (*F. hepactica*/*M. expansa*) or seawater (4 °C) (*D. merlangi*) or river water (4 °C) (*D. sagittata*) to remove any extraneous surface residues.
2. Place a whole worm (*F. hepatica*, *D. merlangi*, *D. sagittata*) or, in the case of *M. expansa,* a small (1 cm) portion, including the scolex, on a microscope slide and cover with 4% (w/v) paraformaldehyde (PFA). Carefully place another microscope slide on top of the parasite.
3. Weight the slides to make sure the parasites are suitably flattened. This can be done by placing a glass beaker on the slide covering the parasite and gradually filling it with water, which still allows a clear view of the worm.
4. Leave the parasites to flat-fix for 1 h at 4 °C making sure they do not dry out. Transfer worms to a container containing 4% PFA (w/v) and leave to fix for another 3 h.
5. Transfer worms to antibody diluent and leave overnight at 4 °C.

6. Transfer specimens to primary antisera at working dilution (1:500 for 5-HT and 1:400 for FMRFamide) and leave to incubate for 48 h at 4 °C.

7. Wash specimens in antibody diluent overnight at 4 °C and then place in FITC-labelled secondary antiserum (swine anti-rabbit IgG; 1:500) for 24–48 h at 4 °C.

8. Wash in antibody diluent overnight (4 °C).

9. Remove the specimens from antibody diluent and mount on a microscope slide in PBS/glycerol containing antifade and view with fluorescence microscope.

10. Green fluorescence will reveal the sites of immunoreactivity, provided the immunostaining is absent in the controls listed below.

11. The slides should be stored (4 °C) horizontally in a slide tray covered with tinfoil to prevent fading.

Cryostat sections

1. Wash freshly-recovered worms in mammalian saline and place whole specimens or segments into a container of 4% PFA (w/v) and leave for 24 h at 4 °C.

2. Specimens should be cryoprotected by transferring them to 5% sucrose in PBS for 24 h at 4 °C, then 30% sucrose in PBS at 4 °C until submerged.

3. Mount specimens in Cryo-M-Bed on a cryostat stub and freeze in the cryostat (−20 °C) for 30 min or snap freeze using liquid nitrogen.

4. Cut sections (5–20 μm in thickness) and collect on subbed slides.

5. Leave the sections to air-dry at room temperature for approximately 30 min.

6. Wash the slides in PBS for 10 min and then carefully remove any excess PBS from the area around the sections using tissue paper.

7. Place slides horizontally in a plastic tray that has been lined with damp tissue paper and add the primary antiserum at the required working dilution to the sections, making sure they are all covered. Cover the tray to prevent the slides drying out and incubate at 4 °C overnight.

8. Wash the slides in PBS for 10 min, then proceed as in 6 above.

9. Place slides as described in 7 above, except replace primary antiserum with secondary antiserum and incubate for 2 h at room temperature.

10. Wash slides in PBS for 10 min, blot off excess and mount on glass slides with PBS/glycerol antifade medium and examine as for whole-mount preparations.
11. Store slides horizontally in slide trays at 4 °C covered with tinfoil until required for examination.

Controls

Suitable controls for staining procedure and specificity of staining must be used for the critical interpretation of immunocytochemistry (ICC) results. The three most commonly used controls in ICC are:

- Omission of primary antiserum;
- Use of non-immune antiserum instead of primary antiserum;
- Liquid-phase preadsorption of the antiserum with a range of concentrations of the appropriate antigen (Fig. 3.7.1B).

All should result in negative staining.
NB The processes of fixation and conjugation may alter the configuration of 5–HT making it very difficult to preadsorb.

Instructions for staff

Check the immunocytochemical media regularly for bacterial growth. Contamination of the media with bacteria gives the sections and whole mounts the appearance of being dirty as they stick to the surface causing specks of non-specific staining. Try to provide the students with specimens of flukes that have regurgitated their stomach contents (appear uniform in colour with no dark pigmented area) as the staining patterns are easier to visualise. It is always a good idea to ring the chosen abattoir in advance to check that you are covered by insurance and are able to gain permission to visit the facility. When using antisera, it is important to check which animal they were raised in as this will dictate the secondary antisera needed, e.g. antisera raised in a rabbit will require an anti-rabbit secondary while antisera raised in guinea pigs will require an anti-guinea pig secondary. Failure to use the correct antisera will result in an absence of staining.

Instructions for students

A series of slides may be set up and the parasites flattened with varying amounts of weight to enable you to determine how much pressure is required to flat-fix the parasite, such that optimum staining is achieved. It is important to make sure that the whole-mount preparations do not dry out during the 1 h period of flat-fixing as this will lead to non-specific staining. To prevent this it is a good idea to place the slide in a Petri dish or similar container and add excess PFA around the slide using a Pasteur pipette. It is also important after washing the subbed slides that the sections contained on them are not allowed to dry out before the addition of primary or secondary antisera. When placing the slides in the plastic tray it may be easier to keep them horizontal if they are placed on each of the four corners of an embryo dish, etc. that prevents the antisera running off onto the tissue. If more than one antiserum is to be used make sure you draw a diagram showing exactly where you have placed each slide with the antisera so as to ensure you keep them in the same order when washing and returning to the slide tray. Once mounted, use a fine-point glass marker to label the slide with the date, tissue type and antisera used.

Ideas for further exploration

- It is possible to use other species of flatworm for both whole-mount and cryostat sections. Nematodes do not lend themselves well to whole-mount immunocytochemistry but can be used successfully for cryostat sections. The intestinal pig nematode, *Ascaris suum*, is readily available from most pig abattoirs in the UK.
- Dual localisation of 5-HT and FaRP immunoreactivities in the nervous system may be attempted by sequential incubation with rabbit anti-5-HT and, for example, guinea pig anti-FMRFamide, followed by washing and sequential incubation in FITC-labelled swine anti-rabbit IgG and tetrarhodamine isothiocyanate (TRITC)-labelled goat anti-guinea pig as secondary antisera, respectively. For further details of the dual labelling procedure in ICC, see Armstrong *et al.* (1997) and Fellowes *et al.* (1999).
- A most informative investigation of the relationship between nerve and muscle in helminths may be made by counterstaining the ICC preparations of 5-HT or FaRP immunostaining with phalloidin, a specific tag for filamentous actin (F-actin) made visible when coupled to a fluorophore (e.g. FITC, TRITC). Further details of phalloidin staining of *F. hepatica* musculature can be found in Mair *et al.* (1998).

- If it is possible to examine the immunostained/phalloidin stained whole-mount preparations of helminth parasites by confocal scanning laser microscopy (CSLM), then high resolution, high contrast internal images can be retrieved from depths up to 200 μm of tissue, depending on the instrument. Moreover, fluorescent images can be collected as an extended focus series from a scan in the Z-axis (optical sectioning) to provide accurate spatial resolution of the nervous and muscle systems in three dimensions. If multiple detectors are available on the confocal microscope, then simultaneous imaging of two or more different fluorophores in the same specimen can be achieved in co-localisation studies.

REFERENCES

Armstrong, E. P., Halton, D. W., Tinsley, R. C., Cable, J., Johnston, R. N., Johnston, C. F. & Shaw, C. (1997). Immunocytochemical evidence for the involvement of a FMRFamide-related peptide in egg production in the flatworm parasite *Polystoma nearcticum*. *Journal of Comparative Neurology* **377**, 41–48.

Coons, Λ. H., Leduc, E. H. & Connolly, J. M. (1955). Studies on antibody production. I. A method for the histochemical demonstration of specific antibody and its application to a study of the hyperimmune rabbit. *Journal of Experimental Medicine* **102**, 49–60.

Day, T. A. & Maule, A. G. (1999). Parasitic peptides! The structure and function of neuropeptides in parasitic worms. *Peptides* **20**, 999–1019.

Fellowes, R. A., Dougan, P. M., Maule, A. G., Marks, N. J. & Halton, D. W. (1999). Neuromusculature of the ovijector of *Ascaris suum* (Ascaroidea, Nematoda): an ultrastructural and immunocytochemical study. *Journal of Comparative Neurology* **415**, 518–528.

Halton, D. W. & Gustafsson, M. K. S. (1996). Functional morphology of the platyhelminth nervous system. *Parasitology* **113**, S47–S72.

Mair, G. R., Maule, A. G., Shaw, C., Johnston, C. F. & Halton, D. W. (1998). Gross anatomy of the muscle systems of *Fasciola hepatica* as visualised by phalloidin-fluorescence and confocal microscopy. *Parasitology* **117**, 75–82.

Maule, A. G., Bowman, J. W., Thompson, D. P., Marks, N. J., Friedman, A. R. & Geary, T. G. (1996). FMRFamide-related peptides (FaRPs) in nematodes: occurrence and neuromuscular physiology. *Parasitology* **113**, S119–S135.

Maule, A. G., Halton, D. W., Day, T. A., Pax, R. A. & Shaw, C. (1997). Parasite neurobiology. In: *Analytical Parasitology* (Ed. M. T. Rogan), pp. 187–226. Heidelberg, Springer-Verlag.

Price, D. A. & Greenberg, M. J. (1977). Structure of a molluscan cardioexcitatory neuropeptide. *Science* **197**, 670–671.

Section 4
Pathology and Immunology

4.1 Encapsulation of foreign matter (not-self) by earthworms

D. WAKELIN, D. I. DE POMERAI & J. M. BEHNKE

Aims and objectives

This exercise is designed to demonstrate:

1. How earthworms respond to not-self matter entering their body cavity, including parasites.
2. How to generate quantitative data for subsequent interpretation by summary statistics, graphs and statistical analysis.

Introduction

All living cells need to be able to distinguish self from not-self molecules. This ability is necessary for activities as diverse as feeding (by phagocytosis/pinocytosis), fertilisation and building multicellular bodies. The more complex an animal becomes the more the integrity of its body is threatened by the invasion of infectious organisms and by the appearance of mutant cells ('cancers'). Self and not-self recognition as a form of adaptive response is seen in all metazoa, the process becoming increasingly more complex in the higher invertebrates, and qualitatively more sophisticated, with the evolution of the immune response, in the vertebrates.

This practical will examine aspects of the ability of a common invertebrate, the earthworm, to recognise not-self molecules. Earthworms have a body cavity, the coelom, and a well-developed blood system. The fluid in the coelom contains free cells, amoebocytes, which carry out a major part of the self/not-self recognition. Amoebocytes can phagocytose material as well as encapsulate objects that are too large to ingest. Aggregations of amoebocytes around foreign bodies form the large 'brown bodies' that accumulate in the tail-end of the coelom.

The major cell involved is the phagocytic coelomocyte, produced from the epithelial lining of the coelom, and present as free cells, amoebocytes, in the coelomic fluid, and as fixed cells, **243** chloragogen cells, covering the intestine and blood vessels. Both cell types take up material from the body cavity; chloragogen cells also act as fixed stores for some excretory products. Amoebocytes scavenge the coelomic fluid and have three means of dealing with not-self particulate material:

(a) *Phagocytosis*: particles are taken into the cytoplasm in vacuoles, after binding to a receptor molecule on the plasma membrane. The vacuoles fuse with lysosomes and the particles are broken down by enzyme action.

(b) *Encapsulation:* larger particles cannot be phagocytosed, the amoebocytes flatten out over the particle surface, often in layers several cells thick. By this means the not-self properties of the particle become hidden; encapsulation, by cutting off nutrient and respiratory sources, can often kill live invaders.

(c) *Cytotoxicity*: in experimental situations, where foreign tissues are transplanted, amoebocytes can actively kill foreign cells, by releasing lytic granules. Graft rejection is well described in earthworms.

Coelomic fluid contains soluble factors that act as agglutinins. These may function against invading bacteria and other organisms. Agglutinins are lectins, molecules that bind to exposed carbohydrates. They probably act as opsonins, i.e. they coat foreign surfaces with self-molecules and enhance phagocytosis by amoebocytes. Earthworms are invaded by a number of protozoan and nematode parasites and details of these can be found in Exercise 1.1.

Laboratory equipment and consumables
(per student or group)

Equipment

Compound microscope
Dissecting dish
50 ml beaker

Consumables

Pasteur pipettes and bulbs	Paper towelling
Slides and coverslips	Invertebrate saline (0.9% NaCl)
Pins (for dissection)	A freshly killed earthworm
Lens tissue	Filter paper

Chloroform   Disposable gloves
Petri dishes
Prepared stained permanent preparations (slides) of transverse
   section of earthworms, particularly the terminal sections.

Sources of parasite material

Earthworms can be found in most types of soil and the larger the
specimen the better for this exercise. A good source is provided
by compost heaps, although species in decomposing matter
tend to be relatively small. See also Additional information in
Exercise 1.1.

Safety

There are few health hazards associated with this practical providing good laboratory practice is followed. As far as is known,
no parasites carried by earthworms in the UK are likely to cause
medical/health problems in humans. However, some soil micro-
organisms adhering to or ingested by the earthworms may be
considered a health hazard, particularly bacteria. This may be
relevant if the worms have been obtained from recently grazed
pastures contaminated by ruminant faeces. Hence disposable
gloves may be necessary. Staff should also take care when handling chloroform.

Instructions for staff

The earthworms *Allolobophora* and *Lumbricus*, but not *Eisenia*,
should be collected a few days or weeks before the class and kept
in suitable containers with soil and organic material, e.g. leaves,
for food, and ventilation. Do not seal in air-tight jars. They
should be killed in a chloroform jar about 15–20 min before the
class commences. A shorter exposure to chloroform may be
insufficient to kill all the worms, particularly the larger specimens and they may show limited movement during subsequent
dissection.

   A table should be prepared on a blackboard, large sheet
of paper, computer spread sheet or database, etc., with
columns/boxes corresponding to the categories listed below so
that students can enter data as they complete the exercise. Each
student should copy down all the data from those participating
in the class, for their own calculations.

Instruction for students

Examination for parasites
You are provided with a freshly killed earthworm.

1. Measure the length of the earthworm and assign to a size class.
2. Dissect the worm, examining both the seminal vesicles and the body cavity as described earlier (See Parasites of the earthworm, Exercise 1.1).
3. Find the 'brown bodies', spherical/oval bodies formed by the accumulation of amoebocytes around foreign bodies in the body cavity. Remove and place in saline.
4. Squash each brown body under a coverslip and examine. What objects are present in the brown bodies?
5. Identify, record and draw. Examine at least 3 or 4 specimens.

Data recording
Comparability of data is most important.

1. For all squash preparations, count the mean number of objects/parasites per low-power ($\times 4$) microscope field. If there are too many to count at this low magnification, use the $\times 10$ objective, but multiply up your mean counts ($\times 100$, then divide by 16) so as to record the number you would have seen at $\times 4$.
2. If you have *Monocystis* present alone (e.g. in seminal vesicles) use the '*Monocystis* only' column and record 0 in the '*Rhabditis* only' column; insert dashes (−) in the twin columns headed 'Both parasites'. Similarly, when *Rhabditis* is present alone (e.g. in body wall), you should enter the number counted in the '*Rhabditis* only' column and zero in the '*Monocystis* only' column, putting dashes (−) in the twin columns for both parasites. When both parasites are present together (e.g. in brown bodies), insert the appropriate numbers in the twin columns headed 'Both parasites', but insert dashes (−) in each of the columns headed '*Rhabditis* only' and '*Monocytis* only'.
3. The total number of counts, zeroes and dashes in each column should equal the total number of worms examined. If you find no parasites at all, enter a zero value in the single parasite columns and dashes in the twin 'Both parasites' columns. Zero entries are useful data! If either brown bodies or seminal vesicles are absent from your worm, record this fact by using dashes rather than zeroes. A zero value for

either site should indicate that the structure was present but that it was devoid of parasites.

4. Within each box of the data table, enter all the data for each individual worm in the same relative position in each box. If everyone does this and copies down the data table accurately, you will be able to trace the data relating to each individual worm examined within the pooled data set. This is most important.

5. For all those worms where you have found one or more brown bodies, calculate a bound:free ratio for each parasite (free = those found in seminal vesicles plus those in the body wall; bound = those encapsulated in brown bodies). This ratio can be compared between different worms to assess variability in the efficiency of the immune response. Good responders should have more bound than free, but poor responders will show the reverse. If you cannot find a brown body, then enter a dash (–) in the bound:free ratio column. It is possible that brown bodies were present but that you missed them. All entries in this column should be entered in the standard form '*n*-bound:1-free', e.g. 0.5:1 would indicate twice as many free compared to bound parasites, while 2:1 would indicate twice as many bound compared to free parasites. Enter these ratios separately for the two parasite species.

6. Enter also the numbers of other objects found in brown bodies. Numbers of chetae can be entered without explanation (they are quite common), but for all other objects include a note of what you think it was. If brown bodies were found inside seminal vesicles (as happens occasionally), enter the data in the brown body columns (not in the seminal vesicle columns), but circle the relevant data entries to indicate this unusual source of brown body material.

7. Select a specimen of brown bodies, measure against graph paper. Record the size of all the brown bodies you can find. From class results, plot length of worm against the number of brown bodies found and the average size of the brown bodies for each worm.

8. Examine the slides, showing sections through brown bodies and draw in detail. Are all the objects not-self? If not, is there any difference in the ways they have been handled by the worm?

9. Write an account to explain how each object may have entered the body cavity, how (and on what basis) it may have been recognised, what its fate will be.

Ideas for further exploration

Use the data you acquire from the practical, and the information provided, to answer the following questions.

- Why does the worm need to recognize not-self molecules in its body cavity?
- What 'invaders' are likely to be encountered?
- What mechanisms are available for
  - detecting not-self, and
  - responding to not-self?
- What is the result of foreign material entering the body cavity?
- Is the fate of not-self different from that of self?
- How can you assess the 'efficiency' of the worms ability to recognize not-self compared with that of the mammalian immune response?
- If you have sufficient specimens, analyse data separately from *Allolobophora* and *Lumbricus*.

REFERENCES

Cox, F. E. G. (1968). Parasites of British earthworms. *Journal of Biological Education* **2**, 151–164.

Ratcliffe, N. A. (1985). Invertebrate immunity – a primer for the non-specialist. *Immunological Letters* **10**, 153–170.

4.2 Opsonisation of trypanosomes

C. M. R. TURNER

Aims and objectives

This exercise is designed to demonstrate:

1. An important humoral immune killing mechanism against *Trypanosoma brucei.*
2. Binding of trypanosomes to macrophages mediated by specific antibodies, the first stage in the process of opsonisation.
3. Antigenic variation by *T. brucei* as a mechanism of evasion of specific antibodies.

Introduction

T. brucei is one of three parasite species that collectively cause Nagana, one of the most important diseases of livestock in sub-Saharan Africa. The other two species are *T. congolense* and *T. vivax*. There are three sub-species of *T. brucei*, two of which, *T. b. gambiense* and *T. b. rhodesiense*, cause sleeping sickness in humans. Currently, sleeping sickness is predicted to cause 300 000–500 000 new cases per year. Infection is normally fatal in the absence of chemotherapy but most infected people will have no access to the relevant drugs.

Vaccination is an obvious alternative to reliance on drugs for prevention of infection and/or disease but the development of potential vaccines has been little explored in African trypanosomiasis because these parasites undergo antigenic variation. Each *T. brucei* parasite is enwrapped in a surface coat that is visible by electron microscopy. The coat is comprised of approximately 10^7 copies of a single molecular species of glycoprotein. It acts as a physical barrier preventing access to the underlying plasma membrane by components of non-specific immune responses that are lethal to uncoated parasites. The coat is, however, strongly immunogenic, and to protect against **249** coat-specific antibody responses trypanosomes regularly undergo antigenic variation whereby they switch from expression of one species of glycoprotein to that of another. Each glycoprotein is immunologically distinct and determines the parasite's variable antigen type (VAT). It is the only portion of the parasite against which effector immune responses can be mounted.

Killing of *T. brucei* parasites is entirely mediated by VAT-specific antibodies and there are two mechanisms by which this is thought to occur: activation of complement by the classical pathway leading to lysis, and opsonisation of parasites leading to phagocytosis by macrophages (mainly Kupffer cells in the liver). This practical illustrates the second of these processes and how trypanosomes evade it.

Equipment and consumables (per student or group)

Equipment

Compound microscope with phase contrast
37 °C incubator

Consumables

Two Petri dishes (35 mm) per student/group, each containing a cover slip (22×22 mm) with adherent macrophages in RPMI medium
Two trypanosome suspensions, to each of which has been added a different antibody

RPMI medium with 5% FCS*	Methanol
Disposable gloves	10% Giemsa's stain in buffer
Microscope slides	Microscope immersion oil
Pasteur pipettes and bulbs	

  *(Foetal calf serum)

Sources of material

The J774 macrophage-like cell line used in this practical is available from a number of commercial suppliers and is easy to maintain and grow. There are numerous laboratories that carry out research on *T. brucei* and one of these should be able to provide a suitable parasite line and the VAT-specific antisera required. Failing this, the author can be contacted for advice. *T. brucei* bloodstream forms are best grown in mice or rats but a few laboratories routinely culture them *in vitro* and if such a source is available, so much the better. In the USA, it may be possible to obtain various species of trypanosomes from the American Type Culture Collection (ATCC); Ward's also supply trypanosomes. Both should be contacted for their latest catalogue. Alternatively, it may be possible to trace potential suppliers through recent publications in one of the biological publications databases and/or through the www.

Safety

Stocks/isolates/lines of *T. brucei* can be functionally divided into three categories. Some lines are sensitive to the lytic action of human serum in all circumstances tested and are considered not infective to humans. Other lines are resistant to killing by human serum or are unstable/intermediate in phenotype. Only lines in the first category should be used. Even then, trypanosomes should be handled as if they are potentially infective. Fortunately, there is no record of transmission by the aerosol route and they can be safely handled on the open bench. Any cuts or abrasions to the hands should be covered using plasters and users should wear lab coats and plastic or latex gloves at all times. Trypanosome suspensions and all materials in contact with trypanosomes should be disposed of into 5% bleach or 3% 'chloros', which kills the parasites instantly. Any spillages should be made safe using one of these solutions or 70% ethanol. As an extra precaution, it is recommended that plastic Pasteur pipettes be used to minimise the risk of broken glass.

Instructions for staff

Cover slips need to be sterilised and added individually to Petri-dishes to which macrophage suspensions are then added, one to four days in advance of the practical. Cell monolayers less then 100% confluent on the day of the practical are ideal.

To grow trypanosome populations you must be aware of the virulence characteristics of your trypanosome/mouse stain combination. Trypanosome suspensions need to be prepared on the same day as the practical. To separate trypanosomes from blood, a differential centrifugation procedure is recommended, which is very robust and maintains very good cellular integrity rather than maximising yield. Exsanguinate mice into anticoagulant and dilute with two volumes of buffer (e.g. RPMI). Centrifuge at 150g for 10 min at 4 °C. Carefully transfer the upper 'straw coloured' layer to a separate tube and centrifuge at 900g for

10 min at 4 °C. Resuspend the pellet in medium at approximately 5×10^6/ml. Split the suspension into two parts, add an appropriately diluted VAT-specific antibody to one of them and an irrelevant antibody to the other. Then dispense each into smaller aliquots for the students. Trypanosomes are very robust cells that will maintain integrity for several hours at room temperature.

Instructions for students

1. Label two Petri dishes A and B and initial.
2. Remove all cell culture medium from both Petri dishes by pipetting.
3. Gently wash the cells by adding RPMI medium (approx. 1 ml) to each Petri dish and immediately removing it.
4. With a clean pipette, to dish A add trypanosome suspension + Antibody A. With a clean pipette, to dish B add trypanosome suspension + Antibody B.
5. Make sure the whole of the cover slip is covered (0.5 ml needed) and that, from now on, the cover slip is not permitted to dry out.
6. With the lid of the Petri dish removed and **with care** observe the macrophages and trypanosomes at $\times 200$ or $\times 400$ with phase-contrast microscopy.
7. Place both cultures in a 37 °C incubator for 15 min. Re-examine. If you do not see any changes re-incubate for another 15 min.
8. Remove the coverslip from each Petri dish and air dry, remove the medium from the dish and return the coverslip, face up.
9. Fix in methanol and stain with Giemsa. To do this, add enough methanol to coat the coverslip and leave for 2 min, then remove the methanol. Add sufficient of the available 10% solution of Giemsa to coat the coverslip and incubate for 15 min, then remove. Add tap water using a Pasteur pipette and remove immediately.
10. Place each coverslip *upside down* on a microscope slide using a drop of microscope immersion oil as a mounting fluid, and observe at $\times 400$ by bright field microscopy.
11. Calculate the proportion of macrophages on each coverslip with one or more trypanosomes attached or engulfed.
12. Explain how trypanosomes have become attached to macro-

phages and why there is a difference between the two cover-slips. What statistical method(s) could you use to analyse your data?

Additional information

The expected outcome is that, in the presence of the VAT-specific antibody, some 20–70% of macrophages should have one or more trypanosomes bound to their surface, whereas with the other antibody less than 5% of macrophages should have adherent trypanosomes. Statistical analysis of data for each group could be by χ^2 test or, alternatively, the class data can be pooled, subjected to an F test and, depending on the outcome, a Mann–Whitney or t-test applied. An observant student might notice numerous smaller structures (1–2 μm) in their preparations. These are platelets that were not separated from the parasites using the differential centrifugation preparation.

Alternative sources of material can be considered. J774 is the only macrophage-like line the author has used but the expectation is that any macrophage-like line would work just as well. Mouse peritoneal exudate cells (PECs) provide a better macrophage source; and the author uses a cell line only out of ethical considerations. If using PECs, they do not need to be elicited before harvesting and should be added to Petri dishes on the day before the practical at approximately 10^6 cells/ml. Procyclic form trypanosomes can be used in place of bloodstream forms. The advantages of doing this are that it obviates the safety requirements, procyclics are easily grown to high density *in vitro*, and antibodies to the major surface protein of procyclic forms are commercially available (e.g. Cedar Lane). The disadvantage is that whilst the practical can illustrate opsonisation, it cannot demonstrate antigenic variation. If employing procyclics, ensure all sera are heat-inactivated before use.

REFERENCE

McLintock, L. M. L., Turner, C. M. R. & Vickerman, K. (1993). Comparison of the effects of immune killing mechanisms on *Trypanosoma brucei* parasites of slender and stumpy morphology. *Parasite Immunology* **15**, 475–480.

4.3 Production and screening of monoclonal antibodies against *Leishmania* promastigotes

M. HOMMEL & M. L. CHANCE

Aims and objectives

This exercise is designed to demonstrate:

1. Hybridoma technology, including the isolation of mouse spleen cells.
2. Hybridization of B-cells and tumour cells.
3. *In vitro* cultivation of fused cells.
4. Screening of monoclonal antibodies produced.
5. Cloning and cryopreservation of cell lines.

Ideally, the whole procedure consists of five 1-day practicals spread out over a period of 5 weeks (when the student actually performs all the steps of the procedure), but this can be shortened to as few as two 1-day practicals (when the student only performs some of the steps and watches a demonstration of others).

Introduction

Monoclonal antibodies have many research and clinical applications (including the detection, localisation, characterisation and purification of antigen). The technique was first described by Köhler and Milstein (1975), but has since been modified and adapted by others (see reviews by Campbell, 1984; Harlow and Lane, 1988). As a practical, the successful production of a few different monoclonal antibodies is considered a very rewarding experience by most students, but even in situations where no antibodies are actually produced during the practical, the students are given the opportunity to learn a variety of useful techniques (sterile handling of cells and centrifugation, tissue culture, cell fusion, preparation of parasite antigens, performance of ELISA and/or IFAT). In addition to the production of monoclonal antibodies itself, the practical provides the demonstrator with an opportunity to discuss (or demonstrate) **255** immunisation, cell cloning and cryopreservation, antibody iso-typing, etc.

As a biological model, *Leishmania* sp. promastigotes are particularly suited for this practical (McMahon-Pratt *et al.*, 1982), because this parasite is easy to grow axenically, is easy to handle and, in view of the fact that promastigotes have a number of characteristic organelles and a flagellum, the antibodies produced by a group of students are often directed against a variety of organelles, making the practical more interesting. However, almost any other parasite may be used in this practical, as long as 'clean' antigen is readily available.

The practical is only suitable for small groups of students (maximum 4 students per demonstrator and for each laminar flow hood).

Laboratory equipment and consumables

Equipment (general)

Laminar flow hood	Centrifuge
Fluorescence microscope	CO_2 incubator
Inverted microscope	Water bath (56 °C)
ELISA reader	Refrigerator (4 °C)
Sonicator	Freezer (-20 °C)

Equipment (per student or pair)

Haemocytometer	Coplin jar
Pipetman P200 and P20	Straight needle (21-guage)
Sterile instruments (forceps, scissors)	

Consumables (per student or pair)

Pipettes (1ml, 5ml, 10ml)	24-well sterile tissue culture plates
Pasteur pipettes (sterile, plugged and unplugged)	Sterile tips for Pipetman
Disposable plastic 1 ml Pasteur pipettes	96-well plates for ELISA (e.g. Immulon-1)
Culture flasks (25 cm², 75 cm²)	12-well Teflon-coated microscope slides
96-well flat-bottom sterile culture plates	Centrifuge tubes (15 ml, 50 ml)
0.22 μm filters	Petri dishes
Whatman #3 filter paper	

Media

1. *Iscove's*: Commercial Iscove's medium (Sigma).
2. *Iscove's without serum*: Includes addition of 2ml of × 100 L-glutamine and 200 μl of stock gentamycin per 100 ml.
3. *Foetal calf serum (FCS)*: Commercial FCS (Sigma). Each batch of serum to be carefully checked before use.
4. *× 100 OPI*: Dissolve 1.5 g of oxaloacetate, 500 mg of sodium pyruvate in 100 ml water, then add 2000 IU of bovine insulin. Sterilise by filtration through a 0.22 μm filter. Aliquot in 2 ml amounts and store at −20 °C until use.
5. *× 100 HT*: Dissolve 136 mg hypoxanthine and 38 mg thymidine in 100 ml water. Heat gently to dissolve. Sterilise by filtration through a 0.22 μm filter. Aliquot in 2 ml amounts and store at −20 °C until use.
6. *× 100 HAT*: Dissolve 136 mg hypoxanthine, 38 mg thymidine and 1.76 mg aminopterin in 100 ml water. Add 0.5 ml 1 N NaOH to dissolve, then bring to neutral pH with 1 N HCl. Sterilise by filtration through a 0.22 μm filter. Aliquot in 2 ml amounts and store at −20 °C until use.
7. *Iscove's plus OPI and HAT*: Consists of 85 ml Iscove's without serum, 15 ml FCS, 1 ml × 100 OPI and 1 ml × 100 HAT. The final mixture is sterilized by filtration through a 0.22 μm filter.
8. *ELISA coating buffer, pH 9.6*: 1.59 g Na_2CO_3, 2.93 g $NaHCO_3$ made up to 1 litre double distilled H_2O.
9. *PBS (phosphate buffered saline), pH 7.2*.
10. *Evans blue*: A stock 1% Evans blue is prepared by mixing 1 g of dry powder with 100 ml double distilled H_2O. The solution is filtered through Whatman 3 paper and stored in a dark bottle until use.
11. *Trypan blue*: Commercial 70% solution (Sigma). Use 1:1 dilution with cell suspension.
12. *Anti-mouse immunoglobulin-FITC*: Commercial antibody (Sigma) used at recommended dilution in PBS (e.g. 1:50).
13. *Anti-mouse immunoglobulin-alkaline phosphatase*: Commercial antibody (Sigma) used at recommended dilution in PBS plus 0.02% Tween 20 (e.g. 1:1000).
14. *Substrate for alkaline phosphatase*: One of the commercial substrate tablets, *p*-nitrophenyl phosphate (Sigma) is dissolved in one of the corresponding buffer tablets in 20 ml double-distilled H_2O.
15. *Polyethylene glycol (PEG 1450)*: Available commercially in 20 ml vials (Sigma); these are heated to 56 °C in a water bath and

5ml PBS plus 0.5 ml DMSO are added. The solution is filtered through a 0.22 μm filter, then aliquoted in 1 ml amounts and stored at −20 °C. The solution is heated to 37 °C before use.

16. *0.85% NH₄Cl in water.*
17. *70% Ethanol.*
18. *DMSO* (Commercial, Sigma).
19. *Gentamycin*: Stock 50 mg/ml gentamycin is prepared in water, filter-sterilised and stored in a dark bottle at +4 °C.
20. *L-glutamine (×100)*: Commercial stock 200 mM L-glutamine (Gibco).
21. *Freund's Complete Adjuvant*: Commercial adjuvant is used (Sigma).
22. *Freund's Incomplete Adjuvant*: Commercial adjuvant (Sigma).
23. *Medium 199 (Sigma) for* Leishmania *culture*: Commercial 199 medium (Sigma).

Sources of biological material

A mouse myeloma cell line selected for the production of monoclonal antibodies (i.e. with a mutation in one of the genes encoding for enzymes of the salvage pathway) is required, e.g. NS-1 line (or equivalent). Such lines are commercially available from most suppliers of tissue culture reagents. A stock of the line should be cryopreserved and a new culture started every 3 months. Any cultured *Leishmania* line will be suitable for the practical, but non-pathogenic parasites are preferred for safety reasons. In the USA, it may be possible to obtain *Leishmania* from the American Type Culture Collection (ATCC) and from Ward's. Both should be contacted for their latest catalogue. Alternatively, it may be possible to trace potential suppliers through recent publications in one of the biological publications databases and/or the www.

Safety

The disadvantage of *Leishmania* sp. is the danger of infection presented by the species that infect mammals. Most *Leishmania* sp. need to be handled in at least category 2 containment facilities. The closely related species that are restricted to reptiles avoid this problem and *Leishmania adleri* is particularly useful since it grows well in culture and shares many features of the mammal-infecting *Leishmania*. Although the myeloma cell lines used are restricted to the mouse and not a risk for humans, it is generally recommended that all the waste generated in the practical be autoclaved.

Performance of monoclonal antibody work

This exercise is complex, inasmuch as it involves advance preparation, the performance of the cell fusion itself, followed by a screening of the culture for hybridoma colonies produced, and a screening of the culture supernatant for the production of antibodies against *Leishmania* promastigotes; the exercise may be (optionally) followed by subsequent activities designed to grow-up, clone and cryopreserve the hybridoma line produced. The normal sequence of events will be:

1. Culture of *Leishmania* promastigotes and preparation of ELISA and IFAT slides (**Part 1**).
2. Immunisation of mice with promastigote extracts (4–6 weeks before the class).
3. Culture of myeloma cells; setting-up cultures of myeloma cells 24 h before fusion (this stage gives the opportunity to introduce students to sterile tissue culture techniques) (**Part 2**).
4. Performance of cell fusion by students (**Part 3**).
5. Screening for hybridoma and preparation of ELISA test (**Part 4**); this exercise must be performed 12–14 days after Part 3.
6. Assay for the presence of anti-*Leishmania* antibodies in culture supernatants, using IFAT and/or ELISA (**Part 5**).

A short version of this exercise can be achieved in two one-day sessions (with 12–14 days interval between the two), where the students perform only the whole of Part 3 and portions of Part 4 and Part 5, with demonstration of the missing steps.

Culture of *Leishmania* promastigotes

Leishmania adleri may be conveniently grown in Medium 199 containing 10% foetal calf serum, 2 mM glutamine and gentamycin 25 μg/ml. A culture should be set up in a 75 cm² plastic culture flask containing 50 ml of medium at a starting density of 5×10^5 promastigotes/ml. After 4–5 days, the promastigote density should be 2–4×10^7 cells/ml.

Once the culture has reached an optimal parasite density, the promastigotes are collected by centrifugation and are thoroughly washed (3 times) with PBS, in order to remove traces of culture medium and serum. After the last wash, the promastigotes are used immediately to prepare IFAT slides or are aliquoted in Eppendorf centrifuge tubes (100 μl of packed promastigotes/tube) and stored at −20 °C until use. Stored promastigotes are suspended in 1 ml PBS and sonicated; the sonicated antigen is either used to immunise Balb/c mice or to coat ELISA plates.

Immunisation of mice

Sonicated promastigotes (100 μl packed promastigotes, stored at −20 °C, resuspended in 1 ml PBS, are sonicated for 2 min in 10 sec bursts) are used to immunise adult female Balb/c mice. After sonication, the suspension of promastigote fragments is emulsified with Freund's adjuvant (1:1 suspension and adjuvant) and 0.2 ml of the stable emulsion is injected subcutaneously in two sites in the previously shaven back of each mouse. After the initial immunisation, the mice are boosted 2–3 times with the same antigen, at 2-weekly intervals; Freund's complete adjuvant is used for the first immunisation, Freund's incomplete adjuvant is used for all subsequent booster injections.

Preparation of IFAT slides

A suspension of washed promastigotes is diluted in PBS (approx. 2×10^4 promastigotes/ml). Drops of 20 μl are placed on each of the wells of a 12-well Teflon-coated microscope slide (Hendley–Essex) and air-dried. The slides are stored at −20 °C in a plastic slide box until use.

Preparation of ELISA antigen

Sonicated promastigotes (100 μl packed promastigotes, stored at −20 °C, resuspended in 1 ml PBS, are sonicated for 2 min in 10 sec bursts) are used as stock antigen. The stock antigen is then diluted in ELISA coating buffer at an optimal concentration to be determined by a checkerboard using normal mouse serum and positive mouse serum (using the serum from one of the immunised mice used for the fusion). A concentration of antigen corresponding to 50 μg protein/ml is generally adequate.

Culture of myeloma (NS-1) cells

Stock NS-1 cells are maintained in Iscove's medium plus 10% foetal calf serum (with 3 ml L-glutamine, 300 μl gentamycin and

1 ml 8-azaguanine per 100 ml of medium). The cells are seeded at a density of 2 to 4×10^4 cells/ml and maintained in 25 cm² flasks (5 ml culture medium per flask) or in 75 cm² flasks (20 ml culture medium per flask). All cell cultures are performed at 37 °C in a CO_2 incubator with a humid atmosphere of 8% CO_2 in air. Cells are transferred every 3–4 days.

Twenty-four hours before fusion, the cells are resuspended (making sure to shake off the cells loosely attached to the culture flask) and carefully counted (a 100 μl sample of the cell suspension is mixed 1:1 with Trypan blue and counted using a Neubauer haemocytometer). For each student, a 40 ml culture is set up at 3×10^5 cells/ml in order to have cells in optimal exponential growth the next day.

Use of haemocytometer

The Neubauer haemocytometer is a counting chamber that has a central area of 1 mm × 1 mm. The central area is divided into 25 large squares each bounded by a triple line. The large squares are subdivided into 16 smaller squares bounded by a single line. When a coverslip is pressed down over the grid, so that interference patterns appear, the depth of the chamber is 0.1 mm. The total volume of the central area is therefore 10^{-4} ml.

To perform a count a small volume of cell suspension is introduced into the well of the haemocytometer by touching the edge of the coverslip with a pipette tip and allowing the chamber to fill by capillary action as the pipette is discharged. When counting promastigotes, buffered 2% formalin should be used to immobilise them. The cells should be allowed to settle down onto the grid to facilitate counting in the same focal plane. Mammalian cells can be counted with a × 10 objective and promastigotes with a × 40. Dead cells take up the trypan blue dye and should not be included in the count. Counting approximately 100 cells will ensure reasonable accuracy. If the whole of the central area is counted then the number per ml is obtained by multiplying the count by 10^4. Remember to include any dilution factor that arises from using typan blue or formalin.

Step-by-step cell fusion

Before the fusion, three different media are prepared: Iscove's without serum, Iscove's with 15 % FCS and Iscove's with OPI and HAT.

1. The immunised Balb/c mouse is killed and exsanguinated (for reference serum).
2. The mouse is pinned out on a board and sprayed with 70% ethanol. A vertical incision is made in the abdomen and the spleen is removed aseptically using sterile dissecting instruments (scissors and forceps). Care must be taken not to injure intestines or to rupture the spleen.
3. The spleen is placed into a sterile Petri dish containing 8 ml Iscove's without serum and transferred to a laminar flow hood.
4. In the hood, the spleen is carefully dissected to remove attached tissues and fat; then it is washed, first by swirling the Petri dish gently, then by transferring it to another Petri dish containing 8 ml of Iscove's without serum.
5. Using a sterile needle, bent at an angle of 90°, to hold the spleen down, carefully tease out the spleen cells with a straight needle (21-gauge).
6. Pipette the cell suspention into a 15 ml sterile conical tube using a 10 ml pipette. Allow large fragments to settle. Transfer the cell suspension into a second 15 ml sterile conical tube, leaving the sedimented fragments behind.
7. Centrifuge at 100 g for 5 min at room temperature.
8. Remove the supernatant and resuspend the cells in 2.5 ml sterile 0.85% NH_4Cl (ammonium chloride) to lyse the red blood cells. Pipette the cells up and down a few times and leave for 2 min. Add 10 ml Iscove's +15% FCS.
9. Centrifuge at 100 g for 5 min at room temperature to wash off the NH_4Cl.
10. Remove supernatant from spleen cells and resuspend in 10 ml Iscove's +15% FCS and count (there should be approximately 10^8 spleen cells in 10 ml).
11. Count an aliquot of the NS-1 cells using 0.4 % aqueous trypan blue and haemocytometer (equal volumes, e.g. 0.1 ml of cells suspension to 0.1 ml trypan blue). Dead NS-1 cells take up the dye and should not be counted.
12. Combine cells in a ratio of 10:1 spleen cells to NS-1 cells in a sterile 50 ml conical tube.
13. Centrifuge at 100 g for 5 min.
14. Remove supernatant and resuspend the cells in 10 ml Iscove's without serum to remove traces of serum.
15. Centrifuge at 100 g for 5 min.
16. Remove supernatant. Resuspend the pellet by tapping the base of the tube.

17. Over a period of 60 sec, add 0.3 ml warm PEG 1450, with constant mixing.
18. Gradually add 15 ml Iscove's without serum, over a period of 90 sec.
19. Incubate the cells at room temperature for 10 min.
20. Centrifuge at 100 g for 5 min.
21. Remove the supernatant and suspend the cells in 30 ml of OPI-HAT medium for 10^8 spleen cells, distribute into 96-well flat bottomed plates, using a 10 ml pipette, and add one drop per well (about 75–100 μl). Each plate has to be labelled (e.g. initials of student/plate number).
22. Incubate in humidified 8% CO_2 incubator at 37 °C.
23. After 24 h incubation add one drop of OPI-HAT medium from a 10 ml pipette per well.
24. Check the plates after 6 days for hybrid clones and re-feed with OPI-HAT.
25. At 10 days incubation the clones are checked and assayed using IFAT or ELISA

Screening for hybridoma colonies

Twelve to 14 days after the fusion, all the wells are examined at × 40 magnification on an inverted microscope for the presence of growing colonies of cells. Positive wells are marked on a scoring sheet. A fusion efficiency is calculated for all the plates in a given fusion (number of hybridoma-containing wells/total number of wells); typically, a fusion efficiency of 30–50% is expected. The supernatant of wells in which the colony covers between 10–20% of the bottom surface of the well is collected for screening. For each selected well, 100 μl of supernatant is aspirated using a Pipetman P200 with sterile tips, changing the tip for each well; the supernatant is transferred into a new 96-well plate or 'test plate'. (Note: careful records have to be kept while this transfer is taking place, in order to ensure that the student is capable of finding which original well was positive.)

Screening for anti-leishmanial antibodies

The supernatants in the test plate are used for both the IFAT (20 μl) and the ELISA (80 μl). Typically, only 10% of hybridoma colonies are expected to contain antibodies to *Leishmania* promastigotes in one or other of the initial tests. A comparison between the results of IFAT and ELISA may provide profitable discussion.

IFAT

The 12-well Teflon-coated slides coated with promastigotes, and stored at −20 °C, are thawed, then air-dried and fixed for 10 min in ice-cold acetone and air-dried again. 20 µl of supernatant is added to each well and incubated in a wet chamber for 1.5 h (a positive and a negative antibody solution should be included routinely). The slides are then washed in a bath of PBS for 5 min. A suitable dilution of anti-mouse immunoglobulin–FITC conjugate is prepared (according to manufacturer's recommendations) in PBS containing 1% Evans blue as a counterstain. Once the slides are washed, 20 µl of conjugate is added to each well and the slides are incubated for 30 min in a wet chamber in the dark. After washing with PBS, the slides are examined on a fluorescence microscope. This part of the exercise needs to be carefully supervised by the demonstrator, particularly if the students have had no previous experience of fluorescence microscopy.

ELISA

Twenty-four hours before the exercise, 96-well Immulon-1 plates must be coated with leishmanial antigen at a suitable dilution (as determined by checkerboard) in ELISA coating buffer; plates are stored in a refrigerator until use. After washing the plates 3 times for 5 min each with PBS containing 0.05% Tween 20, they are shaken dry and 100 µl of supernatant is added to each well (a positive and a negative antibody solution should be included routinely). After 1.5 h incubation at room temperature, the plates are washed in PBS/Tween as above and 100 µl of anti-mouse immunoglobulin–alkaline phosphatase conjugate (diluted in PBS plus 0.02% Tween 20, following manufacturer's recommendations) is added to each well. After 1.5 h incubation, the plates are washed as above and 100 µl of the substrate solution is added and the plates are left to develop until a clear change of colour is detected in the positive control. Optionally, a quantitative reading may be obtained by using an ELISA reader.

Growing up positive hybridoma lines

Following the selection of positive hybridoma lines, the cells have to be transferred into a new plate ('Growing plate') and new medium is added. This starts a laborious growing-up period, where individual lines have to be nurtured to grow, while progressively moving the cells from a 96-well plate, to a 24-well plate, a 25 cm² flask and a 75 cm² flask. This growing-up period is not suitable for a practical, but may be performed with one or two students as part of a specific project. Once the hybridoma line is selected, it needs to be cloned, isotyped and cryopreserved before further characterisation of the antigen recognised. The general principles and methods for cloning, isotyping and cryopreservation can be explained and demonstrated to the students as part of the exercise.

REFERENCES

Campbell, A. M. (1984). *Monoclonal Antibody Technology. The Production and Characterization of Rodent and Human Hybridomas.* Amsterdam, Elsevier.

Harlow, E. & Lane, D. (1988). *Antibodies – A Laboratory Manual.* Cold Spring Harbor Laboratory Publications.

Kohler, G. & Milstein, C. (1975). Continuous culture of fused cells secreting antibody of predefined specificity. *Nature* **256**, 425–497.

McMahon-Pratt, D., Bennett, E. & David, J.R. (1982). Monoclonal antibodies that distinguish subspecies of *Leishmania braziliensis. Journal of Immunology* **129**, 926–927.

Roitt, I., Brostoff, J. & Male, D. (1998). *Immunology*, 4th edn, pp. 28.8–28.10. London, Mosby.

4.4 Pathological effects of *Mesocestoides corti* and *Schistosoma mansoni*

J. CHERNIN

Aims and objectives

This exercise is designed to demonstrate:

1. Qualitative changes in body-organ systems occupied by parasites.
2. Quantitative changes in body-organ systems associated with infection.
3. Histological changes associated with infection, including the direct effects of parasites and the host response.

Introduction

Pathology is the study of the cause and effects of disease as seen in the changes in tissues and organs. The majority of infectious agents, including many parasites, cause some changes in the host that are reflected in alterations in the appearance and function of specific organs. Many parasites cause chronic infections in which pathological changes develop over a protracted period of time, during which the host's health may deteriorate gradually as a consequence of the accumulated damage to organ systems, but seldom sufficiently for the parasite itself to be directly responsible for death. Some of these changes may be sufficiently intense to affect the parasite's route of migration to its preferred site, thereby either protecting the host from further infection or resulting in the spread of the parasites to new sites. In this exercise, the pathological effects of two parasites, commonly maintained in laboratories for medical research, *Mesocestoides corti* and *Schistosoma mansoni*, will be assessed qualitatively and quantitatively. This will be achieved by close comparison of the organs and tissues of infected and uninfected animals, in order to identify changes associated with infection.

Laboratory equipment and consumables
(per student or group)

Equipment

Dissecting instruments
Dissecting board
Binocular dissecting microscope
Petri dishes
Weighing balance

Consumables

Physiological saline (0.9% NaCl solution)
Standard fixative, e.g. buffered formol saline*
Surgical gloves

*Buffered formol saline, pH 7.2: 1 part formaldehyde solution plus 9 parts phosphate-buffered saline (28 ml 2.76% w/v $NaH_2PO_4.H_2O$, 72 ml 2.839% w/v Na_2HPO_4, 100 ml distilled water, 1.7 g NaCl).

Sources of parasite material

Laboratory mice infected with *S. mansoni* and *M. corti* and control non-infected mice of the same strain, age and sex. These would have to be obtained by collaboration with approved commercial suppliers of laboratory animals and a laboratory where these two parasites are routinely maintained. In the UK, an appropriate Home Office licence is required. Workers in the USA should consult Exercise 1.13 for sources of *Schistosoma*. No commercial suppliers of *M. corti* are known to us, so the best route for obtaining infectious material is to try to trace potential suppliers through recent publications in one of the biological publications databases, or through the www.

Safety

It is advisable for students to wear surgical gloves when handling dead animals. The stages of *S. mansoni* encountered during the handling of infected mice, adult worms and eggs, are not infectious to humans. The larval stages of *M. corti*, however, are infectious and appropriate precautions must be taken to ensure that there is no transmission to those handling the material. The post-mortem dissection should be carried out on a tray (or some comparable container) so that parasites or infected tissues do not spill onto the bench. All containers and the bench work area should be cleaned afterwards with 70% alcohol or a suitable detergent to ensure that any infective stages are eliminated.

Instructions for staff

This exercise is best carried out on fresh animal material. If animals cannot be killed in the presence of students, they should be killed, by an approved technique, a few minutes before the start of each class and, in the UK, in a room/facility approved by the Home Office. It is often more convenient for the students to be provided with histological sections of material for examination rather than expecting them to carry out the processing themselves. In this case, standard histological manuals should be consulted for information on the fixation, embedding in wax, sectioning and staining techniques. At least three serial sections of each tissue should be mounted on a single slide. The more sections that can be mounted and stained on a single side the better. These should be stained with haemotoxylin and eosin (H&E), Masson's trichrome, Giemsa, Alcian blue or toluidine blue.

Instructions for students

Wear disposable gloves throughout this exercise.
First weigh the mouse as accurately as you can (preferably to the nearest 0.1 g).

Initial preparation of the mice infected with *M. corti* and their uninfected controls

The stage of this parasite found in the mouse is the metacestode stage or tetrathyridium. Tetrathyridia can be seen with the naked eye as small white organisms about 1 mm in size. In an established infection, they occur both free in the peritoneal cavity and also in organs, mainly the liver and lungs.

1. Place the dead mouse on its abdomen and gently pinch the skin over its back and cut through the skin only. Grip the skin behind the cut end with thumb and forefinger using both hands and gently pull the skin apart. The skin should be pulled back far enough to expose the lower abdomen and neck region.
2. Place the animal on its back so that the abdominal organs

can be seen through the body wall. With the aid of a pair of fine forceps, gently lift the body wall in the mid-abdominal region and cut with a pair of fine sharp scissors. Make only a small incision in the mid-ventral line.

3. Hold the mouse over a beaker, and using a standard wash bottle containing physiological saline, wash the loose parasites out of the peritoneal cavity. After the initial wash, it might be necessary to increase the size of the incision in order to flush out all of the parasites. In most cases the parasites can be decanted into the beaker. In certain strains of mice, however, the tetrathyridia are covered with a viscous mucus-like fluid and require more effort and saline to wash them all out.

4. In order to estimate the total number of parasites, decant the tetrathyridia into a small measuring cylinder or calibrated centrifuge tube to measure the total volume of parasites obtained. Using a pipette, remove a 0.1 ml aliquot of parasites, making certain that the pipette contains only parasites, and empty into a Petri dish; count the parasites. Repeat this process 10 times and calculate the mean number of parasites per ml.

5. Place the mouse, now free of loose parasites, on its back on a dissecting tray to examine the organs. For comparison, a dissected control mouse will be available to the whole class, as a demonstration. Note the colour of the liver of the control mouse and compare it to that of the infected mouse. Note the shape of the individual liver lobes paying particular attention to the edges. Feel the surface of the liver lobes and note if there has been any change to the surface texture. Examine all of the liver lobes for any visible surface lesions. If there are lesions, measure their size and estimate the distribution over the surface of the liver lobe.

6. With a pair of forceps, lift the entire liver thereby exposing the diaphragm and cut the main hepatic artery to free the liver. Weigh the liver and compare to that of the control mouse.

7. Isolate a single liver lobe and with scalpel or razor blade slice through the middle and examine the newly exposed surface for internal lesions and parasites. Place one portion of the cut liver lobe in a Petri dish, wet it with physiological saline, gently squeeze and observe whether or not any parasites emerge.

8. Isolate another lobe, draw an outline of it and then cut into pieces of roughly 1.5 cm^2 and note on the drawing from which part of the lobe each section was derived. Place each section in the fixative to prepare for histological examination.

9. Remove the spleen, weigh, compare colour and size to that of the spleen from the control mouse. Note any changes to the outer capsule, e.g. a layer of fibrous tissue may have formed round the outside of the spleen. If preparing tissues for histological examination, divide the spleen into smaller sections for this purpose.

Mice infected with *S. mansoni* and their uninfected controls

Two stages of this infection can be found in an infected mouse, the adult and the eggs. The adults live mainly in the mesenteric blood vessels and the hepatic portal vein. The eggs pass into the host's gut and are voided from the mouse via the faeces. However, some of the eggs are washed into the portal tracts of the liver, where they become trapped. The more serious pathology is due to the presence of these eggs in the liver and other organs. In general, the extent of obstructive pathology is related to the number of egg producing female worms in the host.

1. When you have exposed the organs in the peritoneal cavity of the mouse, as described above, gently open the abdomen and carefully spread out the small intestine without rupturing the mesenteries or blood vessels. Carefully irrigate the gut with saline so as to prevent desiccation.

2. Observe the blood vessels through a dissecting microscope. In an established infection, live adult worms may be visible within the vessels, depending on how the mouse was killed. Using a fine needle carefully pierce the blood vessel as close to a worm as possible. With a fine Pasteur pipette, aspirate the worms and place them in saline. Note that some worms may have shifted from the mesenteries into the hepatic portal vein. Carefully rupture the vein close to the liver and collect any worms detected.

3. Living worms are very sensitive to fixative and may coil tightly if placed directly into formalin, for example. To avoid this, transfer the worms carefully onto a moistened tongue depressor or similar, using a pipette or fine needle to keep them as straight as possible. Holding the depressor at an

acute angle, slowly drip fixative downwards, from a point above the worms, for a few seconds. Increase the flow of fixative until the worms are washed into a Petri dish. In this way, worms will remain relatively straight to facilitate subsequent examination.

4. Once the worms have been recovered, examine the liver in exactly the same way as described above for *M. corti,* i.e. observe the colour, texture and note any surface markings; record the weight of the liver and note the shape of individual edges of each lobe. Remove one lobe, draw and prepare it for histological examination as described before. Remove the spleen, weigh, examine and prepare for fixation and histology.

Histological examination

You are provided with slides of livers from infected and uninfected mice. Examine an H&E-stained slide of non-infected liver from the control mouse in order to familiarise yourself with the normal structure. Note the shape and size of the hepatic cells and observe the types of cells found surrounding blood vessels, ducts and sinuses. Repeat this procedure with each stain used.

Mice infected with M. corti

1. Examine the slides that have been stained with H&E and observe the distribution of metacestodes in 'host capsules'. Identify the cell types that make up the capsule. Note the changes in cell types from the inner edge of the capsule toward the hepatic cells. Around the sinuses, the capsules and parts of the liver periphery are numerous cells with darkly stained nuclei, which are mainly the leucocytes that make up most of the inflammatory tissue.

2. Examine the slide stained with Masson's trichrome. Scar tissue is formed as a result of the migration of parasites through the liver. Most of the fibrous tissue will stain green and the collagen red.

3. Examine the sections stained with Giemsa and identify the cells that form the inflammatory tissue. Which cell type is most dominant?

4. Examine the sections stained with Alcian blue/toluidine blue and determine whether large granulated cells (mast cells) are present within the fibrous scar tissue. There may be red-staining strands of loose fibres along the inner edges of the host

capsule surrounding the parasite. These represent the early stages of scar tissue formation. Depending upon the fixative used, the spaces between the parasite and the capsule may have stained blue. This indicates the remains of the viscous fluid that normally surrounds the worms.

5. Count the number of parasites within several different sections and estimate their number per cm^2. Estimate the amount of fibrous/scar tissue.
6. Account for the increase in weight of the infected livers.
7. Examine the bile ducts and canals, blood vessels and sinuses for any signs of swellings or blockages.
8. Examine a section of spleen from a control mouse and note the distribution of red and white pulp areas. Apart from lymphocytes try to identify any other cell types present, and pay particular attention to the position and distribution of the giant cells.
9. Measure the size and distribution of the white pulp in infected spleens and compare to that of the spleen from a control mouse. Note the number and distribution of giant cells

Mice infected with S. mansoni

1. Examine the H&E-stained sections using a $\times 20$ objective. Look for the presence of schistosome eggs.
2. Note the shape of the eggs and whether or not they are embryonated. Count the number of eggs per section and the number of granulomas per section and then calculate the number of eggs and granulomas per cm^2.
3. Examine individual eggs under the $\times 40$ objective. Investigate whether the eggs are embryonated and whether there are any host cells attached to the egg surface.
4. Measure the size of individual granulomas. Cells associated with inflammation can be seen surrounding some of the recently deposited eggs. In most cases, the eggs that have been in the liver for several days are surrounded by cells that have formed into a granuloma.
5. Locate the bile ducts, blood vessels and sinuses and compare these with those in the liver of a control mouse. Note if there are any blockages or swellings.
6. Estimate the relative proportions of inflammation and normal liver parenchyma.

Brief account of expected results

Histology. Haemotoxylin and eosin (H&E) is the most commonly used stain. The nuclei stain blue and the cytoplasm pink. The distribution of the cell types and the arrangement of the tissues are clearly visible. Masson's trichrome is good for assessing any damage caused by the parasite. Scar tissue, which is primarily composed of fibre cells, replaces the damaged host tissue. Fibrous tissue stains both red and green, the collagen stains red and elastin green. Giemsa stain is particularly useful for examining the inflammation within the tissues. Leukocytes, in particular lymphocytes, macrophages and eosinophils, are the main cell types that comprise the bulk of inflammatory tissue. The cell nuclei stain dark blue and the cytoplasm pale blue through to pink, depending on the cell type. Alcian blue or toluidine blue are both very useful stains for demonstrating mucus-type secretions, metachromasia and mast cells

Ideas for further exploration

- What do you think are the causes of the changes in the weights of the liver and the spleen?
- What is the reason for the increase in fibrous tissue in the livers from infected animals?.
- Is there any quantifiable change in the proportion of white to red pulp in the spleen of infected compared with control naive animals?
- What are the functions of the eosinophils and mast cells?
- Can you think of alternative staining methods to distinguish between different mast cell types?
- Why do different species of parasites exploit different tissues and organs in their hosts?
- Which of the parasites under study give rise to eggs in host faeces? How would you detect eggs in host faeces and how would you compare the intensity of infection of different hosts?

Information on similar exercises

Although the exercise described herein is based on two helminths, similar classes can be run on other helminths For example, *Heligmosomoides polygyrus* and *Fasciola hepatica*. However, appropriate changes would have to be made to the specific organ systems targeted for examination depending on the biology of the parasite and its location in the host.

REFERENCES

Chernin, J. & McLaren, D. J. (1983). The pathology induced in laboratory rats by the metacestodes of *Taenia crassiceps* and *Mesocestoides corti* (Cestoda). *Parasitology* **87**, 279–287.

Chernin, J., McLaren, D. J., Morinan, A. & Jamieson, B. N. (1988). *Mesocestoides corti:* parameters of infection in CBA/Ca mice and the effect of introducing a concomitant trematode infection. *Parasitology* **97**, 393–402.

Chernin, J., Miller, H. R. P., Newlands, G. F. J. & McLaren, D. J. (1988). Proteinases, phenotypes and fixation properties of rat mast cells in parasitic lesions caused by *Mesocestoides corti*; selective and site specific recruitment of mast cell subsets. *Parasite Immunology* **10**, 433–442.

Riley, S. L. & Chernin, J. (1994). The effect of the tetrathyridia of *Mesocestoides corti* on the livers and peripheral blood of three different strains of mice. *Parasitology* **109**, 291–297.

4.5 Quantification of lymphocyte populations in the spleen and thymus

J. CHERNIN

Aims and objectives

This exercise is designed to demonstrate:

1. Methods for quantifying cellular (lymphoid) changes in body organs.
2. The morphology of the mononuclear cell types.
3. Changes in the lymphoid cell population during the course of an infection.

Introduction

The dominant cell types that are involved in an active immune response are the lymphocytes, of which there are two main classes: B-lymphocytes and T-lymphocytes. Morphologically, they are very similar and can only be distinguished by specialised staining techniques.

Lymphocytes have a limited life-span and are continuously replenished, hence the number of both circulating and organ-based lymphocytes is relatively stable. If an infection becomes established, antigenic molecules derived from the invading pathogen are transported via antigen-presenting cells to the spleen and lymph nodes.

The presentation of the antigen to the T cells with specific antigen receptors (mainly T-helper and to a lesser extent T-cyto-toxic) stimulates the release of a range of cytokines, interleukins and growth factors, which, in turn, promote cell proliferation (cloning) of primed lymphocytes. Hence the numbers of resident lymphocytes in both the T and B cell zones increase.

The total number of viable lymphocytes in various organs can be estimated by a simple extraction technique. This method involves identifying and dissecting out the lymphoid organs (spleen, thymus and lymph nodes) followed by extraction of the lymphocytes, staining and quantification.

277

Laboratory equipment and consumables
(per student or group)

Equipment

Haemocytometer
Retort stands
Safety glasses

Consumables

Petri dishes	Sharp dissecting or syringe
5 ml syringe	needles
Nylon wool	Physiological saline (0.9% NaCl)
Stainless steel fine mesh gauze	Vital stain such as 0.1%
Disposable gloves	Nigrosin
Face mask	Ice

Sources of parasite material

This exercise can be carried out with a spleen and/or thymus removed from a non-infected host and a host infected with any helminth or protozoan parasite and, if possible, staggered infections, i.e. infections from day 0, 7, 14, 28, etc., post-infection. Parasites associated with the blood system (trypanosomes, malaria, etc.) are preferable.

Safety

See Exercise 4.4 or other exercises relating to the species you intend to use. It is also important to wear safety glasses, gloves and face mask when handling the nylon wool.

Instructions for staff

Although this exercise can be carried out using spleen, thymus and lymph nodes, in practice and for first time users it is more practical to use only the spleen and thymus. Ensure that students note the location and orientation of each organ *in situ* before removal. Demonstrate how to distinguish between fatty tissue surrounding the organs and that on associated connective tissues and swollen organs.

Instructions for students

1. Weigh a freshly killed infected and a non-infected mouse. Dissect both mice as described in Exercise 4.4, open the body cavity and remove the spleen. Open the pleural cavity and remove the thymus.

2. Weigh each of the isolated organs carefully (to the nearest 1 mg), place in a Petri dish on ice and keep moist with physiological saline until needed for the next stage. It is important to keep all live tissues cold and moist before teasing them apart.

3. Cut each organ into a number of convenient small pieces, of about 0.5 cm square.

4. Remove a small section of each organ to a separate Petri dish, cover with physiological saline and tease apart using the syringe needles, until no discrete pieces appear to remain. When the tissue has been completely disrupted, pour the contents of the Petri dish through the fine steel gauze to remove any remaining large portions of tissue.

5. Take a 1 ml syringe, remove the plunger, and fill the barrel with nylon or glass wool. Support each 'filled syringe' upright on a retort stand and place above a collecting centrifuge tube. Place a small funnel into the supported syringe barrel and pour in the liquid obtained after the teased organ was passed through the steel gauze. This will filter out any remaining large particles and only single cells should pass through the syringe into the collection tube.

6. Centrifuge the suspension collected at a low speed of about 100 g for at least 5 min. After centrifuging, discard the supernatant and resuspend the pellet into a measured volume of saline (5 ml) and centrifuge again for 10 min.

7. Gently resuspend the pellet and incubate a 0.1 ml aliquot of the suspension in a small tube with an equal volume of nigrosin for 10 min. Remove a small volume with a Pasteur pipette and place under the coverslip of a haemocytometer. All the visible cells should be counted. Viable and non-viable cells may be distinguished because only the former will have taken up the stain.

8. The number of viable cells/ml per g of tissue should be calculated with this information. Smear a suspension of cells onto a microscope slide, dry, fix and stain using the method described in Exercise 4.4. Count 100 cells from each slide and express your results in terms of a percentage of each cell type observed from the total counted.

Interpretation of results

The stained slide should provide a good indication of the cell types that are found within the isolated organs.

Comparing the cell counts per g of tissue obtained from the infected and control mice should give an indication as to whether the mouse has mounted an immune response against the infection.

Brief account of expected results

The population of cells extracted and counted will be mixed but normally most of the cells will be mononuclears, of which the majority will be lymphocytes. There will be some macrophages as well as other leucocytes among these cells.

An increase in the number of lymphocytes is indicative of an active immune response. If there is a significant increase in the number of macrophages and other white blood cell types then this too is indicative of a host response to the infection. However, increases in cells such as neutrophils and monocytes are suggestive of an innate immune response whereas increases in lymphocytes and possibly eosinophils points to an active immune response.

Ideas for further exploration

Take pieces of the spleen, fix in buffered formol saline and prepare the material for histological examination. Inspect the slides for the distribution of red and white pulp and determine whether there is any relationship between the relative size of these areas and the total number of lymphocytes counted by the techniques described above.

- Is there a relationship between the size of each organ and its cellularity?
- Compare the number of giant cells in the histological preparations of each organ.
- What accounts for the increase in the weight of the infected spleens?
- What are the origins and functions of giant cells?
- Where are the B-cells and germinal centres located in the spleen?
- Where are the T-cells found in secondary lymphoid organs?

- Where does antigen presentation take place within secondary lymphoid organs such as the spleen and lymph nodes?

Information on similar exercises

This exercise can be adapted to include lymph nodes such as the mesenteric lymph nodes from rodents.

REFERENCE

Hudson, L. & Hay, F. C. (1989) *Practical Immunology*. London, Blackwell Scientific Publications.

4.6 Use of basic indirect ELISA for the detection of antibodies produced by experimental immunisation

D. A. JOHNSON & C. MCGUIRE

Aims and objectives

This exercise is designed to demonstrate:

1. The use of Enzyme Linked Immunosorbent Assay (ELISA) for the detection of antibody produced by experimental immunisation.
2. The induction of an immune response to a parasite antigen.
3. The change in antibody titre during an immunisation programme.

Introduction

One of the most widely used immunoassays in parasitology is the ELISA. This technique was developed in the 1970s as a quantitative and sensitive assay of antibodies during responses to parasitic infection, but can also be adapted to quantify circulating and faecal antigens. When dealing with parasitic infections, it is important to be able to detect the type of antibody response in terms of class and subclass, which may give some indication of resistance or susceptibility (Hagan *et al.*, 1991). In addition, the ELISA is a useful tool for the detection of antibodies produced by experimental immunisation.

Fig. 4.6.1 shows the various components involved in a standard indirect ELISA. The antigen is bound onto a plastic microassay plate and any free sites are then blocked to prevent non-specific binding. The antiserum is added and, if any antigen-specific antibodies are present in the serum, they will bind to the plate. A second antibody, conjugated to an enzyme that recognises the first class of antibody, is then added. If the primary antibody recognised the antigen in the previous step, the secondary antibody will bind to the primary antibody. A substrate is added that is activated by the enzyme attached to the secondary antibody to produce a colour change. This colour

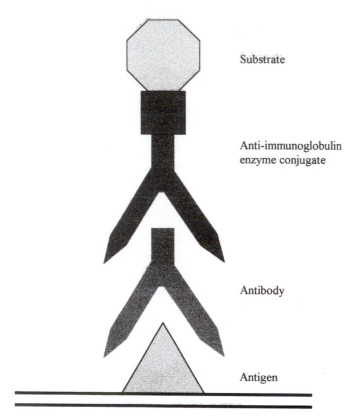

Substrate

Anti-immunoglobulin
enzyme conjugate

Antibody

Antigen

Immunoassay plate

Fig. 4.6.1 The components of a
standard indirect ELISA

change can be measured accurately to detect the level of anti-body present in the serum.

For the purposes of this practical, an antiserum to the homogenate of *Heligmosomoides polygyrus* has been prepared. *H. polygyrus* is a trichostrongyloid parasite which was first introduced as an experimental infection in laboratory mice, and has since been used extensively as an animal–parasite model system. Primary infections are maintained for up to 10 months. *H. polygyrus* is a useful laboratory model as many aspects of its biology have been studied in depth, and the pathology associated with infection can be compared with human hookworm disease and nematode infections of veterinary and agricultural importance. Rabbits are the most frequently used animal for the production of polyclonal antibodies. Typical immunisation procedures involve injection of the antigen together with Freund's adjuvant in order to potentiate the immune response. In this exercise, the ELISA will be used to detect an IgG antibody response following immunisation of rabbits with an injected homogenate antigen of *H. polygyrus*.

Laboratory equipment and consumables
(per student or group)

Equipment

96-well microtitre plate
10–200 µl pipette (plus
  appropriate tips)
1.5 ml Eppendorfs
Centrifuge

Microplate optical density
  reader
Bijoux tubes
Water bath at 37 °C
Wash bottles
Forceps

Consumables

Freund's complete and
  incomplete adjuvant
H. polygyrus homogenate
  antigen
Carbonate–bicarbonate
  coating buffer* (pH 9.6)
Test serum – primary,
  secondary and tertiary
  bleeds (as described in the
  Immunisation procedure)
Goat-anti rabbit IgG alkaline
  phosphatase (or similar)
P-nitrophenyl phosphate
  (pNPP), Sigma Fast, dissolved
  in 5ml (if using alkaline
  phosphatase)

Aluminium foil
Washing buffer (PBS** with
  0.05% Tween 20)
Blocking buffer (3% BSA/PBS,
  0.05% Tween 20)
1:100 dilution of anti-*H.*
  polygyrus tertiary sera in
  PBS Tween
Hanks' saline
Tissue paper; cling film
Disposable gloves

*Carbonate–bicarbonate buffer: Na_2CO_3, 1.59g; $NaHCO_3$, 2.93g; NaN_3, 0.2g (NB: toxic); dissolve in 1 litre distilled water and adjust pH to 9.6.

**Phosphate buffered saline (PBS) NaCl, 8g; KCl, 0.2g; Na_2HPO_4, 1.15g; KH_2PO_4, 0.2g. Make up volume to 1 litre with distilled water and adjust pH to 7.2, if necessary.

Sources of parasite material

Specimens of *H. polygyrus* are widely maintained in research laboratories, and the parasite should be maintained as described in Exercise 1.15. See also Jenkins and Behnke (1977) and the Additional information section of Exercise 1.15.

Safety

There are few health hazards associated with this exercise provided good laboratory practice is followed. Students should be encouraged to keep laboratory coats fastened and to wear gloves throughout the practical. **When using pNPP substrate, students must wear gloves, and use forceps to remove the tablets from the packaging**; if swallowed, medical attention must be sought immediately. As a detergent, Tween may cause skin irritation. Students must wash hands before leaving the laboratory.

Instructions for staff

Immunisation procedure
NB At least six weeks should be allowed for the immunisation procedure before the exercise can be carried out.

1. Remove a 20 ml blood sample from the rabbit prior to immunisation and use as a naive control and to check background antibody levels. Leave the blood samples to clot for 1 h at room temperature. Separate the clot from the wall of the tube by running a Pasteur pipette around the outside of the clot. Leave the blood overnight at 4 °C in order for the clot to retract, then centrifuge for 15 min at 200 g, after which collect and aliquot the serum and store at −80 °C.

2. Emulsify 200 μg of the antigen in a volume of 1 ml PBS with 1 ml Freund's **complete** adjuvant; after 10–15 min, the mixture should be thick and white. Test the preparation by allowing a small drop of the mixture to fall into a beaker of cold water. When the mixture is ready for injection the drop should hold together rather than disperse into the water. Inject 1 ml (in total) subcutaneously into two or three sites on the side and back of the rabbit.

3. Administer primary plus three boosts, 14 days apart, each of 100 μg of antigen, in a volume of 1 ml PBS, mixed with 1 ml Freund's **incomplete** adjuvant, administering 1 ml of the mixture as before. NB:

   naive serum   = pre-immunisation
   primary (1 °)   = 14 days after first immunisation
   secondary (2 °) = 14 days after first boost
   tertiary (3 °)   = 14 days after second boost
   quaternary (4 °) = 14 days after third boost.

Take a blood sample (approximately 10–20 ml, on the basis of the number of students in the class) 1 day before each boost. Ex-sanguinate the rabbit two weeks after the final boost.

Production of homogenate antigen

Kill mice infected with *H. polygyrus* with an overdose of chloroform. Remove the small intestine, cut open longitudinally and place on nylon gauze. Place the gauze containing the intestines into beakers of Hanks' saline and incubate in a water bath at 37 °C. After 2–4 h, remove the gauzes and discard the intestines. Collect the worms from the bottom of the beakers and wash with 10 changes of Hanks' saline, followed by 10 washes with sterile PBS. When no remaining debris is present, place the worms in a small volume of sterile PBS and homogenise on ice for 15 min at 500 g using a polytron Kinematica AG (or similar apparatus). Centrifuge the samples for 10 min at 1000 g, remove the supernatant and discard the sediment (containing debris). Estimate the protein content of the supernatant, using the Bradford method, and store at −80 °C at 100–200 µg/ml in sterile PBS.

Preparation of the ELISA plates

If time is short for the exercise, the incubation periods can be carried out at 37 °C for 30–45 min. Coating of the plate can also be done at 37 °C for 1 h if not done the night before. Coat the 96-well microassay plate the day before the exercise is to take place. Prepare a solution of the *H. polygyrus* homogenate at a concentration of 5 µg/ml in carbonate–bicarbonate buffer (pH 9.6). Add this solution to the plate at 200 µl per well. Cover the plate in cling-film and incubate overnight at 4 °C.

During the exercise

As the secondary antibody to be used should be freshly diluted, and in order to minimise costs, it is suggested that it is prepared by staff during the lunch break. Prepare enough for approximately 1 ml of a 1:2000 dilution for each group.

Instructions for students

Students should work in pairs. You have been provided with a 96-well microassay plate that has been coated with 200 µl *H. polygyrus* homogenate antigen (or excretory/secretory (ES) products) at a concentration of 5 µg/ml. The microassay plate has been stored overnight at 4 °C to allow the binding of the *H. polygyrus* antigen to the plastic. Any excess unbound antigen must now be removed by washing each well.

1. Washing
Holding the plate over a sink, pour the liquid out of the wells. Carefully fill each well with buffer-wash (PBS-Tween) from the wash bottle provided and pour the liquid out again; repeat this washing process twice before refilling the wells. Allow the plate to stand for 2 min before emptying out the wells. Repeat this 2 min wash before tapping the plates upside down onto tissue paper to remove any remaining droplets.

2. Blocking
Add 100 µl of blocking solution (3% BSA in PBS-Tween) from the vial provided to each well, wrap the plate in cling film and incubate at room temperature for 1 h. During this incubation period, you can prepare the appropriate dilutions of the test sera (see below) that you will need for the next stage of the ELISA.

3. Preparation of the serum dilutions
Look at the planned drawing of the ELISA plate (Fig. 4.6.2) and you will see that you will need 4 dilutions of the naive serum: 1:50, 1:100, 1:200 and 1:400. Also note that you will need enough of each dilution to add 100 µl to 3 wells, i.e. 300 µl. Therefore you will need to make up at least 300 µl of each dilution. To prepare these dilutions you will be using a technique regularly adopted in the laboratory, called serial dilution.

3.1. Add 400 µl of PBS-Tween to each Eppendorf tube in the rack provided (there should be 20 arranged in 5 rows of 4). Make sure you label your Eppendorfs correctly, using a waterproof marker pen. Collect the naive serum.

3.2. Add 384 µl of PBS Tween 0.05% to the first Eppendorf (in addition to the 400 µl already there) so there is now a total of 784 µl in the first Eppendorf. Carefully remove 16 µl of the naive serum and add to the first Eppendorf; mix the solution by emptying and filling the pipette several times. You now have a 1:50 dilution of naive serum.

3.3. Transfer 400 µl from this Eppendorf into the second Eppendorf and mix well, as before; you now have a 1:100 dilution of antisera.

Fig. 4.6.2 Plan of ELISA plate and data sheet.

	1	2	3	4	5	6	7	8	9	10	11	12
	<<-	1:50	->>	<<-	1:100	->>	<<-	1:200	->>	<<-	1:400	->>
A												
B												
C												
D												
E												
F												
G												
H												

A = Naïve serum

C= Primary bleed serum

E= Secondary bleed serum

G= Tertiary bleed serum

H= Positive control serum

3.4. Transfer 400 µl of this solution into the next Eppendorf and mix. Continue to dilute the antiserum in this way until you have reached the fourth Eppendorf. Since there are no more dilutions after this, discard 400 µl of the resulting solution, after mixing so as to standardise the volume.

3.5. Change the pipette tip and add 16 µl of primary bleed serum to the first Eppendorf of the second row (prepared as before). Use the same procedure as above to obtain a series of doubling dilutions.

3.6. Repeat for the secondary and tertiary bleed serum. Remember always to use a new pipette tip for each row of Eppendorfs. The blocking buffer can now be emptied out of the wells, and the plate washed with PBS Tween, as described in Section 1.

4. Addition of the antisera to the plate
Refer to Fig. 4.6.2 for assistance with the plate layout.
Naive serum is to be added to row A (100 μl/well):

1:50 to wells 1–3
1:100 to wells 4–6
1:200 to wells 7–9
1:400 to wells 10–12.

Repeat the above procedure adding the primary bleed serum to row C, the secondary bleed serum to row E, the tertiary bleed serum to row G. Add PBS to wells 1–3, and positive control sera (provided) to wells 4–6 in row H.
Wrap the plate in cling film and incubate at room temperature for 1.5 h.

5. Washing
Wash the plate thoroughly with wash buffer as in step 1.

6. Addition of the enzyme conjugated antibody
Add 100 μl of the anti IgG enzyme conjugated antibody, which has been provided at a dilution of 1:2000 in PBS Tween, to each well in rows A, C, E, G and the first 6 wells in row H. The plate should be wrapped in cling film and incubated at room temperature for 1.5 h. When the incubation period is over, the wells of the plate should be emptied and washed as before.

7. Addition of the substrate solution
Make sure you are wearing gloves. Obtain one of each of the two substrate tablets and dissolve in 5 ml of distilled water in the Bijoux provided. The substrate is light-sensitive therefore wrap the Bijoux in aluminium foil. Carefully add 100 μl/well of the enzyme substrate solution to each well in rows A, C, E, G, and the control wells in row H. Wrap the plate in foil and incubate for approximately 30 min, checking regularly for a colour change. The plates are read in a spectrophotometer at a wavelength of 410 nm, using the microplate reader provided.

8. Analysis of results
Results should be presented showing the increase in antibody titre over time and the difference in the signal produced in the ELISA when using different dilutions.

Ideas for further exploration

- What is the purpose of adding blocking solution at stage 2 in the procedure?
- Can you think of a way in which the time to perform this ELISA could be made shorter?
- Why does the antibody titre increase with increased exposure to the antigen?

Information on similar exercises

Various parasite-homogenate antigens or excretory/secretory (ES) products may be used in the same way or used to test for cross reactivity with *H. polygyrus*. Alternatively, serum could be used from animals harbouring a current infection, for instance mice infected with *H. polygyrus* could be tail-bled to obtain an antiserum. However, be aware that the yield of serum from a mouse is much less and so may not be suitable unless the class size is small. The ELISA protocol could be modified to include testing for other antibody isotypes, for example, testing for both IgG and IgM over the course of the immunisation in order to demonstrate the induction of primary and secondary immunity.

REFERENCES

Hagan, P., Blumenthal, U. J., Dunne, D., Simpson, A. J. G. & Wilkins, H. A. (1991). IgE, IgG4 and resistance to reinfection with *Schistosoma haematobium*. *Nature* **349**, 243–246.

Jenkins, S. N. & Behnke, J. M. (1977). Impairment of primary expulsion of *Trichuris muris* in mice concurrently infected with *Nematospiroides dubius*. *Parasitology* **75**, 71–78.

Rogan, M. T. (1997). *Immunological Analysis of Parasite Molecules*. In *Analytical Parasitology* (ed. M. T. Rogan), chapter 10. Berlin, Springer.

Roitt, I., Brostoff, J. & Male, D. (1998). *Immunology*, 4th edn, pp. 28.6–28.7. London, Mosby.

Spurlock, G. M. (1943). Observations on host–parasite relations between laboratory mice and *Nematospiroides dubius*. *Journal of Parasitology* **29**, 303–311.

Venkatesan, P. & Wakelin, D. (1993). ELISAs for Parasitologists: or Lies, Damned Lies and ELISAs. *Parasitology Today* **9**, 228–232.

4.7 SDS PAGE and Western blotting for the detection of antibodies produced by experimental immunisation

C. MCGUIRE & D. A. JOHNSON

Aims and objectives

This exercise is designed to demonstrate:
1. SDS PAGE for the resolution of proteins into their component parts.
2. Western blotting for the detection of immunogenic antigens, using serum from experimentally immunised animals.

Introduction

Sodium dodecyl sulphate polyacrylamide gel electrophoresis (SDS PAGE) and Western blotting are both widely used techniques in parasitology. SDS PAGE is used to separate mixtures of proteins into individual protein bands by dissolving them in SDS (which is a mild detergent) and reducing with 2-mercaptoethanol. The protein solution is then loaded onto an acrylamide gel. When an electric field is applied to the gel, the proteins are separated according to their molecular mass, smaller peptides moving further down the gel. These protein bands can then be stained to visualise their position and therefore their molecular weight can be determined. In this exercise, however, we will not be staining the protein bands but transferring them to nitrocellulose paper where they will be immobilised. These antigens can then be probed by antibodies in order to determine the immunogenic components of the protein.

For the purposes of this practical an anti-serum to the homogenate of *Heligosomoides polygyrus* has been prepared (see Exercise 4.6). *H. polygyrus* is a trichostrongyloid parasite that was first introduced as an experimental infection in laboratory mice over 50 years ago, and has since been used extensively as an animal parasite model system. Primary infections are maintained for up to 10 months. *H. polygyrus* is a useful laboratory model, as many aspects of the biology of this parasite have been studied in depth, and the pathology associated with infection can be compared with human hookworm disease and nematode infections of veterinary and agricultural importance.

Rabbits are the most frequently used animal for the production of polyclonal antibodies. Typical immunisation procedures involve injection of the antigen together with Freund's adjuvant in order to potentiate the immune response.

In this practical SDS PAGE is used to separate, by molecular weight, homogenate antigen into its component parts. Subsequently, Western blotting will be used to determine which parts of the homogenate antigen elicit an antibody response when used to immunise a rabbit.

Laboratory equipment and consumables (per group)

SDS PAGE equipment

Gel apparatus, including glass plates, spacers, casting stand, combs, upper and lower buffer reservoir, e.g. BioRad.
Power pack

SDS PAGE reagents (see Appendix for details)

Acrylamide (NB: toxic)
Resolving gel buffer
  (1 M Tris/HCl pH 8.8)
Stacking gel buffer
  (0.5 M Tris/HCl, pH 6.8)
10% APS
10% SDS
TEMED

H_2O saturated n-butanol
Running buffer (at 4 °C)
Reducing sample buffer
Molecular weight markers
(14 300 – 220 000)

Western blotting equipment

Semi-dry blotting
  apparatus/blotting tank
Power pack
Whatmann No. 3 filter paper
Nitrocellulose sheets
Blotting tray (for blocking)

Multi-channel blotting tray (for probing)
Scalpel blades and cutting board
Rocking platform
Aluminium foil

Western blotting reagents

Transfer buffer
Blocking solution

TBS–Tween, pH 8
Primary antibody

Alkaline phosphatase conjugated secondary antibody

Colour development solution NBT/BCIP

General equipment

1.5 ml Eppendorfs

10–200 µl pipettes and tips

Bijoux tubes

Disposable gloves

Sources of parasite material

Specimens of *H. polygyrus* are widely maintained in research laboratories, and the parasite should be maintained as described in Exercise 1.15. See also Additional information in Exercise 1.15.

Safety

There are few health hazards associated with this practical provided good laboratory practise is followed. Students should be encouraged to keep laboratory coats fastened and to wear gloves throughout the practical and to be extra careful when loading the SDS gels since mercaptoethanol in the reducing sample buffer is thought to be carcinogenic. Acrylamide is highly toxic and should only be handled in a fume hood. Tween-20 is a detergent and may cause skin irritation. Students must wash their hands before leaving the laboratory.

Instructions for staff

Immunisation procedure (see Exercise 4.6 for details)
Production of homogenate antigen (see Exercise 4.6 for details)

Casting the gels

Clean the glass plates with 70% ethanol followed by acetone to remove any debris. Assemble the plates according to the manufacturer's instructions. Make up the 12% resolving gel and pour immediately following the addition of APS and TEMED. Then apply a thin layer of water-saturated butanol on top of the gel and allow 30 min to set at room temperature. When the gel has set wash off the butanol using distilled water and dry the top of the plates using strips of filter paper. Prepare the 4% stacking gel solution and pour immediately, following the addition of APS and TEMED. Carefully insert the comb avoiding trapping air bubbles. Leave the stacking gel to set and then remove the comb and wash the wells with distilled water. The gels can be prepared the day before the practical and stored overnight at 4 °C, wrapped in moist tissue paper and cling film.

Preparation of samples for gel

Remove the homogenate antigen and prestained molecular weight markers from the freezer and dilute 1:1 with reducing sample buffer. Reduce the samples and molecular weight markers by boiling for 5 min. Aliquot the reduced homogenate antigen and molecular weight markers into 200 µl and 15 µl portions respectively; store at -20 °C until required.

Preparation for Western blotting

Prepare six gel-sized pieces of filter paper and one gel-sized piece of nitro-cellulose paper per group of students.

During the exercise

As the secondary antibody needs to be diluted fresh, and in order to cut down on wastage, it is suggested that this is prepared by staff. Make up enough to fill 5 channels of the multi-channel blotting tray at a 1:2000 dilution.

Instructions for students

Students should work in groups of 2–4. You have been provided with a pre-cast 12% acrylamide SDS PAGE gel, which you will load with 75 µg of *H. polygyrus* homogenate antigen. Once the gel has been run, you will be required to blot the gel onto nitro-cellulose paper and probe with serum from the immunised rabbit.

1. Loading and running the gel

1.1. Obtain the pre-cast gel and clamp it into the electrode stand (ask a demonstrator for assistance). Two groups of students should share an electrode stand and tank.
1.2. Place the electrode stand with the plates in place into the electrophoresis tank.

1.3. Fill the centre reservoir and half-fill the outer reservoir with cooled running buffer.

1.4. The prestained molecular weight markers and homogenate antigen have been prepared for you by boiling in reducing sample buffer, in order to reduce the proteins. You must wear gloves from this point onwards! Load 10 µl of the pre-stained molecular weight markers into the small single well, ensuring that the pipette tip is between the two glass plates (if you are unsure or cannot see the well ask a demonstrator for help). Load 200 µl of the *H. polygyrus* homogenate evenly along the long well. Ensure that the level of running buffer is high enough to cover the top of the glass plates.

1.5. Place the electrophoresis tank lid securely in place, matching up the red and black wires.

1.6. Connect to the power pack and run at 20 mA for approximately 1 hour or until the blue dye front is approximately 0.5 cm from the bottom of the glass plates.

1.7. Switch off the power pack and remove the gel by prising apart the glass plates using a spacer. Carefully remove the gel and place in a dish containing transfer buffer. Also place six pieces of gel-sized filter paper and one piece of gel-sized nitro-cellulose paper in the dish and allow to soak for 30 min.

2. Western blotting

2.1. Place three pieces of transfer buffer soaked filter paper on the base (anode) of the semi-dry blotting apparatus, ensuring there are no air bubbles trapped between the layers of paper.

2.2. Place the nitro-cellulose paper on top of the filter paper and cover with the gel, again ensuring that no air bubbles are trapped between the layers.

2.3. Place the final three pieces of filter paper on top to complete the blotting sandwich.

2.4. Position the cathode on top and attach the electrodes to the power supply. Immunoblot the sandwich for 30 min at 10 V or until the prestained markers have been transferred.

3. Blocking

Place the nitro-cellulose paper in the blocking tray and cover with blocking solution. Wrap the blotting tray in aluminium foil and place on the rocking platform for 1 h.

4. Washing

Pour the blocking solution into the sink and cover the nitro-cellulose paper with TBS–Tween. Swirl the TBS–Tween around in the tray and then empty it down the sink. Repeat this action twice and then leave the nitro-cellulose paper submerged in TBS–Tween, on the rocking platform, for 2–3 min. Discard the TBS–Tween and repeat the 2–3 min wash.

5. Cutting up the nitrocellulose

Place the nitrocellulose sheet onto the cutting board and cut four thin strips (thin enough to fit into a lane of the multi-channel blotting tray) using the scalpel blade provided (**TAKE CARE NOT TO CUT YOUR FINGERS!**). Place a strip of nitrocellulose into lanes 1, 3, 5 and 7 of the multi-channel blocking tray.

6. Addition of primary antibody

Each lane in the multi-channel blotting tray can hold 4 ml of liquid. The anti-serum will be added to the nitro-cellulose blots at a dilution of 1 in 200. Therefore we need to make up 4 ml of a 1:200 dilution of each anti-serum provided.

6.1. Add 20 μl of naive serum to a Bijoux tube containing 3.98 ml of blocking solution. You now have a 1:200 dilution of naive serum. Mix well by vortexing for 1 min. Add this to lane 1 of the multi-channel blotting tray.

6.2. Add 20 μl of primary serum to a bijoux tube containing 3.98 ml of blocking solution. You now have a 1:200 dilution of primary serum. Mix well by vortexing for 1 min. Add this to lane 3 of the multi-channel blotting tray.

6.3. Add 20 μl of secondary serum to a Bijoux tube containing 3.98 ml of blocking solution. You now have a 1:200 dilution of secondary serum. Mix well by vortexing for 1 min. Add this to lane 5 of the multi-channel blotting tray.

6.4. Add 20 μl of tertiary serum to a bijoux tube containing 3.98 ml of blocking solution. You now have a 1:200 dilution of tertiary serum. Mix well by vortexing for 1 min. Add this to lane 7 of the multi-channel blotting tray.

Wrap the tray in aluminium foil and place the multi-channel blotting tray on the rocking platform and leave for 1 h.

7. Washing

Remove the primary antibody from each lane by gently tipping the blotting tray out over the sink (take care not to drop the nitro-cellulose strips into the sink). Fill each lane with TBS–Tween, tip out and repeat twice. Refill each lane with TBS–Tween and leave for 5 min on the rocking platform. Tip out the TBS–Tween and repeat the 5 min wash.

8. Addition of the enzyme-conjugated secondary antibody

Add 4 ml of the anti IgG enzyme conjugated antibody, which has been provided at a dilution of 1:2000 in blocking solution, to lanes 1, 3, 5 and 7 of the blotting tray. Wrap the tray in aluminium foil and place on the rocking platform for 1 h.

9. Washing

Wash each strip of nitro-cellulose thoroughly as in step 5.

10. Preparation and addition of the substrate

Add 33 μl of NBT and 44 μl of BCIP to 20 ml of the Colour development solution provided. Mix and add 4 ml of this substrate solution to lanes 1, 3, 5 and 7 of the blotting tray. Wrap the tray in aluminium foil and leave to develop for 30 min.

11. Stopping the reaction

Wash off the substrate as in step 5 and allow the blots to dry in air.

12. Analysis of results

Record the position and intensity of the bands and in particular note any differences between the lanes.

Ideas for further exploration

- Why are the proteins reduced before they are run on the gel?
- What would be the effect of increasing the percentage of acrylamide in the gel?
- When probing the Western blot what is the purpose of adding blocking solution at stage 1?
- Why does the antibody titre increase with increased exposure?

Information on similar exercises

Various parasite homogenate antigens or excretory secretory products may be used in the same way or used to test for cross-reactivity with *H. polygyrus*. Alternatively, serum could be used from animals harbouring a current infection, for instance mice infected with *H. polygyrus* could be tail-bled to obtain an anti-sera. However be aware that the yield of serum from a mouse is much less (maximum 1 ml), therefore may not be suitable unless the class size is small. The Western blots could be probed with other secondary antibodies against specific isotypes, for example, probing with both anti-IgG and anti-IgM in order to demonstrate the induction of primary and secondary immunity over the course of the immunisation.

REFERENCES

Harlow, E. & Lane, D. (1988). Immunoblotting. In *Antibodies: a Laboratory Manual*, pp. 479–504. New York, Cold Spring Harbour Laboratory Press.

Laemmli , U. K. (1970). Cleavage of structural proteins during the assembly of the head of bacteriophage T4. *Nature* **227**, 680–685.

Towbin, H. & Gorden, J. (1984). Immunoblotting and dot immunoblotting–current status and outlook. *Journal of Immunological Methods* **72**, 313–340.

Towbin, H., Staehelin, T. & Gorden, J. (1979). Electrophoretic transfer of proteins from polyacrylamide gels to nitro-cellulose sheets: Procedures and some applications. *Proceedings of the National Academy of Sciences* **76**, 4350–4354.

Appendix

Resolving gel buffer : 4 × Tris–Cl/SDS, pH 8.8

Tris base	18.17 g
10% sodium lauryl sulphate (SDS)	4 ml
Distilled water	40 ml

pH adjusted to 8.8, and made up to 100 ml with distilled water

Stacking gel buffer : 4 × Tris–Cl/SDS, pH 6.8

Tris base	6.05 g
10% SDS	4 ml
Distilled water	40 ml

pH adjusted to 6.8, and made up to 100ml with distilled water

12% Resolving gel

Easi-gel (30% acrylamide/ 0.8% bisacrylamide)	6 ml
4 × Tris–Cl/SDS, pH8.8	3.75 ml
Distilled water	5.25 ml
10% ammonium persulphate (APS)	0.05 ml
TEMED	0.01 ml

4% Stacking gel

Easi-gel	0.65 ml
4 × Tris–Cl/SDS, pH 6.8	1.25 ml
Distilled water	3.05 ml
10% ammonium persulphate	0.025 ml
TEMED	0.005 ml

Running buffer

Glycine	57.6 g
Tris base	12 g
SDS	1 g

Made up to 1 litre with distilled water

Reducing sample buffer

Tris base	1.51 g
Distilled water	25 ml
pH to 6.75 with HCl	
Above added to:	
SDS	40 ml
Glycerol	20 ml
2-mercaptoethanol	10 ml
Bromophenol blue	2 mg

Made up to 100 ml with distilled water

Western blotting transfer buffer

Glycine	57.6 g
Tris Base	12 g
SDS	4 g

Made up to 4 litres with distilled water, then 1 litre of methanol added

TBS Tween 0.05%, pH 8

Tris HCl	100 mM, pH 2.9
NaCl	1 mM

Make up to 1 litre with distilled water, adjust pH to 8, then add 0.5 ml Tween 20 and mix well

Colour development solution (CDS)

Tris HCl	100 mM, pH 9.5
NaCl	10 mM
$MgCl_2$	5 mM

BCIP/NBT

To 20 ml of CDS, add the following, and protect from the light:

5-bromo-4-chloro-3-indolyl phosphate (BCIP) (50 mg/ml in 100% DMF) 33 μl

Nitroblue tetrazolium (NBT) (75 mg/ml in 70% DMF) 44 μl

Section 5
Chemotherapy

5.1 Sensitivity of a coccidial parasite, *Eimeria*, to an ionophore, monensin

M. W. SHIRLEY

Aims and objectives

This exercise is designed to demonstrate the sensitivity of sporozoites, the invasive stages of coccidial parasites, to ionophorous drugs.

Specifically, the practical will examine:

1. How sporozoites may be recovered from transmission stages, the oocysts of the coccidial parasites, by a combination of physical damage and enzymatic digestion (via an intermediate sporocyst stage).
2. The lethality of the ionophore, monensin, to extracellular sporozoites by incubating the mixture of sporozoites, with remaining intact sporocysts and oocysts, in the presence of different concentrations of the drug.

Introduction

Coccidial protozoa are found wherever poultry are reared and an absolute need to prevent or control infections in commercial meat and egg-laying flocks has led to the use of many anticoccidial drugs. The most useful drugs continue to be the ionophorous antibiotics, even though widespread resistance to these drugs has been reported for many years. The drugs act by taking up residence within the cell membrane (pellicle) of the transiently extracellular stages, the sporozoites and merozoites, where they interfere with ion transport across the membrane. The test to be demonstrated in this practical is based on the viability/distortion and/or lysis of sporozoites as the drug interacts with the cell membranes of the parasite and interferes with the functioning of the ATP-dependent pumps that regulate the concentration of intracellular cations. It is thought that the drug acts as a pore in the membrane so that ions, commonly sodium, magnesium or calcium, may move from the outside **305** milieu via the ionophore into the cytoplasm of the parasite. The movement of ions into the parasite is reversed initially by its ion pumps, but as the supply of ATP that drives the pumps becomes exhausted and the concentration of intracellular ions increases sharply, so water enters, the parasites swell and subsequently lyse. Parasites carry the drugs into the host cell and most are destroyed in the first few hours after they become intracellular. However, the effects of the ionophores may be seen conveniently on the extracellular zoite stages if they are incubated in the drugs for a few hours.

Laboratory equipment and consumables
(per student or group)

Equipment

Bench-top centrifuge
Compound microscope with mechanical stage and Kellner eye-
    piece
Counting chambers (a haemocytometer or equivalent)
Vortex mixer (Whirlimix or equivalent)
Water bath (41 °C)
Balance, capable of weighing 50 μg accurately

Consumables

Pipettes and tips
15 ml centrifuge tubes (glass or plastic/round-bottomed or
    conical)
Glass slides
8 glass balls (ballotini) from Jencons
Conical flasks (50 ml)
Methanol
Tissue culture medium (RPMI or equivalent)
Trypsin (Difco 1:250; stored at 4 °C)
Taurocholic acid (Fluka Chemicals)
Phosphate-buffered saline* (PBS) pH 7.6
Monensin sodium (Sigma)
2% Trypan blue solution in PBS pH 7.6 (mix well)
Disposable gloves
Face masks

*PBS pH 7.6

Stock solution A: M/15 KH_2PO_4 (9.073 g/litre); stock solution B: M/15 Na_2HPO_4 (11.87 g/litre).

Mix 12.8 parts A and 87.2 parts B; add 0.9 g NaCl per litre.

Source of parasite material

Preparations of clean, sterile oocysts may be obtained from Dr Martin Shirley, Institute for Animal Health, Compton, Near Newbury, Berks RG20 7NN. In the USA, it may also be possible to obtain various species of *Eimeria* from the American Type Culture Collection (ATCC), which should be contacted for their latest catalogue. Alternatively, it may be possible to trace potential suppliers through recent publications in one of the biological publications databases and/or the www.

Safety

Staff should read the COSSH requirements for handling drug solvents and avoid skin contamination. A mask should be used when weighing trypsin and taurocholic acid.

Monensin sodium is a highly toxic compound (very toxic by inhalation, in contact with skin and if swallowed). Suitable protective clothing, gloves and eye/face protection should be used when handling the powder and appropriate care should be taken when working with the solutions. The Material Safety Data sheet provided by Sigma should be read carefully. *Eimeria tenella* are strictly host specific and will not infect hosts other than the domestic chicken.

Instructions for staff

Parasites should be as fresh as possible (several weeks' advance notice should be given to the Institute for Animal Health). Once received, oocysts should be stored at 4 °C.

The exercise should be started early in the morning, as follows.

Hours 1–2: set up hatching of sporozoites
Hours 2–5: hatching of sporozoites and incubation in monensin
Hour 8: examination of sporozoites after 3 h incubation in monensin
Hour 29: examination of sporozoites after 24 h incubation in monensin
(These timings are approximate and conservative)

The exercise is best fitted around lectures or breaks, so as to avoid the students waiting for the sporozoites to appear. The PBS requires that NaCl is added separately. Addition of NaCl is essential and **must not** be overlooked.

Instructions for students

Procedure 1. Hatching of *Eimeria* sporozoites

Step A. Preparation of sporocysts
Sporocysts will be recovered from oocysts that have been supplied to the laboratory after bleach treatment to remove faecal debris and to sterilise the final culture.

1. Remove 2×10^6 purified oocysts, place in a 15 ml glass or plastic centrifuge tube and add PBS pH 7.6 to a final volume of 15 ml.
2. Centrifuge for 5 min at 1500 g in a bench centrifuge. Discard supernatant. Add 0.5 ml PBS pH 7.6 to pellet.
3. Add 0.5 g glass balls (# 8, Jencons) so they comprise about half of the total resulting volume.
4. Place the tube on a vortex mixer turned to maximum speed and pulse the contents until most of the oocysts have been mechanically fractured to release their sporocysts. Oocysts of *E. tenella* will probably require about 5–20 pulses on the mixer where each pulse is the tube kept on a mixer for a count of 1 sec.
5. Check the progress of the cracking process by examining a very small sample under a microscope at regular intervals, perhaps every 5 pulses or so. If too many strokes are used, a large proportion of the sporocysts will be damaged. As a rule, as soon as free sporozoites are seen in the soup, then STOP! The preparation at this stage will contain unbroken oocysts, free sporocysts and, ideally, very few free sporozoites.
   NB the timings given above are guidelines only and will depend upon the type of vortex mixer used.
6. Recover sporocysts from the glass balls with repeated additions of PBS pH 7.6, so that the final volume is 15 ml.
7. Centrifuge the suspension of parasites at about 1500 g for 5 min and re-suspend the sporocysts in 20 ml of PBS pH 7.6 in a 50 ml conical flask.

Step B. Preparation of sporozoites

1. Separately weigh 50 mg trypsin and 200 mg taurocholic acid into small weighing boats.
2. Add the pre-weighed trypsin and taurocholic acid to the 20 ml of sporocysts, mix by swirling and incubate the flask at 41 °C until most of the sporozoites have excysted from the sporocysts. This process will take about 1.5 h and may be mon-

itored at intervals by microscopic examination: the release of sporozoites will be accompanied by the appearance of empty sporocyst cases.

3. Transfer the preparation of excysted sporozoites (with contaminating oocysts and sporocysts) to a centrifuge tube and wash twice in PBS pH 7.6 (final vol. 15 ml) by centrifugation at 1500 g for 5 min (discard supernatant each time) and re-suspend the pellet of parasites in 20 ml PBS pH 7.6.

4. Invert the tube several times to re-suspend the parasites, count the numbers of sporozoites and adjust to 1×10^5/ml.

5. Spend a few minutes looking at the morphology of the sporozoites as this will be one of the read-outs later.

Procedure 2. Sensitivity of *Eimeria* sporozoites to monensin

Part A: Incubation of sporozoites in monensin

1. Arrange for 1 mg monensin to be weighed (**NB: This must be done by a member of staff**) and dissolved in 1 ml methanol.

2. Dilute with tissue culture medium (RPMI 1640 or equivalent) to a concentration of 2.0 μg/ml and 0.02 μg/ml. (Increase volume to 500 ml initially [2.0 μg/ml] and then dilute 5 ml to 500 ml [0.02 μg/ml]).

3. Add 5 ml of each drug concentration to tubes 1 and 2.

4. Add 5 ml tissue culture medium without drug to tube 3.

5. Invert tube of sporozoites several times and add 5 ml to each of tubes 1, 2 and 3 (to give a monensin concentration of 1.0 μg/ml; 0.01 μg/ml and 0.00 μg/ml in tubes 1, 2, and 3, respectively, and a sporocyst concentration of 5×10^4/ml for each tube)

6. Incubate tubes 1, 2 and 3 at 37–41 °C for 3 and 24 h.

Part B: Evaluation of effect of drug

1. Count the numbers of sporozoites that have a normal or abnormal morphology at both time intervals (3 and 24 h).

2. Determine the viability of the sporozoites.

3. Invert tube several times and remove 100 μl. Place in centrifuge tube and add 10 μl of 2% trypan blue solution. Incubate at room temperature for 1 min.

4. Count blue parasites (dead) and non-staining parasites (viable) during the next 10–15 min.

5. Calculate percentage of viable parasites in the sample.

6. Set out the results as in Table 5.1.1.

Table 5.1.1 *Results and calculations*

Tube no.	Conc. of monensin	Mean % viable parasites 3 h	Mean % viable parasites 24 h	Staining results with trypan blue
1	1.0 µg/ml			
2	0.01 µg/ml			
3	0.001 µg/ml			

Potential sources of failure

- The project is relatively complex and involves many steps.
- Care must be taken with the excystation process to monitor, firstly, the release of sporocysts from the oocysts and, secondly, the release of sporozoites. Too violent vortexing could result in severe damage to the parasites and the unwanted premature physical release and subsequent destruction of large numbers of sporozoites.
- If sporozoites are not observed during the hatching process, it is possible that NaCl has not been added to the PBS. The PBS must be isotonic or the sporozoites will lyse immediately they escape from the sporocysts; the omission of NaCl is an easy error to make.
- Incubation times and temperatures should be monitored carefully.

Expected results

Sporozoites not exposed to monensin will have a normal appearance and a higher percentage will be viable in comparison to those incubated with the drug. Sporozoites exposed to 1 µg/ml of the drug will be affected the most.

Ideas for further exploitation

- A drug-sensitive laboratory strain of *Eimeria tenella* is examined in this study, and an interesting comparison would be between this strain and one isolated more recently from a clinical case of coccidiosis. Any new isolate would almost certainly be characterised by marked resistance to the ionophores and its response to monensin should be significantly different.
- Further intermediate time points and concentrations of monensin could also be used.

REFERENCES

Augustine, P. C., Watkins, K. L. & Danforth, H. D. (1992). Effect of monensin on ultrastructure and cellular invasion by the turkey coccidia *Eimeria adenoeides* and *Eimeria meleagrimitis*. *Poultry Science* **71**, 970–978.

McDougald, L. R. & Galloway, R. B. (1976). Anticoccidial drugs: effects on infectivity and survival intracellularly of *Eimeria tenella* sporozoites. *Experimental Parasitology* **40**, 314–319.

Smith, C. K., Galloway, R. B. & White, S. L. (1981). Effect of ionophores on survival, penetration, and development of *Eimeria tenella* sporozoites *in vitro*. *Journal of Parasitology* **67**, 511–516.

5.2 Egg hatch assay for determination of resistance of nematodes to benzimidazole anthelmintics

F. JACKSON, E. JACKSON & R. L. COOP

Aims and objectives

This exercise is designed to:

Determine the susceptibility of nematode eggs to treatment with a benzimidazole anthelmintic (thiabendazole; TBZ).

Introduction

The widespread use of anthelmintics to control gastrointestinal parasitism in sheep and goats has resulted in the emergence of strains resistant to the benzimidazole, levamisole or avermectin group of drug families. In some countries, multiple resistant strains have been selected. In the UK, the majority of cases of anthelmintic resistance are to the benzimidazole group, and an *in vitro* test for detection of resistant nematodes is based on the fact that benzimidazoles will inhibit the embryonation and hatching of fresh nematode eggs. The assay is based on the ovicidal properties of benzimidazoles and the ability of eggs from resistant strains to develop and hatch at higher concentrations of drug than that of susceptible strains.

The aim is to incubate fresh nematode eggs in serial concentrations of a benzimidazole anthelmintic, normally thiabendazole (TBZ). The percentage of eggs that hatch (or die) at each TBZ concentration is determined and a drug response curve plotted; i.e. drug concentration is plotted against percentage hatch (or death) of eggs, corrected for the normal mortality of untreated eggs. To obtain a linear regression from which to calculate the ED_{50} value (concentration of drug required to kill 50% of eggs) the data are usually transformed using arcsin or log-probit transformation.

Laboratory equipment and consumables
(per student or group)

Procedure 1: Extraction of nematode eggs from faeces

Equipment

1000, 500, 212, 75, 38 μm
    stainless steel sieves
Bench-top centrifuge
Microscope/mechanical
    stage/Kellner eyepiece

Finnpipettes (40–200 μl,
    200–1000 μl), tips
Vacuum suction line
Artery forceps

Consumables

Polythene bags
15 ml Polyallomer centrifuge
    tubes
Saturated sodium chloride solution

4 ml polystyrene macro
    disposable cuvettes
10 ml volumetric flask

Procedure 2: Preparation of thiabendazole solutions

Equipment

Finnpipettes (5–10 ml, 200–1000 μl, 40–200 μl, 5–40 μl), tips
Weighing boats   3-decimal-place top pan
                  balance

Consumables

100 ml volumetric flask
Thiabendazole

Dimethylsulphoxide (DMSO)
Small glass funnel

Procedure 3: Egg-hatch assay for determination of TBZ resistance

Equipment

Finnpipettes (200–1000 μl, 40–200 μl, 5–40 μl), tips
Relative humidity
    container
Incubator set at 25 °C

Stereo microscope
Small Petri dishes

Consumables

Clean suspension of
   nematode ova

Helminthological iodine
(= Lugol's solution)

24-well (2 ml well capacity) culture plate (NUNC)

Source of parasite material

Freshly voided nematode eggs are required. Laboratory animals or a sheep/goat are infected with a gastrointestinal nematode and the eggs extracted from the faeces 21 days later. A high faecal egg count (greater than 300 eggs/g) is essential if sufficient eggs are to be obtained to run the full range of drug concentrations.

Safety

Staff should read the COSHH requirements for handling DMSO and avoid skin contact with undiluted helminthological iodine.

Instructions for staff

It is suggested that the practical is run in the afternoon as this allows for preparation of the following during the morning. Nematode eggs should be extracted from faeces and stored at 4 °C and the stock solution of thiabendazole prepared prior to the exercise. The practical session needs to be planned for two afternoons separated by a period of two days to allow for incubation of the nematode eggs with the different drug concentrations.

Procedure 1: Extraction of nematode eggs from faeces

1. Place the faecal material (amount dependent on egg count) in a polythene bag, add 40 ml water and knead thoroughly using a stomacher or by compression and kneading of the bag by hand.
2. Suspend the macerated faecal material in 1 litre of tapwater and estimate the total number of eggs by placing 200 μl of this suspension into a cuvette, adding 4 ml saturated sodium chloride solution and counting the total number of eggs in the cuvette (dilution factor × 500).
3. Wash the faecal suspension over a series of sieves of decreasing mesh size. The 38 μm retentate will contain fine faecal

debris and nematode eggs. This is centrifuged at 100 g for 1–2 min in a series of polyallomer centrifuge tubes.

4. Remove the supernate using a vacuum line, leaving approximately 1 ml containing faecal debris and eggs, resuspend using 10–12 ml saturated sodium chloride solution, mix thoroughly but gently (violent mixing causes more faecal debris to float); centrifuge at 100 g for 1–2 min.

5. Using artery forceps, clamp off the tube just below the meniscus and wash the contents of the upper chamber (2–3 ml) into a 15 ml polystyrene centrifuge tube using tapwater.

6. Pool extracted eggs and make up to a total volume of 10 ml in a volumetric flask and count the numbers present in 100 μl, again using a cuvette and 4 ml saturated sodium chloride solution. Using this number, resuspend the eggs so that 100 μl contains approximately 100 eggs.

Procedure 2: (a) Preparation of 1000 ppm thiabendazole (TBZ) stock solution

1. Weigh 0.1 g of TBZ into a suitably-sized weighing boat.
2. Carefully transfer the TBZ powder into a 100 ml volumetric flask first by carefully transferring the bulk of the powder using a small glass funnel and then washing the remaining powder into the flask using two washes each containing 10 ml DMSO.
3. Add a further 20 ml of DMSO, mix thoroughly and then make up the total volume to 100 ml with distilled water.
4. This gives a stock solution of 100 ml containing 0.1 g TBZ (i.e. 1 g/1000 ml or 1000 part per million – ppm).
5. A suitable range of dilutions can be prepared from this original stock as required. For investigating ovine and caprine susceptible strains a suitable final range of concentrations would be 0.01–0.1 ppm, whereas for investigating ovine- and caprine-resistant strains a higher final concentration range is required (0.05–1.5 ppm).

(b) Preparation of working TBZ solutions

Using this stock solution, produce a range of working dilutions. Desirable final concentrations (i.e. concentrations in the well) for investigating suspect resistant strains are 0, 0.05, 0.1, 0.3, 0.5, 0.7, 0.9, 1.1, 1.3 and 1.5 ppm. Since 10 μl of working solution will be added to each 2000 μl well the working dilutions should be 200 times more concentrated than the final concentration, i.e. 0, 10, 20, 60, 100, 140, 180, 220, 260 and 300 ppm. Add the

Table 5.2.1 *Preparation of TBZ Solutions*

TBZ stock 1000 ppm (μl)	Dist. water (μl)	Working conc. (ppm)
3000	7000	300
2600	7400	260
2200	7800	220
1800	8200	180
1400	8600	140
1000	9000	100
600	9400	60
200	9800	20
100	9900	10
0	10 000 (= 10 ml)	0

relevant quantity of 1000 ppm stock solution and make up to 10 ml in a volumetric flask using distilled water (Table 5.2.1).

Instructions for students

Procedure 3: Egg-hatch assay for determination of TBZ resistance

1. Add 100 μl containing 100 eggs to each of two wells per drug concentration, i.e. 20 wells for the range above.
2. Add 10 μl of working stock solution to the wells and then add 1890 μl of distilled water to each of the wells to give a total volume of 2000 μl. It is advisable to add the distilled water as two aliquots of 945 μl using the 200–1000 μl Finnpipette as this ensures thorough mixing of the well contents. Use a new tip for each concentration.
3. Label the lid of the culture plate to show the concentration used in each well.
4. Place the culture plate in a 100% relative humidity container (to lessen evaporative losses) and incubate at 25 °C for 48 h.
5. Following incubation, add 1 drop of helminthological iodine to each well. Then count the number of eggs and first-stage larvae using an inverted or stereo microscope (it may be necessary to transfer the contents of each well into a small Petri dish since the edges of wells may be obscured from view depending on the quality of illumination of the microscope).
6. Calculate mean numbers of eggs and larvae at each concentration and the percentage hatch using the formula:

$$\text{Percentage hatch} = \frac{\text{number of larvae}}{\text{number of eggs} + \text{number of larvae}} \times 100$$

7. The percentage of eggs that fail to hatch at each drug concentration are corrected for natural mortality using data from control wells and transformed using arcsin transformation and plotted against drug concentration. The resulting log dose response line enables the ED_{50} to be estimated (i.e. the concentration of TBZ required to prevent 50% of the eggs from hatching).

Ideas for further exploration

Only one species of egg is used in this study. Plot the curves that would have resulted from the introduction of a second species, comprising 25% and 75% of the total number of eggs and assuming that all of the second species were incapable of hatching at concentrations of 0.1 ppm.

Brief details of expected results

Generally, ED_{50} values above 0.1 ppm are indicative of the presence of resistant strains of ovine or caprine nematode. Obviously in mixed infections if a minor species (comprising less than 50% of the eggs) is resistant then using these criteria will not identify resistance. The sensitivity of the technique can be improved by incorporating specific egg identification procedures. Determining the prevalence of different species in pre- and post-incubation samples enables any species hatching at concentrations of greater than 0.1 ppm to be identified. In practice only eggs of susceptible species will remain to be measured and thus resistant species can be identified even when they constitute a minor component of the total population.

REFERENCES

Cawthorne, R. J. G. & Whitehead, J. D. (1983). Isolation of benzimidazole resistant strains of *Ostertagia circumcincta* from British sheep. *Veterinary Record* **112**, 274–277.

Hunt, K. R. & Taylor, M. A. (1989). Use of the egg hatch assay on sheep faecal samples for the detection of benzimidazole resistant nematodes. *Veterinary Record* **125**, 153–154.

Jackson, F. (1993). Anthelmintic resistance – the state of play. *British Veterinary Journal* **149**, 123–137.

Le Jambre, L. F. (1976). Egg hatch as an *in vitro* assay of thiabendazole resistance in nematodes. *Veterinary Parasitology* **2**, 385–391.

Waller, P. J. (1987). Anthelmintic resistance and the future for round-worm control. *Veterinary Parasitology* **25**, 177–191.

Waller, P. J., Dash, K. M., Barger, I. A., Le Jambre, L. F. & Plant, J. (1995). Anthelmintic resistance in nematode parasites of sheep: learning from the Australian experience. *Veterinary Record* **136**, 411–413.

5.3 Larval migration inhibition assay for determination of susceptibility of nematodes to levamisole

F. JACKSON, E. JACKSON & R. L. COOP

Aims and objectives

This exercise is designed to demonstrate:

The susceptibility of nematode larvae to treatment with levamisole. The assay is based on the paralysing property of the drug and the ability of larvae from resistant strains to migrate at a higher concentration of drug than those of susceptible strains.

Introduction

The widespread use of anthelmintics to control gastrointestinal parasitism in sheep and goats has resulted in the emergence of strains of nematodes resistant to the benzimidazole, levamisole or avermectin group of drug families. In some countries, multiple resistant strains have been selected. In the UK, the majority of cases of anthelmintic resistance are to the drugs in the benzimidazole group, although resistance to avermectins and levamisoles has also been reported. An *in vitro* test for resistance to avermectins is based on the fact that the paralysis resulting from treatment with this drug inhibits the migratory behaviour of infective third-stage larvae (L_3).

The aim is to incubate L_3s in serial concentrations of the test drug on a nylon mesh, thus simulating the gut mucosal layer through which the larvae would normally migrate. Paralysis caused by susceptibility to the test drug has the effect of preventing the migration and causing the larvae to be held by the gauze. The percentage of larvae able to migrate at each concentration is determined and a drug response curve plotted (i.e. drug concentration is plotted against percentage migrating through the membrane). This graph is used to estimate an LM_{50} value at which 50% of larvae are paralysed by the drug.

Laboratory equipment and consumables
(per student or group)

Procedure 1 Culture method for ruminant nematodes

Equipment

Culture trays (400 × 200 × 75 mm)

Stainless steel sieve (1.00 mm)

Incubator (22 °C)

Beaker (250 ml)

Filter holder*

Stereomicroscope

Volumetric flasks (50 ml, 10 ml)

Vacuum line

Pipettes and tips (1 ml, 100 μl)

*50 mm diameter tubing with collar to suspend over beaker

Consumables

High-wet-strength paper

Culture flasks (250 ml)

Polythene bags (500 × 300 mm)

Microscope slides

Rubber bands

Procedure 2 Preparation of stock solutions

Equipment

Digital pipettes/tips (5–40, 40–200, 200–1000 μl, 5–10 ml)

Volumetric flasks (100 ml)

Top pan balance (3 decimal place)

Small glass funnel

Consumables

Levamisole hydrochloride (Sigma L9756)

Anthelmintic containing 7.5% w/v levamisole (i.e. Levacide)

Weighing boats

Procedure 3 Exsheathment of ruminant nematodes

Equipment

Microcentrifuge or benchtop centrifuge

Consumables

Eppendorf centrifuge tubes (1.5 ml)

Plastic syringe (1 ml) and needle (20-gauge)

Sodium hypochlorite solution   Microscope slides
   (Milton Sterilising Solution)
0.85% NaCl solution

Procedure 4 Larval migration

Equipment

Tapering plastic tubing (two sizes, one to fit tightly inside the
   other to create fixed collar for mesh, approx. max. diam. 15
   mm)
Nylon mesh (40 mm × 40 mm per well, Nytal nylon 25 μm mesh:
   Lockertex)
24-well (2 ml) culture plate (NUNC)

Consumables

Exsheathed infective larvae
Helminthological iodine (Lugol's solution)
Inverted microscope with suitable stage *or* Stereo microscope
   and Petri dishes (for pipetting well-contents into)

Sources of parasite material

Stored infective nematode larvae of many parasitic nematodes
can be used for the exercise, although small ruminant trichos-
trongyle larvae are particularly suitable. A culture method for
the production of third-stage larvae from monospecifically
infected animals is provided. See also Additional information
section of Exercise 1.15.

Safety

Staff should read the COSHH requirements for handling drug
solvents and avoid skin contact with undiluted helminthological iodine.

Instructions for staff

Stock solutions should be prepared in advance since the larvae
require to be exsheathed and exposed to the drug for a 2 h
period prior to the migration assay. The practical session may be
arranged around a lecture and/or a lunch break to avoid the students waiting during the two 2 h incubation periods.

Procedure 1: Culture of third stage nematode larvae

1. Faeces are collected from a monospecifically infected animal and placed in the culture tray to a maximum depth of 30 mm and incubated inside a loosely sealed polythene bag at 22 °C for ten days.
2. Following incubation the sample is flooded with warm tap-water (22 °C) and left to soak for 1 h.
3. Coarse faecal material is removed using the 1.00 mm sieve and the liquid allowed to sediment for 2 h at 4 °C.
4. The liquid volume is reduced using a vacuum line and the sample cleaned of fine debris using a 'Baermann' apparatus consisting of high-wet-strength paper held over a plastic cylinder and fixed using a rubber band. The liquid suspension is poured through the filter paper, which will temporarily restrain the larvae, inactive following step 3
5. The filter and holder are then immersed in warm tapwater (22 °C). The larvae migrate through the filter and the filter and holder can be removed after 2 h.
6. Cleaned larvae can be stored at 4 °C for up to 6 weeks before use, preferably in a flat-bottomed culture flask with ventilated cap to allow gaseous exchange.

Procedure 2: Preparation of 1000 ppm anthelmintic stock solution

Using pure drug compound from Sigma

1. Weigh 0.1 g of the pure drug into a suitably-sized weighing boat.
2. Carefully transfer into a 100 ml volumetric flask via the funnel, washing any remainder using 2 washes of 10 ml of distilled water.
3. Make up to 100 ml with distilled water.
4. This gives a final stock solution of 100 ml containing 0.1 g anthelmintic (i.e. 1 g/1000 ml or 1000 ppm).
5. A suitable range of working dilutions can be made up from this stock solution by the students in the class. (Note pre-experimentation is recommended in order to establish a suitable range of working concentrations of anthelmintic for each nematode species/isolate).

Using a commercial anthelmintic product
Levacide low-volume – 7.5% levamisole hydrochloride (Norbrook).
A 1 in 75 dilution of this product in distilled water gives a suit-

Fig. 5.3.1 Improvised filter apparatus. A filter of Nytal mesh can be held in place between an outer collar made from the barrel of a 2 ml disposable syringe, cut off 1.5 cm from the end, and an inner collar that is a plastic pipette tip trimmed to a suitable size.

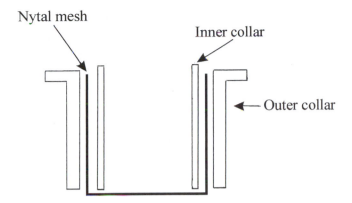

Nytal mesh

Inner collar

Outer collar

able 1000 ppm stock solution, i.e. 1 ml of Levacide plus 74 ml water.

Procedure 3: Exsheathment of third stage larvae

1. Place larvae into centrifuge tubes and add 80 µl sodium hypochlorite solution per ml larval suspension.
2. Take a small subsample of treated larvae, place on a slide and observe under a microscope. When all larvae have exsheathed (usually after 1–2 min) centrifuge (microcentrifuge short burst at lowest speed or benchtop centrifuge 2 min 100 *g*) and then wash centrifugally three times with 0.85% NaCl.
3. Larvae can then be counted by taking 100 µl subsamples and counting on a slide and an estimate made of total numbers of larvae. Adjust volume to approximately 100 larvae per 100 µl sample.

Filter apparatus

The filters should be held by two concentric rings as shown in Fig. 5.3.1.

Instructions for students

1. Make up working dilutions of levamisole using Table 5.3.1. Use 10 ml volumetric flasks and distilled water as the diluent.
2. Count the exsheathed larvae and adjust the concentration so that 100 µl contains approximately 100 larvae. Dispense 100 µl containing the larvae into a series of labelled Eppendorf tubes and add 1 ml of the working stock solutions.

Table 5.3.1 *Preparation of levamisole solutions*

Stock 1000 ppm levamisole (μl)	Dist. water (μl)	Working conc. (working stock solution) (ppm)
400	9600	40
200	9800	20
100	9900	10
50	9950	5
20	9980	2
10	9990	1
0	10000 (= 10 ml)	0

Centrifuge, remove the supernatant and resuspend using 1 ml of the working stock solution. Repeat the centrifugation/resuspension process to ensure that the larvae are incubating in the appropriate drug concentration and incubate for 2 h at 37 °C.

3. Following a 2 h incubation, centrifuge and reduce the volume to 200 μl.

4. Add 1800 μl of each drug dilution to the wells on the test plate and place a filter into the well ensuring that the mesh is fully submerged and that there are no air bubbles trapped beneath the mesh.

5. Mix the larval suspension thoroughly and add the larvae in 200 μl of fluid to each filter by pipetting gently down the inside of the inner collar. NOTE the suspension must be mixed for every well as larvae settle rapidly in water.

6. Place a cover over the plates and incubate for two hours at 37 °C.

7. Carefully remove the filters and wash any remaining larvae into individual labelled Petri dishes and stain with a few drops of helminthological iodine.

8. Add a few drops of helminthological iodine to each well on the culture plate.

9. Using an inverted microscope at × 100 magnification count the number of larvae in each well and in each Petri dish.

10. The percentage migration is calculated for each concentration using the formula:

$$\% \text{ migration} = \frac{(N_m)}{(N_m + N_r)} \times 100$$

Where: N_m = number of larvae migrating through mesh (i.e. found in the well)

N_r = number of larvae retained by the mesh
(i.e. washed off from mesh)

11. Plot a graph of drug concentration against percentage migration and from this estimate the LM_{50} value (concentration at which 50% of larvae fail to migrate)

Ideas for further exploration

• The use of two or more strains, one resistant and one susceptible, is recommended to demonstrate the application of this technique.

REFERENCES

Condor, G. A. & Campbell, W. C. (1995). Chemotherapy of nematode infections of veterinary importance with special reference to drug resistance. *Advances in Parasitology* **35**, 1–84.

Hunt, K. R. & Taylor, M. A. (1989). Use of the egg hatch assay on sheep faecal samples for the detection of benzimidazole resistant nematodes. *Veterinary Record* **125**, 153–154.

Jackson, F. (1993). Anthelmintic resistance – the state of play. *British Veterinary Journal* **149**, 123–137.

Le Jambre, L. F. (1976). Egg hatch as an *in vitro* assay of thiabendazole resistance in nematodes. *Veterinary Parasitology* **2**, 385–391.

Waller, P. L (1987). Anthelmintic resistance and the future for roundworm control. *Veterinary Parasitology* **25**, 177–191.

Waller, P. J., Dash, K. M., Barger, I. A., Le Jambre, L. F. & Plant, J. (1995). Anthelmintic resistance in nematode parasites of sheep: learning from the Australian experience. *Veterinary Record* **136**, 411–413.

5.4 Effect of anthelmintics on nematodes

D. L. LEE & J. E. SMITH

Aims and objectives

This exercise is designed to:
  Study the effect of levamisole on *Heligmosomoides polygyrus*, *in vivo*.

Introduction

Mice infected with the parasitic nematode *H. polygyrus* will be dosed with the human and veterinary anthelmintic, levamisole. The effect of the treatment after different periods of exposure to the drug will be assessed.

Laboratory equipment and consumables
(per student or group)

Equipment

Dissecting board and pins	Marker pen
Scissors	Binocular dissecting
Petri dishes	microscope
Physiological saline	Disposable gloves
(0.85% NaCl)	Graph paper
Measuring scale	Tally counters

Consumables

Disinfectant for dissection instruments
Levamisole hydrochloride (0.4% solution in distilled water)

Source of parasitic material

Infective stage larvae (juveniles) of *H. polygyrus* are required. Our source was the School of Biology, Leeds University, UK. See also Additional information section of Exercise 1.15.

Safety

Students should ensure that all equipment and instruments are placed in the dishes of disinfectant at the end of the practical class. Personal dissecting instruments should be immersed in the disinfectant then thoroughly washed before leaving the laboratory. Students should also wash their hands thoroughly before leaving the laboratory.

Instructions for staff

Mice will need to be infected with *H. polygyrus* 20 days before the exercise. Each mouse should receive 100 infective stage larvae (juveniles) administered orally in water by means of a blunt-ended syringe. Levamisole (0.25 ml of a 0.4% solution of levamisole hydrochloride per mouse) will be given orally by means of a blunt-ended syringe 1, 2 and 3 h before the start of the practical. One group of mice will receive water instead of levamisole. The mice should be killed according to approved procedure immediately prior to dissection. Please note that this practical exercise requires appropriate licences for experiments on animals.

Instructions for students

Each group will receive one mouse infected 20 days previously with 100 infective stage juveniles (larvae) of *H. polygyrus*. Adult males and females will be found in the intestine.
The mice have been divided into four groups:

Group 1 were dosed with levamisole 1 h before killing
Group 2 were dosed with levamisole 2 h before killing
Group 3 were dosed with levamisole 3 h before killing
Group 4 are infected controls which have not been drug treated

Make a note of which group your mouse is from.

1. As soon as possible dissect out the whole of the small intestine, caecum and large intestine. Try not to break the alimentary tract but if this happens then keep the sections in their correct sequence.
2. Separate the caecum and the large intestine and place each in a Petri dish of warm saline.
3. Measure the length of the small intestine, using the scale provided, and cut it into five equal sections. Place each section in a separate Petri dish containing saline. Keep the Petri dishes

in order and label them with the anterior small intestine in the first dish and the large intestine in the last dish.

4. Open each section along its length and examine for the presence of *H. polygyrus*. Using a binocular dissecting microscope note the number of worms in each section. The mice may harbour natural infections, so note the number and position of any other helminths present. These may include the nematode *Aspiculuris tetraptera*, or the cestode *Hymenolepis* (= *Rodentolepis*) *nana*, or possibly other helminths.

5. Enter the results in your group and then in a class table. Plot the percentage distribution of the worms in the alimentary tract. If other helminths are present record their position.

Expected results

This exercise should illustrate the effect of levamisole on worm position in the alimentary tract, which should vary according to the period of drug exposure. Why are the positions of the nematodes different in the different treatments? Speculate on the effect the anthelmintic has had on the physiology/biochemistry of the nematodes. What effect, if any, has the anthelmintic had on other helminths, if they are present? If there are differences in the effect on different helminths speculate on the reasons for these differences. How would you improve this experiment?

Ideas for further exploration

- Find out if the sexes of the nematode respond differently to the treatment.
- Include a different helminth if no others were present. *Hymenolepis nana* would be an ideal choice.
- Find out if there are statistical differences in the position of the nematodes between the groups of mice.
- Note activity or inactivity of the helminths present.

REFERENCES

Aceves, J., Erliji, D. & Martinez-Marnon, R. (1970). The mechanism of the paralysing action of tetramisole on *Ascaris* somatic muscle. *British Journal of Pharmacology* **38**, 332–344.

Coles, G. C., East, J. M. & Jenkins, S. N. (1975). The mechanism of action of the anthelmintic levamisole. *General Pharmacology* **6**, 309–13.

Martin, R. J., Pennington, A. J., Duittoz, A. H., Robertson, S. & Kusel, J. R. (1991). The physiology and pharmacology of neuromuscular transmission in the nematode parasite, *Ascaris suum. Parasitology* **102**, S41–S58.

Thienpont, D., Vanparijis, O. F. J., Raeymaekers, A. H. M., Vandenberk, J., Demoen, P. J. A., Allewijn, F. T. N., Marsboom, R. P. H., Niemegeers, C. J. E., Schellekens, K. H. L. & Janssen, P. A. J. (1966). Tetramisole (R8299), a new, potent broad-spectrum anthelmintic. *Nature* **209**, 1084–1086.

Section 6
Molecular Parasitology

6.1 Purification of DNA

P. A. BATES, T. KNAPP & J. M. CRAMPTON

Aims and objectives

This exercise is designed to demonstrate the isolation of genomic DNA; it can, in principle, be adapted for use with any organism, including parasites. Specifically, the objectives are:

1. To perform bacterial cell lysis.
2. To extract the cell lysate with phenol/chloroform.
3. To isolate the DNA by ethanol precipitation.

Introduction

DNA encodes genetic information for all known living organisms. A basic method in molecular biology is the isolation of DNA. The version given here is for the preparation of DNA from *Escherichia coli*. In this context the practical is useful for demonstrating some basic methods in molecular biology as well as introducing students to the use of bacteria as hosts for gene cloning. This is a prerequisite for more complex molecular biological methods including the isolation of parasite genes and analysis of their expression products.

Laboratory equipment and consumables
(per student or group)

Equipment

Safety glasses
P20, P200, P1000 Gilson pipetmen or equivalents
Waste beaker for chemically contaminated plasticware
Autoclavable waste bags for bacterially contaminated waste
Waste container for bacterial culture supernatant
Bench top centrifuge capable of producing 1000 g
Water bath (37 °C) with suitable racks for Universal tubes
Centrifuge capable of producing 8000 g at 10 °C
Use of cold room/refrigerator and freezer

Consumables

Latex gloves in a range of sizes

Sterile tips for pipetmen

Universal tube containing a 25 ml culture of *E.coli* grown overnight at 37 °C in L-broth

3 ml of bacterial cell lysis buffer

300 μl of 10% (w/v) sodium dodecyl sulphate (SDS) in a microfuge
tube

20 μl of 10mg/ml (in distilled water) Proteinase K in a microfuge
tube

3 ml of chloroform/isoamyl alcohol in a glass bottle

6 ml of phenol/chloroform in a dark glass bottle (or wrap in
silver foil)

Access to polyethylene glycol (PEG) in case of phenol contamination (e.g. 500 ml bottle)

6 ml of absolute ethanol

300 μl of sodium acetate in a microfuge tube

1 ml of 70% ethanol

1 ml of TE buffer

10 ml polypropylene capped test-tubes (sterile)

Sterile 1.5 ml microfuge tubes

Sterile disposable plastic Pasteur pipettes with a wide bore

Glass Pasteur pipette, heated in a flame and bent to produce a
small hook

Sources of biological material

Any of the disabled strains of *E. coli* K12 commonly used in
molecular cloning work are suitable and have the advantage of
rapid growth. Cultured parasitic protozoa such as *Leishmania*,
Trypanosoma, *Plasmodium*, etc. can also be used although it will
be necessary to conduct preliminary trials to establish the
amount of material required and a longer period for growth of
the organisms will be necessary. Similarly, DNA can also be
extracted from helminth parasites. Here other preparatory steps
will be required to disrupt the organisms, such as grinding with
a pestle and mortar (see below). If using parasite sources the
protocol can be used, but starting with frozen cell pellets, which
is more convenient and avoids exposure in the case of pathogenic organisms.

Safety

In addition to the following, local safety rules and regulations should be followed.

- **Phenol is the most dangerous organic solvent used in molecular biology,** but can be used safely provided certain precautions are taken. Disposable latex gloves should be worn whenever phenol is being handled as contact with skin can cause burning. Make sure that the gloves used are an effective barrier against phenol. Eye protection is essential and safety spectacles should always be worn when using phenol or chloroform. Avoid unnecessary exposure to phenol vapours by keeping the stock bottles closed when not in use. Pipettes and tips that have been in contact with phenol should be collected separately and destroyed by incineration. Phenol waste should be disposed of according to local safety rules for the disposal of organic solvents.

  In the event of phenol exposure to skin, remove contaminated clothing immediately and flush off excess chemical with water. Wash the affected area with polyethylene glycol (PEG), molecular weight 300, for 30 min wearing latex gloves, followed by rinsing with ethanol. In the event of exposure to the eyes wash with flowing water – **do not use PEG** –and seek hospital attention.
- Avoid unnecessary exposure to chloroform vapours by keeping the stock bottles closed when not in use. Chloroform waste should be disposed of according to local safety rules.
- Ensure that all tubes for centrifugation are properly balanced.
- *E. coli* K12 is not considered a pathogenic organism, but caution dictates that culture waste should be sterilised by autoclaving or chemical treatment and disposed of according to good microbiological practice.
- SDS is supplied as light flakes that are easily disturbed. Care should be taken when weighing as SDS is irritating to eyes and skin. Use of a face mask or a balance in a containment hood is recommended.

Instructions for staff

1. Bacterial cell lysis buffer is 50 mM glucose, 25 mM Tris-HCl pH 8, 10 mM EDTA, 5 mg/ml lysozyme. The buffer can be

prepared and autoclaved in advance and stored at room temperature. However, the lysozyme should be added only just prior to use, weighing out an appropriate amount of lyophilised enzyme.

2. 'Chloroform' is a 24:1 (v/v) mixture of chloroform/isoamyl alcohol, stored in a capped bottle at room temperature.

3. Phenol/chloroform is 1:1 (v/v) of TE-saturated phenol and chloroform/isoamyl alcohol prepared as follows. Water-saturated phenol can be purchased directly and should be stored at 4 °C. The aqueous layer should be replaced with 1 M Tris-HCl pH 8 and the mixture agitated. This aqueous layer should also be discarded and replaced with TE buffer (10 mM Tris/HCl, 1 mM EDTA pH 8). Addition of 8-hydroxyquinoline to 0.1% (w/v) is useful – this is an antioxidant but it also imparts a yellow colour to the phenol phase, particularly helpful during the extraction procedure. Again, agitate the mixture and allow to separate into organic and aqueous phases. The TE-saturated phenol can now be removed from the lower organic phase and mixed 1:1 with chloroform/isoamyl alcohol in a separate bottle. Phenol/chloroform should be stored at 4 °C until use.

4. A solution of 3 M sodium acetate is prepared by dissolving sodium acetate at 3 M, adjusting the pH to 5.5 with acetic acid and autoclaving.

5. TE buffer should be autoclaved and stored at room temperature.

6. Aliquots of 10 mg/ml (in distilled water) proteinase K, weighed out while wearing a face mask, should be stored at −20 °C.

Instructions for students

1. Grow a 25 ml culture of *E. coli* in culture broth overnight at 37 °C in a sterile plastic Universal tube (already provided).

2. Spin down the cells in a bench top centrifuge, 1000 *g* for 10 min at room temperature.

3. Remove the supernatant by carefully pouring off into a waste container, and resuspend the cell pellet in 3 ml of lysis buffer (50 mM glucose, 25 mM Tris-HCl pH 8, 10 mM EDTA, 5 mg/ml lysozyme). Then add 300 μl of 10% SDS and leave at room temperature for 5 min. These additions are made with a 1 ml pipetman.

4. Add 17 µl of proteinase K to give a final concentration of 50 µg/ml and incubate in a 37 °C waterbath for 1 h.

5. Transfer the cell lysate into a polypropylene test tube with a sterile wide-bore plastic Pasteur pipette. The lysate should be viscous and sticky from the release of DNA from ruptured cells into solution. Extract the lysate with phenol/chloroform. Add 3 ml of phenol/chloroform, cap the tube and mix thoroughly but gently by repeatedly inverting the tube for 5 min. Vigorous shaking is not required and will shear the DNA. Centrifuge at 8000 *g*, 10 min, 10 °C to separate the aqueous and phenol phases.

6. Remove upper, aqueous phase into a new tube using a sterile plastic Pasteur pipette. The upper aqueous phase is readily distinguished from the lower yellow-coloured phenol phase. Take as much of the upper phase as possible without removing material from the interface. Add another 3 ml of phenol/chloroform to the aqueous phase and repeat the phenol extraction.

7. Centrifuge at 8000 *g* for 10 min at 10 °C. Remove upper aqueous phase into a new tube as before. This time extract with chloroform alone, using the same method as above.

8. After centrifugation, remove upper aqueous phase into another tube, add 300 µl of 3 M sodium acetate and mix with the aqueous phase. Then carefully overlay with 6 ml (2 volumes) of absolute ethanol.

9. Mix gently to precipitate the DNA. This is best achieved by stirring the interface with a bent glass Pasteur pipette. As the ethanol mixes with the aqueous phase the DNA is precipitated and can be spooled out as a whitish precipitate. Hook the DNA out with the pipette, wash by dipping briefly in 70% ethanol and then slip the DNA off the Pasteur pipette into a microfuge tube containing 1 ml of TE buffer. Leave to dissolve overnight at 4 °C.

Expected results

Lysis of the bacterial cells (steps 1–4) should result in a sticky, viscous solution. This is caused by the presence of high molecular weight DNA. Use of a wide-bore pipette helps to avoid shearing the DNA.

After phenol extraction (step 5) protein contamination will be concentrated at the interface between the aqueous and organic phases. When removing the upper aqueous phase some of the interface material may be carried over. A small amount of carryover is acceptable but should be minimised. The main effect is that the DNA will not be as clean, unless extra extractions are performed. If all that is required is to demonstrate DNA extraction then this is no problem. However, if it is desired to continue on the following or another day with endonuclease restriction using the DNA prepared in this exercise, then it is essential that the DNA is as clean as possible. Extra phenol/chloroform extractions may be advised.

Spooling out the DNA (step 9) is interesting and usually works very well. Low yields can result if there is poor recovery of the aqueous phase following phenol extractions. In this event the spooling may not work as well. The same effect can be produced by vigorous handling of the sample, which can shear the DNA. In either event there will be some DNA precipitate after mixing in the ethanol. If necessary this can be recovered by centrifugation ($10\,000\,g$, $10\,\mathrm{min}$, $4\,^\circ\mathrm{C}$) instead of spooling.

Ideas for further exploration

- The DNA produced can be used in the next exercise: restriction endonuclease digestion and agarose gel electrophoresis.
- The DNA can also be quantified by measuring the A_{260} in a UV spectrophotometer. Since the DNA needs to be left at least overnight to dissolve, this is most conveniently done in a subsequent practical session (see Exercise 6.2).
- Alternatively it can be performed as a short follow-up to this exercise. At 260 nm in a 1 ml quartz cuvette, a $50\,\mu\mathrm{g/ml}$ solution of double-stranded DNA produces an A_{260} of 1.0. The purity can also be assessed by determining the A_{260}/A_{280} ratio. For clean DNA, this should be 1.9 or higher, but will be lower if there is protein contamination.

Information on similar exercises

Variations on this practical can be made with virtually any biological source. Once a cell lysate has been made the procedure given above will work in most cases. Preliminary runs are strongly recommended to determine the amount of material required. Depending on the source, additional phenol/chloroform extractions may be needed to remove protein contaminants, i.e. to produce a clean interface between the aqueous and phenol phases. The major differences will lie in the initial preparation of the material. Single-cell suspensions of eukaryotic cells will work as above, but lysozyme will not be required – here this is used to help in digestion of the bacterial cell wall. However, certain metazoan organisms, including helminth parasites, will require some form of homogenisation before cell lysis. This is most easily achieved by placing the worm material into a pestle and mortar cooled on dry ice/methanol, adding a small quantity of liquid nitrogen, and grinding to a powder. Since such material is normally only available in small amounts and the procedure involves use of liquid nitrogen, it is not generally recommended for large practical classes and should only be used with small groups of advanced students that can be closely supervised.

REFERENCES

Brown, T. A. (1991). *Essential Molecular Biology, A Practical Approach*, vol. 1 (ed. T. A. Brown). Oxford, IRL Press.

Heath, S. (1997). Molecular techniques in analytical parasitology. In: *Analytical Parasitology* (ed. M. T. Rogan). Chapter 3. Berlin, Springer.

Hyde, J. E. (1993). *Protocols in Molecular Parasitology* (ed. J. E. Hyde). Methods in Molecular Biology Volume 21. New Jersey, Humana Press.

Maizels, R. M., Blaxter, M. L., Robertson, B. D. & Selkirk, M. E. (1992). *Parasite Antigens, Parasite Genes: A Laboratory Manual for Molecular Parasitology*, Cambridge, Cambridge University Press.

Sambrook, J., Fritsch, E. F. & Maniatis, T. (1989). *Molecular Cloning: A Laboratory Manual*. New York, Cold Spring Harbour Laboratory Press.

6.2 DNA digestion and gel electrophoresis

P. A. BATES, T. KNAPP & J. M. CRAMPTON

Aims and objectives

This exercise aims to perform spectrophotometric analysis of *Escherichia coli* DNA, followed by endonucleolytic digestion of this bacterial genome and DNA from a variety of other sources with the restriction enzyme, *Eco*RI.

The specific objectives of this practical are:

1. To measure the A_{260} and A_{280} of *E. coli* DNA.
2. To perform restriction digests with *Eco*RI.
3. To prepare an agarose gel.
4. To analyse the restriction digests by agarose gel electrophoresis.

Introduction

The yield and purity of isolated DNA can be estimated relatively easily using ultraviolet (UV) spectrophotometry. A further test of DNA purity is to assess the extent that it can be cut with restriction endonucleases, enzymes that recognise and make a double-stranded cut at specific nucleotide sequences. This is an essential technique used in the construction of gene libraries and the analysis of cloned genes. The digestion products are separated by size on an agarose gel and visualised under UV illumination, after staining with the fluorescent dye ethidium bromide.

Laboratory equipment and consumables
(per student or group)

Equipment

UV spectrophotometer and quartz cuvettes
P20, P200 and P1000 Gilson pipetmen or equivalents
37 °C waterbath; 65 °C waterbath

Heater/stirrer, magnetic stirrer bar
Heat-resistant gloves
Agarose gel electrophoresis equipment and combs
Power supply for running gels
UV transilluminator and photographic system

Consumables

Latex gloves in a range of sizes
Sterile tips for pipetmen
Ice bucket and wet ice
*Eco*RI restriction enzyme (5 units/μl)
Sterile distilled water and TE buffer
15 μl 10 $\times$ *Eco*RI reaction buffer
0.5 ml and 1.5 ml sterile microfuge tubes
12 μl Phage λ DNA at 50 μg/ml
12 μl *E. coli* DNA at 100 μg/ml*
12 μl *Plasmodium falciparum* DNA at 100 μg/ml
12 μl mosquito (*Aedes aegypti*) DNA at 100 μg/ml
12 μl human DNA at 100 μg/ml
Flask containing weighed amount of agarose for 0.8% gel**
Masking tape for sealing gel trays
Sufficient 1 $\times$ TBE gel buffer containing ethidium bromide to
    prepare and run gel**
Gel loading buffer
λ *Hind*III DNA markers
Whatman No. 1 filter paper
Wrapping foil

   *Either use DNA prepared by the students themselves or provide.

   **Amounts required depend on specific make of electrophoresis equipment used.

Sources of biological material

As with Exercise 6.1, a wide variety of DNA sources can be used. Further, this practical can be run independently or as a continuation of Exercise 6.1. In the latter case, it is most satisfying for the students if they can use the DNA they have prepared themselves, i.e. from *E. coli* or another source. Depending on the purity and concentration, this may or may not be suitable for illustrating agarose gel electrophoresis. Therefore, as suggested here, a number of additional DNA sources should be provided. These will need to be purchased or prepared in advance. In the examples used here phage λ and human DNA are available commercially, whereas *Plasmodium* and mosquito DNA are prepared in-house.

Safety

Ethidium bromide is toxic, carcinogenic in mice and a powerful mutagen and, after use, all solutions should be decontaminated. Latex gloves or an equivalent known to be an effective barrier should be worn at all times when handling solutions or gels containing ethidium bromide. Dilute solutions of ethidium bromide such as those being used here (i.e. 0.5 μg/ml or less) can be decontaminated after use by adding Amberlite XAD-16 resin or equivalent. Mix 3 g of resin per 10 ml of solution and shake periodically at room temperature for 12 h to overnight. The resin absorbs the ethidium bromide. To dispose, filter the solution through a Whatman No.1 filter. The liquid waste that flows through can then be poured down the sink. The resin and filter paper should be retained and sealed in a plastic bag for incineration.

A perhaps safer and less costly means of decontamination of ethidium bromide, especially for disposing of mini-gel tank volumes of ethidium bromide contaminated buffer, is as follows: add one volume of 0.5 M potassium permanganate and mix; add one volume of 2.5 M hydrochloric acid, mix, and leave at room temperature for at least 3–4 h. Add one volume of 2.5 M sodium hydroxide and mix. The solution is now safe to pour down the drain.

This exercise calls for the use of electrical equipment that should be regularly serviced by a qualified electrician. Care should be taken to prevent contact between solutions and power packs – electricity and water do not mix!

A UV transilluminator designed for viewing ethidium bromide stained gels should be used. The gel should be viewed through a UV safety screen and/or UV-blocking face shields should be worn.

Instructions for staff

DNA from various sources (those above or alternatives) should be prepared in advance, and preferably should also be checked for digestion by *Eco*RI. Other restriction enzymes can be substituted if desired (see information on similar exercises). Restriction enzymes are normally supplied with an accompanying 10×

reaction buffer. Agarose of a grade suitable for gel electrophoresis should be used. $1 \times$ TBE buffer is 89 mM Tris borate pH 8.3, 2 mM EDTA, with ethidium bromide added to a final concentration of 0.5 µg/ml. Stock solutions of ethidium bromide at 10 mg/ml should be stored at 4 °C and protected from light, e.g. by wrapping in foil. A variety of gel-loading buffers can be used, which is largely a matter of personal preference. A simple mixture that works well is 0.05% bromophenol blue, 40% (w/v) sucrose, 0.1 M EDTA pH 8.0, 0.5% (w/v) SDS. Addition of the gel-loading buffer will terminate the digestion and protect the DNA fragments. The sucrose increases the density and the dye provides colour, both of which make loading the gel easier. Once running, the dye provides a front enabling the progress of the gel to be followed easily. λ *Hind*III DNA markers are commercially available and should be diluted to 50 µg/ml in a form ready to load on to the gel, i.e. including loading buffer.

Instructions for students

Dilute 10 µl of the *E. coli* DNA you have prepared 1/100 by mixing with 990 µl TE buffer, transfer the diluted DNA to a quartz cuvette and use the spectrophotometer to ascertain the quantity and quality of DNA produced. At 260 nm in a 1 ml quartz cuvette, a 50 µg/ml solution of double-stranded DNA produces an A_{260} of 1.0. The purity can also be assessed by determining the A_{260}/A_{280} ratio. For clean DNA, this should be 1.9 or higher, but will be lower if there is protein contamination. From the A_{260} reading estimate the original stock DNA concentration, and then dilute this DNA to 100 µg/ml. Use this DNA in the reactions detailed below.

1. Mix the reactants in order on ice (see Table 6.2.1) and use a new yellow tip for dispensing the enzyme into each reaction tube. Mix by gently flicking each tube and incubate at 37 °C for 2 h.
2. While the DNA is digesting, mix 2.4 g of agarose with 300 ml of $1 \times$ TBE buffer containing ethidium bromide (care) in a 500 ml flask. Cap with foil and heat with mixing on a heater/stirrer until the agarose is completely dissolved. Allow to cool to 45 °C (hand hot), tape the ends of the gel tray with masking tape, insert the slot-former provided and pour the gel. Allow to set, remove the tape, cover the gel with running buffer and carefully remove the slot-former.

Table 6.2.1 *DNA digestion – reaction mixtures*

Reaction no.	DNA	Water (μl)	10 × Buffer (μl)	Enzyme (μl)
1	10 μl λ	7	2	1
2	10 μl *E. coli*	7	2	1
3	10 μl *Plasmodium*	7	2	1
4	10 μl *Aedes*	7	2	1
5	10 μl human	7	2	1

3. Once the 2h incubation period is past, add 5 μl of the gel-loading buffer (blue coloured) to each of reactions 1–5. Heat the samples and the λ *Hind*III markers for 5 min at 65 °C. Load the samples onto the gel and 10 μl of the markers provided. Put the cover on the gel box and run the gel at 35 volts overnight (or at 70 V for 1.5 h) or until the dye front has migrated ¾ of the way down the gel. On the next day, view the gel on the transilluminator and take a photograph.

Expected results

The result of the gel electrophoresis will depend on the purity of the DNA and the combination of DNA source(s) and enzyme(s) used. It is useful to include a simple viral DNA such as phage λ as this will generate a small number of easily visible discrete bands when viewed on the UV transilluminator. Most prokaryotic or eukaryotic genomic DNA will generate a large number of fragments (depending on the enzyme used), which will be visible as a smear under UV illumination, i.e. most individual fragments cannot be discriminated. Sometimes banding can be seen in small genomes containing repeated sequences. Often there is a band of high molecular weight DNA, representing large fragments that are not resolved by standard gel electrophoresis.

Ideas for further exploration

A calibration curve can be drawn in which the sizes of the λ *Hind*III markers are plotted on the y-axis using a log scale (23.1 kb, 9.4 kb, 6.6 kb, 4.4 kb, 2.3 kb, 2.0 kb, 564 bp and 125 bp) and the distance migrated by each marker fragment plotted on the x-axis using a linear scale. This can be used to estimate the sizes of fragments produced when the λ DNA was cut with *Eco*RI. Theoretically this should produce six fragments of 21.2, 7.4, 5.8, 5.6, 4.9 and 3.5 kb. The practical can also be used to lead in to a discussion of Southern blotting or more sophisticated forms of electrophoresis such as pulsed-field gels.

Information on similar exercises

- In the present exercise, a variety of DNA sources are digested with a single enzyme.
- Alternatively, one or a single DNA source can be digested with a range of enzymes. This could be used to illustrate the different sizes of fragments generated by a 4-base-pair cutter (e.g. *Sau*3A) versus a 6-base-pair cutter (e.g. *Eco*RI) versus an 8-base-pair cutter (e.g. *Not*I).
- Another idea is to use a variety of, say, 6-base-pair cutters that vary in their recognition sequences and therefore the range of fragments generated. This can be particularly effective when using certain parasite DNAs with extreme base compositions, for example *Plasmodium falciparum*, which is very AT rich, or *Leishmania*, which is GC rich. The protocol suggested calls for running the agarose gel overnight, in which case a medium-to-large-sized gel apparatus is most convenient.
- It is also possible to conduct this exercise within one day if mini-gels are used, enabling the electrophoresis run time to be reduced to 1 h. In this case, the restriction digests can be set up in the morning and incubated until the afternoon, when the electrophoresis can be performed.

REFERENCES

Brown, T. A. (1991). *Essential Molecular Biology, A Practical Approach*, vol. 1 (ed. T. A. Brown), Oxford, IRL Press.
Heath, S. (1997). Molecular techniques in analytical parasitology. In: *Analytical Parasitology* (ed. M. T. Rogan). Chapter 3. Berlin, Springer.
Sambrook, J., Fritsch, E. F. & Maniatis, T. (1989). *Molecular Cloning: A Laboratory Manual*. New York, Cold Spring Harbour Laboratory Press.

6.3 Restriction enzyme mapping

P. A. BATES, T. KNAPP & J. M. CRAMPTON

Aims and objectives

This exercise is designed to demonstrate the construction of a restriction map from a set of data provided.

The specific objectives of the exercise are:

1. To construct a calibration curve using DNA markers.
2. To determine the sizes of fragments produced by digestion of a plasmid with different restriction enzymes, singly and in combination.
3. To use this information to construct a restriction map of the plasmid.
4. To determine the sizes of fragments detected by Southern blotting of genomic DNA cut with different restriction enzymes, singly and in combination.
5. To use this information to construct a restriction map of genomic DNA.

Introduction

One of the main aims of molecular biology is to understand the organisation and regulation of genes. An essential element of this process is the physical mapping of genes. In this exercise restriction maps of (i) a circular plasmid and (ii) a linear portion of a chromosome, are constructed from the data provided.

Laboratory equipment and consumables
(per student or group)

Equipment

Ruler	Protractor
Compass	Pencil
Calculator	

Consumables

Semi-log graph paper

Instructions for staff

This practical class is a 'dry lab' in which the purpose of the exercise is to illustrate the construction of two simple restriction site maps. Either or both can be used or adapted as required. Each has two parts, in which students first determine the fragment sizes, and second use this information to construct a simple restriction map. The latter is essentially an exercise in logic, and is fairly straightforward if approached correctly. However, it is useful if the staff demonstrating have been through the exercise themselves beforehand, preferably 'blind', as this will help them to assist students as they work through.

Instructions for students

Map 1

A circular, plasmid DNA molecule was digested with five restriction enzymes, singly and in combination. The digested DNAs were fractionated on a 1.5% agarose gel using λ DNA digested with *Hind*III as a marker. The gel was stained with ethidium bromide and viewed on a UV transilluminator to visualise the DNA. A diagram of the resulting gel is shown in Fig. 6.3.1. The samples loaded onto each track are listed in Table 6.3.1. The sizes of the λ *Hind*III marker bands are 23130 bp, 9416 bp, 6557 bp, 4361 bp, 2322 bp, 2027 bp, 564 bp and 125 bp.

Construct a calibration curve of fragment size against distance migrated on semi-log graph paper using the λ markers and determine the size of the DNA fragments resulting from each of the digests. Estimate to the nearest 0.1 kb. Using this data, construct a restriction map of the plasmid DNA placing the single *Eco*RI site at map position 0. Remember that the map will be circular.

Map 2

Genomic DNA was prepared from *Leishmania mexicana* and digested with five restriction enzymes, singly and in various combinations. The digested DNAs were fractionated on a 0.8% agarose gel, transferred to a nylon membrane by Southern blotting, and the resulting blot probed for the presence of a single copy gene, gene X. The probe used was derived from a cDNA

Fig. 6.3.1 Agarose gel electrophoresis of plasmid DNA cut with restriction endonucleases. The distance fragments have migrated should be measured from the top of the gel. See Table 6.3.1 for full explanation of lanes.

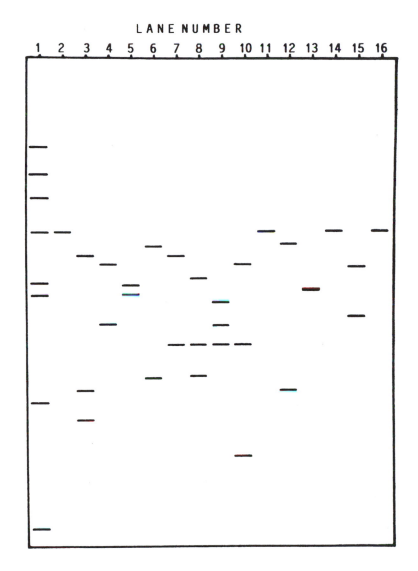

clone containing the complete 3 kb open reading frame, which from sequencing is known to contain an *Ava*I site at the 5′ end, an internal *Hind*III site 1 kb from the 5′ end and a *Pst*I site at the 3′ end. Gene X contains no intervening sequences. The pattern of bands obtained in the Southern blot is shown drawn to scale, with the positions of molecular weight markers as indicated in Fig. 6.3.2. The samples loaded onto each track are listed in Table 6.3.2. By direct inspection of the blot determine the sizes of fragments detected in each digest to the nearest 0.5 kb (a calibration curve is not necessary). Use this information to construct a linear restriction map of the 20 kb portion

Table 6.3.1 *Restriction endonuclease digested plasmid DNA samples for agarose electrophoresis (Fig. 6.3.1)*

Lane	Sample
1	λ DNA digested with *Hind*III
2	Plasmid DNA digested with *Eco*RI
3	Plasmid DNA digested with *Eco*RI and *Hind*III
4	Plasmid DNA digested with *Eco*RI and *Ava*I
5	Plasmid DNA digested with *Eco*RI and *Pvu*II
6	Plasmid DNA digested with *Eco*RI and *Pst*I
7	Plasmid DNA digested with *Hind*III
8	Plasmid DNA digested with *Hind*III and *Ava*I
9	Plasmid DNA digested with *Hind*III and *Pvu*II
10	Plasmid DNA digested with *Hind*III and *Pst*I
11	Plasmid DNA digested with *Ava*I
12	Plasmid DNA digested with *Ava*I and *Pvu*II
13	Plasmid DNA digested with *Ava*I and *Pst*I
14	Plasmid DNA digested with *Pvu*II
15	Plasmid DNA digested with *Pvu*II and *Pst* I
16	Plasmid DNA digested with *Pst*I

of the chromosome containing gene X. The positions where each of the restriction enzymes cut should be indicated and drawn to scale, with the 5′ end of the open reading frame to the left of the 3′ end.

Expected results

Map 1
Students should begin by measuring the distances migrated in cm by each of the λ markers in lane 1. The calibration curve should be drawn with the size of the fragments on the the *y*-axis (log scale) and the distance migrated on the *x*-axis (linear scale). This should produce a straight line plot for all but the largest marker. However, the linear portion covers all the fragment sizes that need to be determined. Students should then measure the distances migrated and use their calibration curves to determine the fragment sizes in lanes 2–16. The data in Table 6.3.3 should be obtained.

From this information students should be able to determine that there is one *Eco*RI site, one *Ava*I site, one *Pvu*II site, one *Pst*I site and two *Hind*III sites that need to be placed on the map. Starting with the *Eco*RI site at position 0 (Fig. 6.3.3A), it is easiest to fix one of the other single sites first. Which one does not

Fig. 6.3.2 Southern blot of genomic DNA hybridised with a specific gene probe.

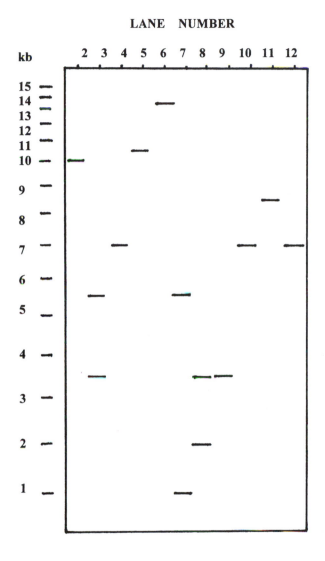

matter and a choice must be made over the position of the first site. Ultimately, this does not matter: two alternative maps are possible that are mirror images of each other, i.e. essentially the same maps but one viewed from behind. Here the *Ava*I site has been fixed next (Fig. 6.3.3B) using the fragments generated by the *Eco*RI/*Ava*I double digest. Following through the various fragments should ultimately yield the map in Fig. 6.3.3C (or its mirror image).

Map 2
Students should begin by determining the fragment sizes for each of the digests, which should be as listed in Table 6.3.4. To

Table 6.3.2 *Restriction endonuclease digested genomic DNA samples for Southern blot (Fig. 6.3.2)*

Lane	Sample
1	1 kb ladder of molecular weight markers
2	Genomic DNA digested with *Ava*I
3	Genomic DNA digested with *Hind*III
4	Genomic DNA digested with *Pst*I
5	Genomic DNA digested with *Bam*HI
6	Genomic DNA digested with *Eco*RI
7	Genomic DNA digested with *Ava*I and *Hind*III
8	Genomic DNA digested with *Pst*I and *Hind*III
9	Genomic DNA digested with *Bam*HI and *Ava*I
10	Genomic DNA digested with *Bam*HI and *Pst*I
11	Genomic DNA digested with *Eco*RI and *Ava*I
12	Genomic DNA digested with *Eco*RI and *Pst*I

begin construction of the restriction map, students should use the information in the opening paragraph to produce a map as in Fig. 6.3.4A. The precise order in which the remaining sites are positioned is not important, but ultimately a map as in Fig. 6.3.4B should be produced. This is the only solution possible providing the students begin with the orientation of the 5' and 3' ends as instructed. Otherwise the map will be inverted (but topologically identical to that shown).

Ideas for further exploration

- The first exercise familiarises students with the concepts of circular plasmid maps. Since plasmid cloning and/or subcloning is ubiquitous in molecular biology this is a useful concept to understand.
- The second exercise can be used to introduce principles of Southern blotting, use of gene probes and gene organisation in the linear chromosomes found in eukaryotes, including parasitic protozoa and helminths.

Information on similar exercises

The examples given here are theoretical examples of the two commonest types of restriction mapping. They can be simplified or made more complex as desired.

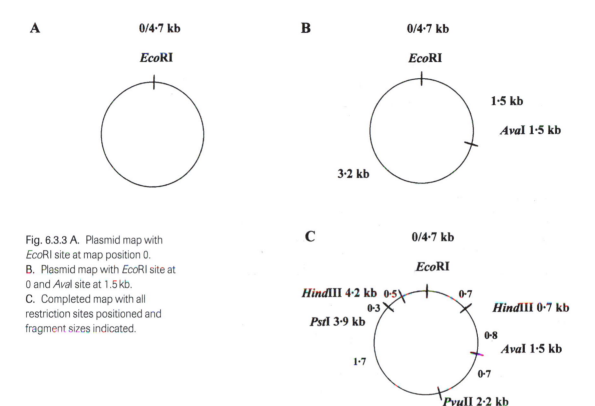

Fig. 6.3.3 A. Plasmid map with *Eco*RI site at map position 0.
B. Plasmid map with *Eco*RI site at 0 and *Ava*I site at 1.5 kb.
C. Completed map with all restriction sites positioned and fragment sizes indicated.

Table 6.3.3 *Fragment sizes of the restriction enzyme digested plasmid DNA samples*

Lane	Sample	Fragments (kb)
2	*Eco*RI	4.7
3	*Eco*RI and *Hind*III	3.5, 0.7, 0.5
4	*Eco*RI and *Ava*I	3.2, 1.5
5	*Eco*RI and *Pvu*II	2.5, 2.2
6	*Eco*RI and *Pst*I	3.9, 0.8
7	*Hind*III	3.5, 1.2
8	*Hind*III and *Ava*I	2.7, 1.2, 0.8
9	*Hind*III and *Pvu*II	2.0, 1.5, 1.2
10	*Hind*III and *Pst*I	3.2, 1.2, 0.3
11	*Ava*I	4.7
12	*Ava*I and *Pvu*II	4.0, 0.7
13	*Ava*I and *Pst*I	2.35×2
14	*Pvu*II	4.7
15	*Pvu*II and *Pst*I	3.0, 1.7
16	*Pst*I	4.7

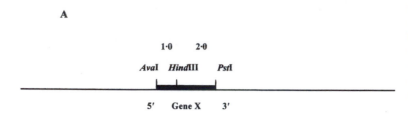

A

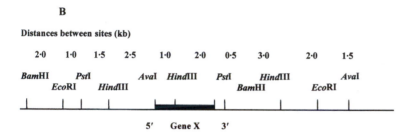

B

Fig. 6.3.4 A. Initial map showing restriction sites of gene X.
B. Completed restriction map showing all sites in correct relationship to gene X and distances between sites as indicated.

Table 6.3.4 *Fragment sizes of the restriction enzyme digested genomic DNA samples*

Lane	Sample	Fragments (kb)
2	*Ava*I	10
3	*Hind*III	3.5, 5.5
4	*Pst*I	7
5	*Bam*HI	10.5
6	*Eco*RI	13.5
7	*Ava*I and *Hind*III	1, 5.5
8	*Pst*I and *Hind*III	2, 3.5
9	*Bam*HI and *Ava*I	3.5
10	*Bam*HI and *Pst*I	7
11	*Eco*RI and *Ava*I	8.5
12	*Eco*RI and *Pst*I	7

REFERENCES

No references exist as such for the exercises given here. However, it is useful to point out that a variety of computer software is available that can perform the same tasks, e.g. Lasergene by DNASTAR Inc. As with all computer programmes, it is the quality of the data input that is crucial to the outcome!

6.4 Construction of a genomic library

P. A. BATES, T. KNAPP & J. M. CRAMPTON

Aims and objectives

This exercise is designed to demonstrate the construction of a simple genomic DNA library. It can be adapted for use with any organism including parasites. The version given here is for the construction of a *Plasmodium falciparum* library.

The specific objectives of this exercise are:

1. To perform a ligation reaction.
2. To prepare competent *E. coli* cells.
3. To transform and plate these competent cells.
4. To grow small-scale cultures from recombinant colonies.
5. To prepare plasmid DNA from recombinants.
6. To analyse recombinant plasmids by agarose gel electrophoresis.

Introduction

The work to be described is aimed at the construction of a genomic library from an isolate of *P. falciparum* such as might be undertaken to analyse a specific gene. This type of cloning experiment is fundamental to a whole range of analyses in molecular biology.

Laboratory equipment and consumables

(per student or group)

Equipment

P20, P200 and P1000 Gilson pipetmen or equivalents
Waste beaker for chemically contaminated plasticware
Autoclavable waste bags for bacterially contaminated waste
37 °C, 42 °C and 65 °C waterbaths
Bench-top centrifuge capable of producing 1000 g
37 °C incubator, static and orbital

Vortex machine
'Camping gaz' and lighter
Vacuum pump and desiccator
Eppendorf microfuge or equivalent
Heater/stirrer and magnetic stirring bar
Agarose gel electrophoresis apparatus and combs
Power supply for running gels
Photographic system for recording results
Safety glasses

Consumables

Latex gloves in a range of sizes
Sterile tips for pipetmen
Ice bucket and wet ice
0.5 and 1.5 ml sterile microfuge tubes and racks
Disposable gloves

Day 1

5 μl or more of *Eco*RI digested *P. falciparum* DNA at 200 μg/ml
5 μl or more of *Eco*RI digested, dephosphorylated pUC 18 DNA at
    200 μg/ml
1.5 μl or more of 10× ligase buffer
1.5 μl or more of DNA ligase enzyme
2.5 μl or more of sterile water
20 ml log phase culture of *E. coli* strain DH5α or equivalent
5 ml L-broth
20 ml of CPG solution (60 mM $CaCl_2$/10 mM PIPES/15% glycerol)
5 μl or more of supercoiled pUC18 at 1 μg/ml
Three L-agar plates containing ampicillin at 50 μg/ml
Plate spreader

Day 2

3 × 20 ml L-broth containing ampicillin at 50 μg/ml
Inoculating loops

Day 3

2 ml of plasmid prep solution I (50 mM glucose/25 mM Tris/10 mM
    EDTA, pH 8)
3 ml of plasmid prep solution II (0.2 M NaOH, 1% SDS)
2 ml of plasmid prep solution III (5 M potassium acetate)
Phenol/chloroform mixture

Polyethylene glycol (PEG) in case of phenol contamination
Chloroform/isoamyl alcohol mixture
5 ml isopropanol
300 µl of TE buffer containing 50 µg/ml RNAse
Flask containing weighed amount of agarose for 0.8% gel
Tape for sealing gel trays
Sufficient 1 × TBE gel buffer containing ethidium bromide to run gel
Gel-loading buffer
λ *Hind*III DNA markers
Aluminium foil

Sources of parasite material

Any convenient source of parasite DNA can be used in this practical. See Exercise 6.1 for a suggested protocol.

Safety

- **Phenol is one of the most dangerous organic solvents used in molecular biology**, but can be safely used provided certain precautions are taken. Disposable latex gloves and safety spectacles should be worn whenever phenol is being handled as contact with skin and eyes causes burning. Make sure that the gloves used are an effective barrier against phenol. Avoid unnecessary exposure to phenol vapours by keeping the stock bottles closed when not in use. Pipettes and tips that have been in contact with phenol should be collected separately and destroyed by incineration. Phenol waste should be disposed of according to local safety rules for the disposal of organic solvents.

  In the event of phenol exposure to skin, remove contaminated clothing immediately and flush off excess chemical with copious amounts of water. Wash the affected area with PEG, molecular weight 300, for 30 min wearing latex gloves followed by rinsing with ethanol. In the event of exposure to the eyes, wash with flowing water – **do not use PEG** – and seek hospital attention.
- Avoid unnecessary exposure to chloroform vapours by keeping the stock bottles closed when not in use; wear safety spectacles when using chloroform.
- Ensure that all tubes for centrifugation are properly balanced.
- *E. coli* K12 is not considered a pathogenic organism, but

caution dictates that culture waste should be sterilised by autoclaving or chemical treatment and disposed of according to good microbiological practice.

Instructions for staff

1. *Chloroform* is a 24:1 (v/v) mixture of chloroform/isoamyl alcohol, stored in a capped bottle at room temperature.
2. *Phenol/chloroform* is 1:1 (v/v) of TE-saturated phenol and chloroform/isoamyl alcohol prepared as follows. Water-saturated phenol can be purchased directly and should be stored at 4 °C. The aqueous layer should be replaced with 1 M Tris–HCl pH 8 and the mixture agitated. This aqueous layer should also be discarded and replaced with TE buffer (10 mM Tris/HCl, 1 mM EDTA pH 8). Addition of 8-hydroxyquinoline to 0.1% (w/v) is useful – this is an antioxidant but it also imparts a yellow colour to the phenol phase, particularly useful during the extraction procedure. Again agitate the mixture and allow to separate into organic and aqueous phases. The TE-saturated phenol can now be removed from the lower organic phase and mixed 1:1 with chloroform/isoamyl alcohol in a separate bottle. Phenol/chloroform should be stored at 4 °C until use.
3. *TE buffer* is 10 mM Tris/HCl, 1 mM EDTA pH 8, and should be autoclaved and stored at room temperature.
4. *Plasmid solution I* is 50 mM glucose, 25 mM Tris pH 8, 10 mM EDTA pH 8 and should be autoclaved and stored at 4 °C.
5. *Plasmid solution II* is 0.2 M NaOH, freshly diluted from a 10 M stock, 1% (w/v) SDS.
6. *Plasmid solution III* is prepared by mixing 60 ml of 5 M potassium acetate, 11.5 ml of glacial acetic acid (dispense in a fume hood), and 28.5 ml of sterile distilled water. See Exercise 6.2 for details of agarose gel electrophoresis.

Instructions for students

A. Ligation of DNA and transformation of *E. coli* (day 1)

Set up a ligation of digested plasmid DNA vector with digested *P. falciparum* DNA to generate recombinant DNAs ready for introduction into bacteria.

Prepare the following reaction on ice in a 0.5 ml microfuge tube:

pUC DNA	5 μl of 200 μg/ml, *Eco*RI digested, dephosphorylated
P. falciparum DNA	5 μl of 200 μg/ml, *Eco*RI digested
water	2.5 μl
10× ligase buffer	1.5 μl
Ligase	1 μl
TOTAL VOLUME	**15 μl**

Incubate at 37 °C for 1–2 h.

Prepare competent *E. coli* cells ready for transformation with recombinant DNA; set up transformation and plate transformed cells.

1. Remove the log phase culture of *E. coli* from the incubator. These cultures were prepared by inoculating 20 ml of L-broth with 200 μl from an overnight culture of DH5α and incubated with shaking at 37 °C for 2.5 h (A_{600} should be approx 0.5).
2. Chill cells on ice for 10 min. Centrifuge in a bench-top centrifuge at 1000 g for 10 min (brake off). Gently pour off supernatant and resuspend cells **VERY GENTLY** in 1 ml CPG solution. When thoroughly resuspended, add a further 10 ml of CPG solution and leave on ice for 20 min.
3. Centrifuge the cells as above. Gently pour off the supernatant and **VERY GENTLY** resuspend the cells in 2 ml CPG solution. Keep on ice until required.
4. Once the ligation reaction is complete, set up three transformations on ice in 1.5 ml microfuge tubes. Use all the ligation reaction you set up in the first transformation. Put 1 μl of supercoiled pUC18 into the second tube; do not add DNA to the third tube. Add 200 μl of prepared cells to each. Leave for 20 min on ice then transfer to a 42 °C water bath for 3 min. Remove from the water bath and add 1 ml of fresh L-broth to each tube. Cap the tube and incubate at 37 °C in the air incubator for between 30 min and 1 h.
5. Remove 200 μl from transformation 1, apply it to one of the agar plates provided and spread the liquid evenly over the plate with the spreader. Sterilise the spreader between each plating by immersing in ethanol and flaming. Repeat this process for the other transformation cultures, so that you end up with three plates with the transformed bacteria spread over the surface of the plates. Any transformed cells containing the plasmid DNA will grow in the presence of ampicillin and form a colony. Incubate the inverted plates overnight at 37 °C.

B. Growth of recombinant plasmids (day 2)

Analyse transformation results. Pick recombinant clones to grow small-scale cultures for analysis of the recombinant DNA.

1. Count the colonies on each of the plates. Calculate the transformation efficiency for each of the transformations and express this as the number of colonies obtained per μg of DNA in the transformation. There should be no colonies on plate 3 and many colonies on plates 1 and 2.
2. The colonies on plate 1 should contain bacteria with recombinant plasmid DNAs, i.e. that contain pieces of DNA from *P. falciparum*. The next part of the experiment is to pick individual clones and grow small-scale cultures of each so that the plasmid DNA they contain can be analysed.
3. Each person should set up 20 ml overnight cultures of three clones using L-broth containing ampicillin. Grow overnight at 37 °C in an orbital shaker.

C. Isolation and analysis of plasmid DNA (day 3)

Prepare purified plasmid DNA from the bacterial cultures and analyse the DNA on agarose gels.

1. Spin the bacterial culture at 250 g in the bench top centrifuge for 10 min in the Universal tubes.
2. Pour off the L-broth and resuspend pellet in 400 μl of Soln I. Leave at room temperature for 10 min.
3. Add 800 μl Soln II, mix well but gently and leave for 5–10 min on ice.
4. Add 400 μl Soln III, vortex and leave for 30 min on ice. While you are waiting go to step 9 below.
5. Transfer solution to 1.5 ml microfuge tube, and spin in the microfuge for 3 min.
6. Transfer 700 μl of the supernatant to a new microfuge tube and extract it twice with an equal volume of phenol/chloroform and twice with an equal volume chloroform/isoamyl alcohol.

CARE! – PHENOL CAUSES BURNS – USE GLOVES AND
WEAR SAFETY GLASSES

7. To the extracted aqueous phase add 450 μl of isopropanol, mix and leave at least 5 min at room temperature. Spin down the DNA at 250 g in the microfuge for 10 min.
8. Dry the pellet under vacuum and take up in 50 μl TE/RNase

($50\,\mu g/ml$). Remove $25\,\mu l$, add $5\,\mu l$ gel-loading buffer and run on a gel (see step 10 below).

9. While you wait for step 4 to be completed, mix 2.4 g agarose with 300 ml of $1\times$ TBE buffer containing ethidium bromide (care) in a 500 ml flask. Cap with foil and heat on heater/stirrer until agarose is completely dissolved. Allow to cool to 45 °C (hand hot), tape ends of gel tray with masking tape, insert slot former provided and pour gel. Allow to set, remove tape, cover gel with running buffer and remove slot former.

10. Heat the samples and the λ *Hind*III markers for 5 min at 65 °C. Load the samples or $10\,\mu l$ of the markers provided onto the gel and also include a control sample with super-coiled pUC DNA. Put the cover on the gel box and run the gel at 100 volts until the dye front has moved three-quarters of the way down the gel. View the gel on the transilluminator and take a photograph.

Expected results

This is a relatively ambitious exercise and requires staff and/or demonstrators with a practical knowledge of gene cloning if it is to work well. On day 2, typically there should be in the order of 60 colonies on plate 1, 600 colonies on plate 2, and none on plate 3. Experience dictates that such results are sometimes obtained by students, but not always, so it is advisable to run a demonstration in parallel, as well as trying out the practical in advance. The plasmids viewed on the gel will be undigested, so accurate determination of insert size is not possible, but they should be seen to be larger than the pUC control.

Ideas for further exploration

• The recombinant plasmids could be cut with a restriction enzyme to linearise the DNA and run on another agarose gel. In this case size determination of the inserts will be possible.

Information on similar exercises

For larger classes, or where some guarantee of success is important, the exercise can be designed to produce results as follows. A 'library' is constructed in advance, made with DNA in a range of sizes inserted into *Eco*RI digested pUC18. A glycerol stock is prepared, which is used to grow a culture overnight. At the same time a suitable host strain (DH5α) containing pUC18 is prepared. Prior to the practical class, the 'library' and the host containing pUC18 are both diluted 1 µl in 1 ml of L-broth. These are used to spike the transformations, unknown to the students. This is done after 1 ml of L-broth has been added to the transformation and the culture is recovering at 37 °C. 10 µl of the pUC18 dilution is added to transformation reaction 2; 1 µl of the 'library' dilution is added to transformation reaction 1. The components of the ligation reactions and the tube of supercoiled pUC DNA provided to the students actually only contain distilled water. This should produce results that approximate to a real transformation.

REFERENCES

Brown, T. A. (1991). *Essential Molecular Biology, A Practical Approach*, vol. 1 (ed. T. A. Brown). Oxford, IRL Press.

Sambrook, J., Fritsch, E. F. & Maniatis, T. (1989). *Molecular Cloning: A Laboratory Manual*. New York, Cold Spring Harbour Laboratory Press.

6.5 Detection and differentiation of *Entamoeba histolytica* and *E. dispar* by PCR

J. E. WILLIAMS & D. BRITTEN

Aims and objectives

This exercise aims to:

1. Differentiate *Entamoeba histolytica* from *E. dispar*.
2. Introduce students to PCR.

Introduction

Diagnosis of infectious diseases by the detection of nucleic acids specific for the infectious agent may offer considerable advantages over traditional laboratory diagnostic methods such as microscopy and culture. One such area where this may prove very useful is in the detection and differentiation of the two morphologically similar human parasitic amoebae, *E. histolytica* and *E. dispar*. Recent work has shown that *E. histolytica*, the causative organism of amoebic dysentery in humans, is in fact two separate species. Prior to this work it was thought that *E. histolytica* generally existed in humans without disease and, in about 10% of cases, gave rise to clinical disease. It has now been shown using a combination of methods (isoenzymes, biochemical, immunological and molecular) that infection with *E. histolytica*, *sensu stricto*, is likely to lead to disease whilst infection with *E. dispar* results in no symptoms.

The problem in trying to diagnose which of these two organisms is present in a patient is that both organisms are morphologically identical, both in the trophozoite stage and in the cystic form. This PCR (polymerase chain reaction) method uses two pairs of discrete primers that are able to recognise and differentiate the two species. PCR provides differential diagnosis of DNA extracted from faeces. The result is detected colorimetrically in a microtitre plate, which is simpler and safer than using radio-labelled probes.

Two specific primers are used, one of each pair is labelled **365** with biotin. One set is specific for a 145 bp (base pair) repetitive region from extrachromosomal circular DNA in *E.histolytica*:

5′ . . . Biotin-TCA AAA TGG TCG TCG TCT AGG C . . . 3′
5′ . . . CAG TTA GAA ATT ATT GTA CTT TGT A . . . 3′

The second set is specific for a 135 bp repeat in *E. dispar*:

5′ . . . TCG GAT CCT CCA AAA AAT AAA GTT T . . . 3′
5′ . . . Biotin- ACA GAA CGA TAT TGG ATA CCT AGT A . . . 3′

After amplification in the PCR two labelled specific DNA probes are added:

5′ . . . digoxygenin-CCC GAG GTT CTT AGG AAA TGG A . . . 3′ (*E. histolytica*)
5′ . . . fluorescein-AAT GAG GTT GIA GCA GAG CCC C . . . 3′ (*E. dispar*)

Following denaturation at 99 °C they are allowed to hybridise to the PCR product at 37 °C.

If the original sample contained *E. histolytica* or *E. dispar* DNA then the PCR will have produced amplicons labelled with biotin. Once the specific probes have hybridised to their target the amplicon will have become labelled with a second label provided by the probe. The PCR product can be 'captured' via the biotin in duplicate avidin-coated wells of a microtitre plate. The specific amplicon is then identified with enzyme-labelled anti-digoxygenin or anti-fluorescein antibodies as in a simple ELISA.

Laboratory equipment and consumables
(per student or group)

Equipment

PCR thermal cycler machine
Microcentrifuge
Automatic pipettes (e.g. P1000, P100 and P20)
Vortex mixer
Water bath
Microplate reader
Safety glasses
Access to 37 °C incubator
Access to cold room/refrigerator and deep freeze

Consumables

Aerosol pipette tips (Anachem: RI10-f3, RT96-fl, RT96-f4 and RT1000-f3)

0.5 ml microcentrifuge tubes (e.g. Scotlab SL-7011)

1.5 ml screw-capped microfuge tubes and caps (e.g. Alpha CP5514 and CP552R)

Sterile 1 ml pastettes (plastic disposable pipettes)

8-well microtitre strips (e.g. Greiner 2 × 8-well strips, cat. 756061)

Disposable gloves

PCR wax discs – Lambwax, melting temp 57–58 °C

Diatomaceous earth (acid washed, calcined: Sigma D5384)

Dried skimmed milk powder

Reagents

Ethanol 100% and 70%

Acetone (propanone)

Sterile double distilled water (DDW)

Guanidine thiocyanate (Sigma G6639)

WARNING: Thiocyanates liberate hydrogen cyanide on contact with acid. Discard all solutions containing thiocyanates into excess 10 M sodium hydroxide solution, allow to stand for 30 min and then dilute with a large volume of water before discarding.

EDTA disodium salt

Sodium chloride

Disodium hydrogen orthophosphate, anhydrous

Potassium dihydrogen orthophosphate, anhydrous

Concentrated hydrochloric acid

Sodium hydroxide 10 M

Sodium carbonate, anhydrous

Sodium bicarbonate, anhydrous

Tris base – Sigma TRIZMA

Tris–HCl – Sigma T3149

Citric acid monohydrate

Potassium hydroxide

Hydrogen peroxide 30% w/w

Triton X100

Tween 20

Solid carbon dioxide

Primers Eh1, Eh2, Ed1, Ed2 from R&D Systems Europe Ltd, 4–10 The Quadrant, Barton Lane, Abingdon, Oxon OX14 3YS. UK Technical Services Tel 01235 531074

Probes EH & ED from R&D Systems Europe Ltd.

Taq polymerase (Promega *Taq* DNA polymerase in Buffer A; M1861)

Promega 'Buffer A' kit

Deoxynucleotide triphosphates – Pharmacia ultrapure dNTP set 27-2035-01. Stock solution prepared by mixing 10 μl of each dNTP and adding 360 μl distilled water. Store at −20 °C.

Peroxidase-labelled anti-digoxigenin (anti-digoxigenin-POD Fab fragments: Boehringer 1207 733)

Peroxidase-labelled anti-fluorescein (Boehringer 1426 346)

Peroxidase substrate, TMB (Sigma T2885)

Preparation of solutions

1. *Hydrochloric acid*, about 1 M (and 2 M if final ELISA colour is to be measured in a microplate reader). Add carefully 8.6 ml of concentrated hydrochloric acid to approximately 80 ml of double-distilled water (DDW) and make up to 100 ml with DDW. **Remember to add acid to water. Eye protection is essential when working with concentrated acids**.

2. *Sodium hydroxide* solution, approximately 10 M. To make 100 ml add 40 g of sodium hydroxide pellets to 60 ml DDW; allow to cool and make up to 100 ml. Alternatively, this can be purchased as a ready-made 10 M solution. **Eye protection is essential when working with sodium hydroxide**.

3. *Sodium hydroxide* solution, approximately 1 M. Use 4.0 g sodium hydroxide per 100 ml and proceed as above. **Precautions as above.**

4. *Potassium hydroxide solution*, 4 M. To make 100 ml add 22.4 g of pellets of potassium hydroxide to about 70 ml of DDW, allow to cool and make up to 100 ml. **Precautions as above**.

5. *Phosphate-buffered isotonic saline pH 7.2 (PBS)*. This is easily prepared from Sigma PBS tablets.

6. *0.1 M Tris–HCl buffer, pH 6.4*. Dissolve 15.76 g Tris–HCl in about 950 ml DDW, adjust pH to 6.4 and then make up to 1 litre with DDW.

7. *0.5 M EDTA*. Dissolve 18.61 g disodium EDTA dihydrate in about 80 ml DDW. Adjust pH to 8.0 and make up to 100 ml.

8. *Diatomaceous earth suspension*. Add 2 g diatomaceous earth to 10 ml DDW and add 100 μl concentrated hydrochloric acid. Vortex mix, aliquot into approximately 1 ml amounts in small glass bottles and cap tightly; sterilize by autoclaving. Store at 4 °C.

9. *Lysis buffer*. **(Prepare all solutions containing thiocyanates in a fume hood)**. Dissolve 120 g guanidine thiocyanate in

111.2 ml 0.1 M Tris–HCl buffer, pH 6.4, add 8.8 ml 0.5 M EDTA and 1.3 g Triton X100.

10. *DNA wash buffer.* **(Prepare all solutions containing thiocyanates in a fume hood.)** Dissolve 120 g guanidine thiocyanate in 100 ml of 0.1 M Tris–HCl buffer, pH 6.4.

11. *TE buffer.* Dissolve 1.21 g Tris base and 0.372 g disodium EDTA dihydrate in about 900 ml DDW. Adjust pH to 7.5 and make up to 1 litre with DDW.

12. *Primer mix.* Dilute and mix the four primers so that 18 μl contains 25 pmoles each of Eh1, Eh2, Ed1 and Ed2. To prepare 18 μl mix 2.5 μl of the stock solutions of each primer and add 8 μl sterile DDW. Store at −20 °C.

13. *Probe mix.* Mix equal volumes of the 10 pmol/μl stock solutions of EH and ED. Store in small aliquots at −20 °C.

14. *Tris-buffered saline (TBS).* To prepare one litre, dissolve 2.54 Tris–HCl, 0.47 g Tris base and 8.76 g sodium chloride in about 900 ml DDW. Make up to 1 litre with DDW.

15. *TBS/Tween (TBST).* Add 5 ml Tween 20 to 1 litre of TBS.

16. *TBST/milk (TBSTM).* Always prepare fresh. Add 3 g dried skim milk to 100 ml TBST.

17. *Plate coating buffer.* Dissolve 0.79 g sodium carbonate anhydrous and 1.46 g sodium bicarbonate anhydrous in about 400 ml DDW. Make up to 500 ml with DDW.

18. *Working dilution of peroxidase-labelled anti-digoxigenin.* Dilute 1:5000 in TBSTM. Do not store.

19. *Working dilution of peroxidase-labelled anti-fluorescein.* Dilute 1:5000 in TBSTM. Do not store.

20. *Peroxidase substrate (TMB) solution:*
    (i) TMB stock solution. Dissolve 480 mg TMB in 10 ml acetone. Add 90 ml of 100% ethanol, shield from light and store at 4 °C.
    (ii) Substrate buffer. Dissolve 42 g citric acid in 800 ml DDW. Adjust pH to 3.6 with pellets of potassium hydroxide. Leave to cool, then adjust to pH 3.95 with 4 M potassium hydroxide. Make up to 1 litre with DDW and add 275 μl 30% hydrogen peroxide. Store at 4 °C for up to 6 months.

21. *Avidin stock solution.* Prepare a 1 mg/ml solution in water.

Sources of parasite material

Cultured *E. histolytica* and *E. dispar* can be 'seeded' into a sample of faeces to provide the samples for DNA extraction. These organisms are available from the Diagnostic Parasitology Laboratory, London School of Hygiene and Tropical Medicine, Keppel St., London WC1E 7HT UK or from National Collection of Typed Cultures (NCTC), 61 Colindale Avenue, London NW9 5HT. In the USA, it may be possible to obtain various species of *Entamoeba* from the American Type Culture Collection (ATCC), which should be contacted for their latest catalogue, or through the www.

Safety

Trophozoites of *E. histolytica* and *E. dispar* are not infective by ingestion, however it is possible that cysts, which are infective, are present so gloves should be worn when handling initial specimens before extraction. Gloves should also be worn for general safety reasons and to help prevent PCR contamination. A number of the chemicals used are toxic and a full hazard assessment should be carried out before starting the work.

Instructions for staff

This exercise can be divided into two sections depending on requirements.

(a) DNA extraction can be carried out prior to the practical, either by the students or by teaching staff and the prepared DNA given to the students for use.
(b) The PCR and hybridisation followed by the colorimetric detection can be carried out as a single practical on ready-prepared DNA.

NB The microtitre plates need to be coated overnight with avidin; the students can do the final wash of the wells while the probes are hybridising.

The colour reactions can be read by eye or the reactions can be stopped with 2 M hydrochloric acid and read at 450 nm in a microplate reader.

Although this method is detecting both *E. histolytica* and *E dispar*, the exercise can be done using only one organism and the relevant primers and probe. This makes the handling very simple and cuts down on the cost of reagents. A 'negative' sample containing no amoebae can be run as a control in this case.

DNA extraction

Appropriate precautions for handling possible pathogenic material must be observed.

1. Mix one gram of faeces with sufficient PBS to make the sample semi-fluid.
2. Lyse with three cycles of freeze–thaw using a solid carbon dioxide/ethanol freezing mixture and a water bath at room temperature to thaw.
3. Centrifuge for 5 min at 500 g to remove debris.
4. Remove 250 μl of supernatant to a clean tube.
5. Add 40 μl of diatomaceous earth suspension and 900 μl of lysis solution. Mix very thoroughly and allow to stand for 2 min.
6. Microfuge at 13 000 g for 30 sec.
7. Remove and discard supernatant.
8. Add 1 ml DNA wash buffer, mix thoroughly, re-spin and discard supernatant. Repeat wash step once.
9. Add 1 ml 70% ethanol, mix thoroughly, re-spin and discard supernatant. Repeat ethanol wash once.
10. Wash with 1 ml acetone, mix thoroughly, re-spin and discard supernatant.
11. Leave tube open to allow residual acetone to evaporate.
13. Add 100 μl TE buffer. Mix thoroughly and place in 56 °C waterbath for 10 min. Mix again before centrifuging at 13 000 g for 5 min.
14. Remove supernatant containing DNA to a clean tube and store at −20 °C until required for use.

Instructions for students

Polymerase chain reaction (PCR)

PCR is a highly sensitive technique. Take great care to avoid contamination of reagents. Wear gloves and always use a clean pipette tip.

You have been given four tubes labelled A, B, C and D containing either a test sample or a positive and negative control.

1. Label four reaction tubes (0.5 ml tubes) A, B, C and D and mark them with your initials.
2. Place 18 μl of primer mix from primer mix tube in each of your reaction tubes. The primer mix contains 25 pmol of the four primers. Spin briefly in the microcentrifuge to ensure all the liquid is at the bottom of the tube.

3. Add a wax sealing disc. Gently close the caps and place the tubes at 80 °C for 5 min to allow the wax to melt.
4. Leave at room temperature for a few min for the wax to solidify, forming an unpenetrable barrier above the primer mix.
5. Add 20 µl of reaction mix to each of your reaction tubes. This mix contains the *Taq* enzyme and nucleotides.
6. Add 2 ml from each of the samples A, B, C and D to the appropriate reaction tube.
7. Firmly close caps and spin for 30 sec in the microfuge to ensure that the reagents are on top of the wax layer.
8. Place in the thermal cycler machine. Once all samples are ready the PCR can be run using the following amplification cycle:

94 °C	2 min	1 cycle
94 °C	30 sec	
58 °C	1 min	35 cycles
72 °C	2 min	
72 °C	10 min	1 cycle

Hold at 80 °C

Hybridisation

Hybridisation follows PCR, the tubes having been held at 80 °C.

1. To each of your reaction tubes add 2 µl of the probe mix provided. This contains 10 pmol of both probes.
2. Heat the tubes in the thermal cycler at 99 °C for 10 min and then incubate at 37 °C for 15 min.
3. During this incubation prepare your microtitre strips.
   (i) Wash the plate by tipping the avidin into a sink or bucket.
   (ii) Fill the wells from a wash bottle of TBS/Tween taking care to avoid bubbles.
   (iii) Empty wash solution into sink AND repeat twice.
   (iv) Finally remove excess liquid from the wells by knocking the strip upside down on a paper towel.
   (v) Label one strip ER and the other ED.
   The wells are now ready to receive the samples.
4. Following the 15 min hybridisation add 450 µl of sample diluent, which must be prewarmed to 37 °C. This will rest on top of the wax that has resolidified.
5. Working quickly but carefully:

     (i) Taking one tube at a time pierce the wax plug with a yellow pipette tip.

     (ii) Taking a fresh tip mix the contents of the tube.

     (iii) Transfer 100 μl aliquots to 2 wells in each strip (i.e. 4 × 100 μl in total).

6. Continue doing the same with the other 3 tubes and then place the strips in the 37 °C incubator for 10 min.

7. Following the incubation wash the strips (× 3) as in step 3 above.

8. (i) To the EH strip add 100 μl of anti-digoxygenin antibody.

     (ii) To the ED strip add 100 μl of anti-fluorescein antibody.

9. Allow plate to stand at room temperature for 30 min.

10. Wash to remove unbound antibody as in step 3.

11. To every well add 100 μl of TMB substrate.

12. Wait for colour to develop.

Interpretation of results

Blue colour in the EH wells indicates the sample was *positive* for *E. histolytica* while a blue colour in the ED wells indicates the sample was positive for *E. dispar.* No colour indicates a negative sample for both.

REFERENCES

Aguirre, A., Warhurst, D. C., Guhl, F. & Frame, I. A. (1995). Polymerase chain reaction-solution hybridisation enzyme-linked immunoassay (PCR-SHELA) for the differentiation of pathogenic and non-pathogenic *E. histolytica. Transactions of the Royal Society for Tropical Medicine and Hygiene* **89**, 187–188.

Diamond, L. S. & Clarke, O. C. (1993). A redescription of *Entamoeba histolytica* Schaudin, 1903 separating it from *Entamoeba dispar* Brumpt, 1925. *Journal of European Microbiology* **40**, 340–344.

Heath, S. (1997). Molecular techniques in analytical parasitology. In: *Analytical Parasitology* (ed. M. T. Rogan). Chapter 3. Berlin, Springer.

Sargeaunt, P. G., Williams, J. E. & Greene, J. D. (1976). The differentiation of invasive and non-invasive *Entamoeba histolytica* by isoenzyme electrophoresis. *Transactions of the Royal Society for Tropical Medicine and Hygiene* **72**, 519–521.

6.6 Differentiation between parasite species by agglutination and detection of parasite surface carbohydrates, using non-conjugated lectins

G. A. INGRAM

Aims and objectives

Using commercial lectins, these two exercises are designed to:

1. Distinguish between different species of parasites and two strains of the same parasite species.
2. Determine the type of carbohydrate present on the parasite surface membrane.

Introduction

Lectins (often referred to as agglutinins) are proteins or glycoproteins that bind specifically to carbohydrates, usually present as glycoconjugates, on cell or tissue surfaces or in cell tissue fluids (Sharon & Lis, 1989; Jacobson, 1994). Lectin reactivity is generally designated according to the monosaccharides or oligosaccharides that cause inhibition of lectin-mediated agglutination of cells or adherence to cell membranes.

Lectins have been widely used in parasitology to detect and determine the types of saccharide moieties on the surface of trypanosomatids, e.g. *Crithidia* (Petry *et al.*, 1987), *Leishmania* (Schottelius & Aisen, 1994) and *Trypanosoma* species (Maraghi *et al.*, 1989), to demonstrate differences between parasite growth and stationary phases (Jacobson & Schnur, 1990) and to distinguish between the different stages of the trypanosomatid life cycle (Rudin *et al.*, 1989). In addition, lectins have been employed to differentiate between parasite species (Schottelius & Aisen, 1994) and in the identification of various strains (Schnur & Jacobson, 1989) and stocks (Schottelius, 1987) of these flagellates. The following methods are applied to kinetoplastid flagellates but can be adapted to study other unicellular parasites.

375

Laboratory equipment and consumables
(per student or group)

Equipment

Bench centrifuge	100 ml glass beaker
Adjustable pipette (1–10 μl) and tips	Incubators (20 °C, 25 °C)
Neubauer haemocytometer	1 and 5 ml glass/plastic pipettes
Microscope	

Consumables

Appropriate culture(s) of parasites

10 ml screw-capped conical-base test tubes

Microtitre (Terasaki) plates (approx. 12 μl vol; 60 wells)

Phosphate buffered saline (PBS) containing divalent cations pH 7.3

(8.5 g NaCl; 1.07 g Na_2HPO_4; 0.40 g $NaH_2PO_4.2H_2O$; 0.060 g $CaCl_2.2H_2O$ and 1.30 g $MgC1_2.6H_2O$ all dissolved and made up to 1 litre distilled water)

Locke's solution

Filter paper and scissors

Solutions of commercial lectins; Sigma-Aldrich Company Ltd (stock 1000 μg/ml in PBS; initial molarities cited in Table 6.6.1)

Stock solutions (each 300 mM) of specific and non-specific sugars (for Part 2 only)

Pasteur pipette and bulb

Disposable gloves

Sources of parasite material

The parasites used in these exercises are obtained from previously isolated samples (Liverpool School of Tropical Medicine) and frozen in liquid nitrogen. *Trypanosoma brucei brucei* procyclic forms are cultured in Cunningham's medium (Taylor & Baker, 1978). *Crithidia fasciculata* and *C. luciliae choanomastigotes* and the promastigotes of two *Leishmania hertigi hertigi* strains (designated LV42 and LV43) have been cultivated using a nutrient agar–blood medium with Locke's solution (Taylor & Baker, 1978) as overlay. Workers in the USA should consult Exercise 4.2 on advice for trypanosomes. It may be possible to obtain *Leishmania hertigi* from the American Type Culture Collection (ATCC), which should be contacted for their latest catalogue. *Crithridia fasciculata* cultures can be obtained from Carolina Biological Supply Co. and from Ward's.

Laboratory safety

Good laboratory practice should be performed at all times for these experiments. Some lectins interact with cells/tissues in potentially damaging ways not yet fully understood and should always be regarded as potential biological hazards. Wear disposable gloves when handling solutions of non-conjugated lectins. Toxic lectins must **NEVER** be used for laboratory practical purposes. The parasites used here are non-infectious to humans but nevertheless, care should be taken when handling and working with any culture forms (wear gloves). Spillages of either lectin or parasite on the bench surface must be wiped up immediately using bleach or similar substances.

Instructions for staff

Appropriate parasite cultures should be available for the class and lectins (see below) prepared in phosphate buffered saline (PBS) at a stock concentration of 1000 μg/ml or known molarity (Table 6.6.1). Depending upon the length of the practical class, the students may be provided with pre-washed parasites in suspension in PBS at a concentration of $4-5 \times 10^6$ cells/ml. PBS should be available for parasite washings, suspension adjustment, agglutination plate and lectin dilution preparation. The parasites should initially be at high density in the culture medium.

Instructions for students

Materials

You are provided with stock solutions of lectins prepared in PBS pH 7.3 and 1 ml of each culture form *(T.b. brucei, L.h. hertigi* strains LV42 and LV43, *Crithidia fasciculata* and *Crithidia luciliae)*. For Part 2, you will be provided with stock solutions of specific sugar inhibitors (each 300 mM in PBS) and a non-specific sugar, fructose, also at the above initial concentration. The lectins used in these exercises, together with their sugar-binding specificities and chosen specific sugar inhibitors of lectin activity, are given in the Table 6.6.1.

Table 6.6.1 *Lectins commonly used in agglutination together with their initial molarities (M), metal ion requirement, sugar binding specificities and carbohydrates used for binding inhibition*

Lectin (source and abbreviation)	$M \times 10^{-6}$	Metal ion	Specificity	Specific inhibitor
Canavalia ensiformis (Jack bean; Con A)	9.8	Ca^{2+} Mn^{2+}	α D-mannose, α D-glucose	α-CH_3-mannopyranoside, α-CH_3-D-glucopyranoside, D-mannose, and D-glucose
Codium fragile (a green seaweed; CFA)	0.17	–	D-GalNAc*	D-GalNAc, D-GlcNAc*
Limulus polyphemus (Horseshoe crab; LPA)	2.5	Ca^{2+}	NeuNAc*	NeuNAc, Fetuin
Triticum vulgaris (Wheat germ; WGA)	0.28	–	$(D\text{-GlcNAc})_2$	N,N¹-diacetylchitobiose GlcNAc
Glycine max (Soybean; SBA)	9.1	–	D-GalNAc	GalNAc-D-galactose
Sophora japonica (SJA) (Japanese pagoda tree)	7.5	–	β-D-GalNAc	α-Lactose, GalNAc
Lens culinaris (Lentil; LCA)	0.2	–	α-D-glucose, α-D-mannose	As for Con A
Arachis hypogaea (Peanut; PNA)	8.3	–	β-D-gal(1–3)GalNAc	GalNAc, D-galactose
Tetragonolobus purpureas (Lotus; TPA)	3.4	–	α-L-fucose	α-L-fucose
Dolichos biflorus (Horse grain; DBA)	7.1	–	α-D-GalNAc	GalNAc
Bandeiraea simplicifolia (Bandeiraea; BS-II)	8.8	Ca^{2+} Mg^{2+} Mn^{2++}	D-GlcNAc	D-GlcNAc
Vicia villosa (Hairy vetch; VVA)	7.2	–	D-GalNAc	D-GalNAc

Notes:
*NeuNAc, *N*-acetylneuraminic acid; GlcNAc, *N*-acetyl-D-glucosamine; GalNAc, *N*-acetyl-D-galactosamine. –, no metal ion requirement. M, molarities equivalent to 1000 μg/ml intial lectin concentration.

Lectins
In Exercises 1 and 2, commercially available LPA, PNA, WGA, LCA, BS-II, VVA and CFA are used.

Part 1: Lectin-mediated parasite agglutination

The following direct agglutination technique is applied to *Trypanosoma brucei brucei*, *Leishmania hertigi hertigi* and *Crithidia fasciculata/C. luciliae* cultured procyclic, promastigote and choanomastigote forms respectively.

(i) Parasite suspension preparation

1. Centrifuge Cunningham's medium containing *T.b. brucei* and Locke's solution containing *Leishmania* or *Crithidia* at $200\,g$ for 10 min in a conical-base tube.
2. Resuspend the parasite pellet obtained by agitation in PBS, wash in the same buffer and recentrifuge the parasite suspension as above.
3. Repeat the process a further three times with washing followed by centrifugation.
4. Enumerate the cells using an improved Neubauer haemocytometer and finally adjust the parasite suspensions with PBS to approximately 4–5×10^6 cells/ml.
   NB More concentrated suspensions (i.e. $>10^8$ cells/ml) may lead to non-specific agglutination and false positive results.

(ii) Agglutination test

1. Line the internal margin of each microtitre plate with four filter paper strips. Keep the plates internally humid (to avoid drying out of the well mixtures) by moistening the filter paper strips with distilled water.
2. Add to each well $5\,\mu l$ PBS pH 7.3.
3. Prepare two-fold serial $5\,\mu l$ dilutions of each lectin, initial concentration $1000\,\mu g/ml$ in PBS (with appropriate divalent cations added), i.e. 12 samples ranging from 500 to 0.49 $\mu g/ml$, with thorough mixing.
   NB Many lectins require the presence of certain metal ions in the buffer for activity purposes (see Table 6.6.1). BS-II requires Mn^{2+} as well as Ca^{2+}.
4. Dispense into all wells $5\,\mu l$ adjusted parasite suspension.
5. Control wells comprise $5\,\mu l$ parasite suspension plus $5\,\mu l$ buffer alone (to check for the presence of non-specific parasite agglutination occasionally associated with cultured forms of parasites).
6. Cover the plates and incubate at 25 °C for 30 min.
7. The degree of parasite agglutination can be assessed (especially in Part 2) using a relative scale of 4+ (100% agglutination), 3+ (75%), 2+ (50%), 1+ (25%), tr (trace) and 0 (negative). However, the end point agglutination titre is usually

expressed as the concentration of lectin that just gives visible agglutination.

Part 2: Carbohydrate inhibition of agglutination

Sugar inhibition is performed to determine more accurately the reactivities of non-conjugated lectins for specific parasite surface membrane carbohydrate moieties, especially where lectins are reactive against more than one sugar. Stock solutions of various sugars, each of equivalent molalities (300 mM), are available in PBS pH 7.3. A non-specific sugar(s) is usually included in the inhibition assay; fructose or sucrose are commonly chosen. Inhibition of agglutination is achieved by the sugar that is present on the parasite surface.

(i) Competitive saccharide inhibition

1. Prepare 2 μl volume two-fold dilution series of each positive lectin determined in Part 1.
2. To the wells containing each lectin dilution add an equal volume of either non-specific or specific sugar(s), final concentrations 150 mM, with mixing.
3. Prepare a two-fold dilution series of positive lectin (each 2 μl) and add 2 μl PBS to each well; these are the control plates.
4. Incubate the plates for 30 min at 25 °C before use in the agglutination test.
5. Dispense 4 μl adjusted parasite suspension to all wells, re-incubate at 25 °C for 30 min after which time determine the agglutination end-point titres.
6. Inhibition of agglutination is considered as a reduction in end-point titre, upon comparison to the controls, by at least two well dilutions and can be given a relative score of 1+. A reduction in titre by three wells is designated 2+, four wells as 3+, five wells as 4+, six or more wells, 5+ and total inhibition as 6+. Inhibition of one well or no inhibition is designated ± and − respectively. Alternatively, the results can be expressed as endpoint dilution obtained in the presence of sugar inhibitor, i.e. change in titre when compared to the control end-point values.

(ii) Mininium sugar inhibitor concentration determination
This experiment is performed to ascertain the most predominant carbohydrates present on the parasite surface against

which the lectin is the most specific. Only the carbohydrate inhibitors that cause inhibition are used in this experiment. The agglutination titres for positive lectins against the parasite cells (i.e. controls) have been evaluated in (i) and the lectin solution is diluted with PBS to a concentration that gives just 100% agglutination.

1. Prepare two-fold serial dilutions (each 2 μl) of appropriate sugar inhibitor(s) in PBS to give final concentrations ranging from 150 to 0.073 mM.
2. Add to each sugar dilution adjusted lectin (2 μl) and incubate all plates as previously stated.
3. Finally, add 4 μl appropriate parasite suspension to all the plates and incubate as before.
4. Control wells comprise lectin or sugar plus PBS and parasites, or PBS and parasites alone.
5. Assess the degree of agglutination and determine sugar concentration at the end-point agglutination titre.
6. The minimum inhibitor concentration is regarded as that which causes a reduction in agglutination from 100% to 10% or less as compared to the appropriate controls.
7. The sugar(s) that causes the highest degree of agglutination inhibition (note the millimolarity) is the predominant carbohydrate on the parasite surface against which the lectin is the most specific.

Expected results

LPA agglutinates only *C. luciliae*; CFA agglutinates *C. fasciculata* at a much higher titre than *L.h.hertigi* (LV43) and neither WGA nor LCA cause the agglutination of *L.h. hertigi* (LV42). PNA and WGA strongly agglutinate *T.b. brucei*; the former weakly agglutinates *C. fasciculata* and LV42 whilst the latter lectin also agglutinates LV43. The remaining three lectins give varying end-point titre values to enable distinction between the parasites used. LCA agglutinates *C. fasciculata* > LV43; BSII *C. luciliae* > *T.b. brucei* > LV42 and VVA *T.b. brucei* > LV42 > *C. luciliae*. These results should enable the investigators to distinguish between the four parasite species and two *Leishmania* strains.

The sugar inhibition experiment would indicate the occurrence of glucosyl > galactosyl moieties on the *T.b. brucei* surface. In the case of *C. luciliae* neuraminyl > (GlucNAc)$_2$; *C. fasciculata* (mannosyl > galactosyl); LV42 (galactosyl > glucosyl) and LV43 glucose > mannose. The results would give an indication as to the sugars present on the parasite surface membrane that would be potential binding sites for lectins.

Ideas for further exploration

- How could the parasite surface be modified to enable better binding of lectins?
- Give other examples of how lectins could be used in parasitology.
- Does the use of lectins enable successful identification of different parasite species of the same genus or different strains of the same species?
- What are the roles of naturally occurring lectins (agglutinins) in insect vector–parasite interactions?

Additional information

Although seven lectins of varying sugar specificities have been used in these experiments, other lectins could be employed in cases where negative results have been found. In some instances, a given lectin may be negative whereas another lectin of the same or similar specificity might produce a positive result. Furthermore, a wider range of sugars may be examined. It may be feasible to divide the student class into several groups to allow the use of more lectins/parasite species/carbohydrates and thus obtain additional data. An alternative source of material may be the use of viable parasites obtained directly from infected blood (following isolation on an ion exchange column). Different stages of the parasite life cycle may be examined and also different stages of growth to determine any variations in membrane surface carbohydrates.

REFERENCES

Jacobson, R. L. (1994). Lectin–*Leishmania* interaction. In *Lectin– Microorganism Interactions* (eds. R. J. Doyle & M. Slificin), pp. 191–223. New York, M. Dekker.

Jacobson, R. L. & Schnur, L. F. (1990). Changing surface carbohydrate configurations during the growth of *Leishmania major*. *Journal of Parasitology* **76**, 218–224.

Maraghi, S., Molyneux, D. H. & Wallbanks, K. R. (1989). Differentiation of rodent trypanosomes of the subgenus Herpetosoma. *Tropical Medicine and Parasitology* **40**, 273–278.

Petry, K., Schottelius, J. & Dollet, M. (1987). Differentiation of Phytomonas sp. and lower trypanosomatids (Herpetomonas, Crithidia) by agglutination tests with lectins. *Parasitology Research* **74** , 1–4.

Rudin, W., Schwarzenbach, M. & Hecker, H. (1989). Lectin-bound sites in the midgut of the mosquitoes *Anopheles stephensi* Liston and *Aedes aegypti* L. (Diptera: Culicidae). *Journal of Protozoology*, **36**, 532–538.

Schnur, L.F. & Jacobson, R. L. (1989). Surface reaction of *Leishmania* IV. Variation in the surface membrane carbohydrates of different strains of *Leishmania major*. *Annals of Tropical Medicine and Parasitology* **83**, 455–463.

Schottelius, J. (1987). *Parasitology Research* **73** 1–8.

Schottelius, J. & Aisien, M. (1994). Trypanosome–lectin interactions. In *Lectin–Microorganism Interactions* (eds. R. J. Doyle & M. Slifkin), pp. 225–248. New York, M. Dekker.

Sharon, N. & Lis, H. (Eds.) (1989). *The Lectins*. London, Chapman & Hall.

Taylor, A. E. R. & Baker, J. R. (Eds.) (1978). *Methods of Cultivating Parasites In Vitro*. London, Academic Press.

Appendix (See also Taylor & Baker, 1978)

Cunningham's medium (mg/100 ml)

NaH_2PO_4	53
$MgCl_2.6H_2O$	304
$MgSO_4.7H_2O$	370
KCl	298
$CaCl_2.2H_2O$	15
Glucose	70
Fructose	40
Sucrose	40
Malic acid	67
Ketoglutaric acid	37
Fumaric acid	5.5
β-alanine	200
DL-alanine	44
L-asparagine	24
L-aspartic acid	11
L-cysteine HCl	8
L-cysteine	3
L-glutamic acid	25
L-glutamine	164
Glycine	12
L-histidine	16
DL-isoleucine	9
L-leucine	9

L-lysine	15
DL-methionine	20
L-phenylalanine	20
L-proline	690
DL-serine	27
DL-threonine	10
L-tryptophan	10
L-tyrosine	20
DL-valine	21
BME×100 (vitamin mixture)	0.2 ml
Phenol red (0.4% w/v)	0.4 ml

Dissolve in double distilled water and adjust to pH 7.4 with 2 M NaOH. Sterilise by filtration through a 0.22 μm filter and add 20% foetal bovine serum prior to parasite culture.

6.7 Tentative identification of parasite and tissue surface carbohydrates by conjugated lectins

G. A. INGRAM

Aims and objectives

Using commercial conjugated lectins, these two exercises aim to:

1. Detect and ascertain surface membrane carbohydrates on trypanosomes and insect vector salivary glands.
2. Determine differences in salivary gland surface carbohydrates between *Glossina* species and between different regions of *Anopheles* salivary glands.

Introduction

Lectins are proteins, usually glycoproteins, derived from both animal and plant material, that recognise and bind to specific sugar moieties (or certain glycosidic linkages) of polysaccharides, glycoproteins and glycolipids present on cell membrane surfaces or in biological fluids (Lis & Sharon, 1986; Sharon & Lis, 1989). Lectin specificity is usually defined in terms of the monosaccharides or simple oligosaccharides that inhibit precipitation or binding reactions.

In order to detect membrane-bound carbohydrates, commercial lectins have been employed either non-conjugated (for use in agglutination/precipitation reactions) or conjugated with a marker (for use in cell or tissue surface membrane studies). The labels used include fluorescein isothiocyanate (FITC) and enzymes (e.g. peroxidase). These lectin-based techniques encompass the use of light and ultraviolet microscopy (Jacobson & Doyle, 1996).

The surface of intact vector tissues to which parasites may attach in order to undergo morphogenesis during transformation from one life cycle stage to the next, or attachment sites prior to transmission, have been studied (Ingram & Molyneux, 1991). These investigations, in part, enabled detection of carbohydrate **385** moieties on a membrane surface that might act as binding sites to facilitate parasite adhesion. For example, tsetse fly (*Glossina* spp.) and mosquito (*Anopheles* spp.) salivary glands and midguts have been used with FITC-conjugated lectins (Rudin & Hecker, 1989; Molyneux *et al.* 1990; Okolo *et al.* 1990; Mohamed *et al.* 1991) and/or peroxidase-labelled lectins (Mohamed & Ingram, 1993).

Laboratory equipment and consumables (per student or group)

Equipment

Microscope slides and
  coverslips
Dissection kit
Adjustable pipette (1–10 μl)
  with disposable tips
Pasteur pipettes plus
  bulbs

Disposable gloves
Microscope (light/UV
  fluorescent, with camera)
Incubator (25 °C)
Cavity microscope slides
Safety glasses

Consumables

1. *Trypanosoma brucei brucei*–infected blood smears (for Part 1, Experiment A). Tissue samples (e.g. salivary glands) for Part 1, Experiment B and Part 2. Glutaraldehyde (0.25 M) in 100 mM sodium cacodylate buffer, final pH 7.3 (Parts 1 and 2).
2. 10 mM Tris/HCl buffer pH 8.0 (Parts 1 and 2).
3. Phosphate buffered saline (PBS) containing divalent cations, pH 7.3. (8.5 g NaCl; 1.07 g Na_2HPO_4; 0.40 g $NaH_2PO_4.2H_2O$; 0.060 g $CaCl_2.2H_2O$ and 1.30 g $MgCl_2.6H_2O$ all dissolved in 1 L distilled water).
4. Tris–HCl/glycerol (1:1 v/v) pH 8.0 (Part 1, Experiments A & B).
5. PBS/glycerol (1:1 v/v) pH 7.3 (Part 1, Experiment A).
6. Solutions of FITC-conjugated commercial lectins (stock 1000 μg/ml in PBS) for Part 1, Experiment B).
7. Solutions of peroxidase-labelled commercial lectins (stock 1000 μg/ml in PBS) for Part 2).
8. FITC solution (50 μg/ml in PBS) for Part 1, Experiment B.
9. Specific sugar inhibitors (150 mM in PBS).
10. Substrate (3-amino-9-ethylcarbazole) tablets (Sigma). Part 2.
11. Dimethylformamide (Part 2).

12. Acetate buffer (50 mM; pH 5). 27.62 g sodium acetate dissolved in 850 ml distilled water, adjusted to pH 5.0 with glacial acetic acid and made up to a final volume of 1 L (Part 2).
13. Hydrogen peroxide (30%) (Part 2).

Sources of material

Trypanosoma brucei brucei or other trypanosomatid flagellate parasites can be obtained from living organisms and after isolation either tested immediately, placed into culture or frozen down in liquid nitrogen and stored until use. Parasites for blood-smear purposes are obtained from infected animals. Salivary glands may be dissected, usually from clean, laboratory bred colonies of tsetse flies, *Glossina* species (e.g. *G.m. morsitans; G. palpalis gambiensis* and *G. palpalis palpalis*) and anopheline mosquitoes (e.g. *Anopheles gambiae* Nigeria strain and *An. arabiensis* Swaziland strain) or suitable insect vectors. Supply of the above material may be possible from: London School of Hygiene and Tropical Medicine, Keppel Street, London; Liverpool School of Tropical Medicine, Pembroke Place, Liverpool. Workers in the USA should consult the National Centre for Disease Control, Atlanta, GA, for the latest advice on the availability of anopheline mosquitoes.

Laboratory safety

Good laboratory practice should be performed at all times for these experiments. Lectins should always be regarded as toxic and possible biohazards since some interact with living cells/tissues in potentially damaging ways not yet fully understood. Always wear gloves when handling conjugated lectin solutions. Safety glasses should be worn when working with dimethylformamide, fluorescein isothiocyanate (FITC) and glutaraldehye in sodium cacodylate buffer because these substances are skin and eye irritants. Always regard most organic compounds as potentially toxic materials and therefore wear gloves! Salivary glands should be obtained from uninfected insect vectors. Spillages of any chemicals or biological materials on the bench must be wiped up immediately. Relevant COSHH (Control of Substances Hazardous to Health) forms must be completed and made available for inspection.

Table 6.7.1 *FITC- and peroxidase-labelled lectins commonly used in parasitology, together with their sugar-binding specificities and carbohydrates used for binding inhibition*

Lectin	Abbreviation	Specificity	Inhibitors
Canavalia ensiformis (Jack bean)	Con A	α D-mannose, α D-glucose	α-CH$_3$-mannopyranoside, α-CH$_3$-D-glucopyranoside, D-mannose, D-glucose
Triticum vulgaris (Wheat germ)	WGA	(D-GlcNAc)$_2$	N, N^1-diacetylchitobiose, GlcNAc
Glycine max (Soybean)	SBA	D-GalNAc	GalNAc, D-galactose
Arachis hypogaea (Peanut)	PNA	β-D-gal(1–3)GalNac	GalNAc, D-galactose
Tetragonolobus purpureas (Lotus)	TPA	α-L-fucose	α-L-fucose
Dolichos biflorus (Horse grain)	DBA	α-D-GalNAc	GalNAc

Instructions for staff

Smears of infected rat or mouse blood with a high level of parasitaemia could be prepared and fixed beforehand. Alternatively, students can be given infected blood from which to prepare their own smears at the start of the laboratory class. Stock solutions of conjugated lectins and sugars (see materials and consumables section and Table 6.7.1) should be prepared and be ready to use at the start of the experiments. Living specimens of tsetse flies or mosquitoes should be available to enable students to obtain salivary glands, especially with *Glossina* species. However, it may be worthwhile giving the students a practical lesson on tsetse fly or mosquito salivary gland dissection to enable glands to be stored for this exercise. This is advisable in the case of mosquitoes but tsetse salivary gland dissection is quick and easy to achieve. In view of time limitations for practical classes, it is suggested that FITC-conjugated lectins be used in one laboratory session and enzyme-labelled lectins in another.

Materials

You are provided with either blood infected with *Trypanosoma brucei brucei* trypomastigotes or prepared infected blood smears.

Tsetse flies and/or mosquitoes will be available for dissection and removal of salivary glands. Conjugated lectins, specific sugar inhibitors (see Table 6.7.1) and all necessary solutions will have been prepared in advance. The FITC-conjugated lectins used on the blood smears and tsetse fly tissues are WGA, Con A, PNA, TPA and SBA (Part 1; Experiments A and B), whilst peroxidase-labelled WGA, Con A, PNA, DBA, TPA and SBA are applied to mosquito salivary glands (Part 2).

Part 1: Fluorescein isothiocyanate (FITC)-conjugated lectins

Experiment A: Blood parasite smears

1. Thin blood smears from rats or mice infected with *Trypanosoma brucei brucei* trypomastigotes have been prepared (a high level of parasitaemia is important). The smears have been fixed in 95% methanol for 10 min and the slides dried.
2. Treat the slides with chosen diluted FITC–lectin conjugate (75 µg/ml in PBS) or FITC alone (50 µg/ml). See later section on Control samples.
3. Incubate for 30 min at 25 °C.
4. Wash the microscope slide smears in PBS pH 7.3 (containing divalent cations) three times, each for 3 min.
5. Re-wash for 3 min with PBS/glycerol (1:1 v/v) and finally wash with Tris-HCl buffer pH 8.0 for 3 min.
6. Mount the smear sample in Tris–HCl/glycerol buffer pH 8.0 under a coverslip and examine for fluorescent activity using a fluoresence microscope. Oil immersion is usually required.

The fluorescent intensity of FITC–lectin binding to the parasite cell surface is designated as either 3 + (intense), 2 + (intermediate), 1 + (weak to trace) and 0 (no fluorescence). A simple quantification can be obtained by the exposure times of a camera attached to the microscope.

Experiment B: Salivary glands

1. With the aid of a binocular microscope, slowly withdraw tsetse fly salivary glands from the thorax and abdomen by carefully pulling the head away from the thoracic region with fine forceps whilst placing a seeker needle, using light pressure, on the abdominal surface. Remove the head away from the thorax very slowly to allow the paired salivary glands to be exposed. Remove them from the body and place

in PBS on a microscope slide. (Mosquito salivary glands should have been obtained previously or provided.)

2. Place the tissues on a cavity slide and fix in 250 mM glutaraldehyde in 100 mM sodium cacodylate buffer pH 7.3 for 15 min at 25 °C. Alternative fixation can be achieved by treatment with a mixture of 5% ethanoic acid: 95% ethanol (v/v) at 5 °C for 5 min.

3. Wash the glands three times, each 3 min, **WITH CARE** using PBS pH 7.3 and a Pasteur pipette.

4. Incubate the samples with chosen FITC-conjugated lectins at concentrations of 50 μg/ml in PBS or FITC alone (50 μg/ml in PSB) for 30 min at 25 °C.

5. Following incubation, wash the tissues five times in PBS, each 3 min, and finally wash in Tris–HCl buffer pH 8.0.
   NB Some salivary glands should be incubated only in PBS following fixation as an additional control (see section on Control samples).

6. Mount the samples in a drop of Tris–HCl/glycerol buffer (1:1 v/v) pH 8.0, cover with a coverslip and examine for degree of fluorescence.

The fluorescent activity is assessed as Experiment A.

Part 2: Peroxidase-labelled lectins

1. Fix and wash all the tissues exactly as for Part 1: Experiment B. Treat the salivary glands with 50 μg/ml chosen peroxidase-labelled lectins prepared in PBS pH 7.3. As controls, some glands should be incubated in peroxidase alone at 50 μg/ml and others in PBS (see section on Control samples).

2. Incubate at 25 °C for 30 min after which time wash the tissues in PBS three times, each 4 min.

3. *Freshly* prepare the substrate (3-amino-9-ethylcarbazole) prior to use by dissolution of a tablet (Sigma) with stirring in 1 vol. dimethylformamide mixed with 19 vols 50 mM acetate buffer pH 5.0. To this mixture add 30 μl fresh 30% H_2O_2.

4. Immediately incubate the tissues in the substrate solution for 30 min at 25 °C in the dark and then wash thoroughly as before with PBS.

5. Mount the salivary glands in PBS/glycerol buffer pH 7.3 and visually assess the degree of colour intensity of the red-brown insoluble end product formed by light microscopy. A dark red-brown colour is regarded as a 3 + relative score; intermediate

brown colour 2 +; light brown 1 +; yellow-brown trace (±) and yellow alone as negative.

Control samples

Several controls (see below) can be employed in the FITC-conjugated and peroxidase-labelled lectin assays and should be treated under exactly the same conditions as the lectin-conjugate exposed tissue samples (e.g. salivary glands) or infected blood smears. Selected controls may vary according to each experiment but it is suggested that at least (v), (vi) and (vii) be included in the experiments.

(i)   Unfixed, isolated tissues or blood smears that are washed and incubated only in appropriate buffer.
(ii) Samples fixed with glutaraldehyde and buffer washed only.
(iii) Samples unfixed, washed, incubated with either FITC, substrate solution or peroxidase and then rewashed.
(iv) Unfixed samples washed and reacted with a previously incubated mixture (30 min at 25 °C in the dark) of either FITC- or peroxidase-labelled lectin plus specific sugar (used at 150 mM).
(v)   As per (iv) except that a non-specific sugar (at 150 mM) is used.
(vi) As per (iii), but using fixed samples.
(vii) Similar to (iv) but with fixed samples.
(viii) Similar to (v) but with fixed samples.

Expected results

The surface membrane of *T.b. brucei* in the blood smears should display 3 + fluorescence for Con A and WGA with 2 + and 1 + fluorescent activity for PNA and SBA, respectively. The use of sugar inhibitors (especially glucose, mannose and their derivatives with Con A) indicate a preponderance of glucosyl moieties with limited galactosyl residues on the trypanosome surface. Degrees of fluorescence of 2 + to 3 + obtained with Con A and WGA on the surface of *G.m. morsitans* salivary glands suggest a predominance of glucosyl moieties but no galactosyl or fucosyl residues were detected in view of negative results for PNA, SBA and TPA. Strong fluorescence (3 +) would be found using WGA on both *G.p. palpalis* and *G.p. gambiensis* salivary glands (*N*-acetyl-D-glucosamine prevalence) but Con A (glucosyl moieties) only gives fluorescent activity (3 +) with *G.p. gambiensis* (negative for

G.p. palpalis). Trace amounts of galactosyl residues are present on both *palpalis* species salivary gland membrane as indicated by PNA and SBA weak binding whilst TPA (fucose specificity) displayed 1 + fluorescence with *G.p. palpalis* and was negative for the other two tsetse species examined.

WGA binds strongly to both the proximal and distal portions of the lateral lobes (3 +) of *An. gambiae* Nigeria strain salivary glands but does not bind at all to the median lobe (indicating lack of presence of N-acetyl-D-glucosamine here). Con A, however, reacts only against the lateral lobes with less binding (1 + and 2 + for the distal and proximal lobes, respectively). By comparison, both WGA and Con A bind to *Anopheles arabiensis* lateral and median lobes. Furthermore, N-acetylgalactosamine is present on surfaces of all *An. arabiensis* salivary gland lobes as indicated by binding of SBA and DBA (2 +) although galactose is demonstrated only on the lateral lobes when PNA (2 +) is used.

The lectins used would enable the participant to tentatively ascertain the types of sugars present on both parasite and tissue surfaces in conjunction with the use of sugar inhibitor to confirm binding specificity.

Ideas for further exploration

- What is the function(s) and importance of carbohydrates present on the salivary gland surface membrane in relation to the parasite–insect vector association?
- What other methods are available for detection of carbohydrates on surface membranes?
- Would you choose FITC- or enzyme-labelled lectins to detect surface carbohydrates? Comment on the merits/limitations of your choice!
- Could other enzymes (e.g. glycoside hydrolases) be used to confirm further the specificity of lectin binding or alternative treatments of the cell/tissue surface to allow improved binding? Comment!

Additional information

The lectins used are only a few from a wide range of commercially available types; other lectins could be utilised. Whilst tsetse flies and mosquitoes have been used in these practicals, different insect vectors, e.g. sandflies or blackflies, may be examined along with different tissues, e.g. peritrophic membrane and midgut. These exercises may be conducted as a class (group) exercise combining results. A non-vector such as *Calliphora eryth-rocephala* (blowfly) larvae could easily be used to obtain salivary gland material. Moreover, nematode, cestode or trematode parasites may be another good source of material for staining with conjugated lectins.

REFERENCES

Ingram , G. A. & Molyneux, D. M. (1991). Insect lectins: role in parasite-vector interactions. *Lectin Review* **1**, 103–127.

Jacobson, R. L. & Doyle, R. J. (1996). Lectin–parasite interactions. *Parasitology Today* **12**, 55–61.

Lis, M. & Sharon, N. (1986). Lectins as molecules and tools. *Annual Review of Biochemistry* **55**, 35–67.

Mohamed, M. A., & Ingram, G. A. (1993). Salivary gland surface carbohydrate variation in three species of the *Anopheles gambiae* complex. *Annales de la Société belge de Médicine Tropicale* **73**, 197–207.

Mohamed, M. A., Ingram, G. A., Molyneux, D. M. & Sawyer, B. V. (1991). Use of fluorescein-labelled lectin binding of salivary glands to distinguish between *Anopheles stephensi* and *An. albimanus* species and strains. *Insect Biochemistry* **21**, 767–773.

Molyneux, D. M., Okolo, C. J., Lines, J. D. & Kamhawi, M. (1990). Variations in fluorescein-labelled lectin staining of salivary glands in the *Anopheles gambiae* complex. *Medical Veterinary Entomology* **4**, 459–462.

Okolo, C. J., Jenni, L., Molyneux, D. M. & Wallbanks, K. R. (1990). Surface differences of Glossina salivary glands and infectivity of *Trypanosoma brucei gambiense* to Glossina. *Annales de la Société Belge de Médicine Tropicale* **70**, 39–47.

Rudin, W. & Mecker, M. (1989). Lectin-binding sites in the midgut of the mosquitoes *Anopheles stephensi* Liston and *Aedes aegypti* L. (Diptera: Culicidae). *Parasitology Research* **75**, 268–279.

Sharon, N. & Lis, M. (Eds.) (1989). *The Lectins*. London, Chapman & Hall.

Section 7
Behaviour

7.1 Behaviour of the miracidia of *Fasciola hepatica* and demonstration of other larval stages

C. E. BENNETT

Aims and objectives

This exercise is designed to demonstrate:

1. The mechanism of hatching of miracidia from typical platy-helminth eggs.
2. Styles of movement of platyhelminths, using *Fasciola hepatica* miracidia, namely:
   (i) vermiform peristalsis using circular and longitudinal muscle on a hydrostatic skeleton to facilitate emergence of the miracidium from the egg;
   (ii) co-ordinated beating of surface cilia for swimming of the miracidium;
   (iii) beating of flagella in the paired flame cells in the miracidium as the primary propulsive force in the movement of excess water and excretory waste.
3. Behavioural responses of miracidia that facilitate host-snail location for continuation of the life cycle.

Introduction

This exercise pays particular attention to the eggs (Fig. 7.1.1) and miracidia of the common liver fluke, *Fasciola hepatica*. It focuses on the behaviour of the miracidia in relation to movement and location of sites where an encounter with the intermediate host is most likely. It also demonstrates response to stimuli of host origin to ensure location of snails and hence successful transmission. The life cycle of *F. hepatica* is indirect and includes freshwater snails as intermediate hosts; in the UK this is the mud snail, *Lymnaea truncatula* and in North America *L. columella* and *L. cubensis*. **NB** *F. halli* and *F. californica* are also to be found in cattle and sheep in the USA. Adult *F. hepatica* inhabit the bile ducts of sheep, cattle and humans and their unembryonated eggs are voided in host faeces. Development of the miracidium is governed largely

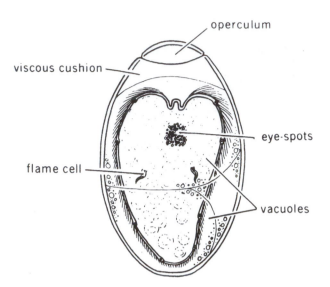

viscous cushion

operculum

eye-spots

flame cell

vacuoles

Fig. 7.1.1 Embryonated egg of *F. hepatica* (after Smyth & Halton, 1983).

by external factors, particularly temperature, and after hatching the larvae must infect *Lymnaea* within hours of release, otherwise they will die.

Movement in platyhelminths can take the form of:

1. *Vermiform peristalsis.* A hydrostatic skeleton operates a vermiform peristalsis with a basket-like arrangement of mainly circular and longitudinal muscles operating on the tissue-filled mesoderm. This peristalsis is easy to observe during the emergence of the first larval stage, the miracidium, from the egg capsule as it squeezes out through the operculum. This type of peristalsis occurs in all platyhelminth classes [Cestoda (tapeworms), Digenea and Monogenea (flukes), and the free-living Turbellaria].

2. *Co-ordinated beating of surface cilia.* Miracidia demonstrate the alternative locomotory system of platyhelminths via the beating of surface cilia: many adult Turbellaria make use of surface cilia to glide along. A variety of larval stages of Cestoda, Digenea and Monogenea also use cilia.

3. *Flagella in the excretory system:* As a flatworm, *Fasciola* provides a useful example of the excretory system that evolved under selective pressure to remove excess water and excretory products. The basic functional unit of the excretory system in flatworms is the flame cell, or protonephridium (Fig. 7.1.2). This consists of a single cell bearing a tuft of flagella or 'flame'

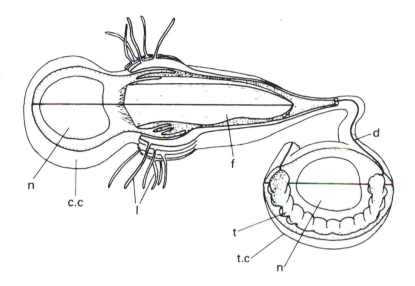

Fig. 7.1.2 Protonephridial flame cell and tubule of *F. hepatica* (after Wilson, 1969).c.c, cap cell; d, desmosome; f, "flame"; l, leptotriches; n, nucleus; t, tubule; t.c, tubule cell.

that extends into a tube leading to a system of excretory ducts. The ducts finally connect with an excretory bladder and pore. Flagella beat more or less continously and in doing so draw excess water containing nitrogenous wastes from the surrounding tissues into the tubule for voidance via the excretory pore. A pair of flame cells can be seen in miracidia, with phase-contrast microscopy, while they remain constrained within the translucent egg in the few minutes immediately prior to hatching.

Behavioural responses of miracidia facilitating host location

The motile miracidium is a non-feeding stage of the life cycle. It is entirely dependent on stored energy reserves and has a limited life-span, depending on temperature, of 12–48 h. Unless a suitable snail is located within this period, the miracidium will die. Host location is facilitated by patterns of response to environmental stimuli that result in miracidia swimming into micro-environments where the snails are most likely to be located.

Location of the snail intermediate host, *Lymnaea truncatula*, is most likely at the surface and perimeters of a body of water. Miracidia have negative geotaxis, positive phototaxis and respond to the presence of snails by an increased rate of turning. This taxis is called chemo-klinokinesis and increases the likelihood of contacting a snail, resulting in attachment and penetration. **NB**: the response is not a chemotaxis, which is a directional and concentration-dependent response. The former taxes will be demonstrated in the practical.

Laboratory equipment and consumables
(per student or group)

Equipment

Phase-contrast microscope
Binocular microscope

Consumables

Solid watch-glasses
Microscope slides (plain or cavity)
Lymnaeid snails, e.g. *Lymnaea truncatula*, *L. stagnalis* or other available species
Fine-drawn glass rod
Oxygenated bottled still water, pH adjusted to 6.5
Bleach solution (10%)

Sources of parasite material

Eggs of *F. hepatica*
The recovery of *F. hepatica* from abattoirs is unpredictable, but the best opportunities are from cattle and sheep that have been reared in the wetter Western regions of the UK. Adult infections are to be found in sheep and cattle in February–March and are greatest after high levels of infection of fluke in snails. These are at their highest when there has been wet weather in June/July/August of a previous year. A good working relationship with a meat inspector at a local abattoir is important. Do not expect to predict when fluke will be available. Rely on a telephone call and always respond! Infected livers from sheep and cattle will provide adult fluke and, more importantly, the gall bladders from infected animals are a ready source of thousands of eggs. See also the Additional information section in Exercise 1.6.

Larval stages
Prepared slides of sporocysts, rediae and cercariae are available from Philip Harris Education. They can provide useful additional demonstration material to the practical.

Safety

The miracidium of *F. hepatica* is non-infectious to vertebrates. As such, the exercise is entirely safe. Note, however, that if *Lymnaea* species are infected during the course of the exercise they will produce cercariae in approximately 35 days, which can transform into infective metacercariae. Snails that are exposed will need to be disposed of safely by immersion in 10% bleach and not released into the environment.

Instructions to staff

Handling of eggs and storage
Eggs are washed clean of bile and debris through sieves: coarse (150–300 μm) and fine (38 μm), the latter retaining the eggs. Eggs are washed with tap water and allowed to settle. They may be stored in a non-embryonated state for up to 6 months at 4 °C in the dark in tap water.

Embryonation of eggs prior to use
Samples of eggs are placed in a maximum depth of 1 cm of tap water in a flat-bottomed flask closed only with non-absorbent cotton wool to allow an adequate oxygen supply. They are made ready for embryonation by wrapping the flask in aluminium foil to maintain the eggs in the dark (to prevent premature hatching) and incubated at 22 °C for 14 days or 25 °C for 11 days. Eggs brought into the light will hatch with a high percentage of success even up to a week after these dates.

The first stages of hatching (carried out by the class supervisor)
The aluminium foil wrapping is removed from the flat-bottomed flask and the eggs are quickly separated from the water in which they have been kept, by centrifugation: the entire volume being quickly spun down at 150 *g* for 30 sec and the supernatant removed. The remaining eggs are pipetted into fresh oxygenated water, at which point they can be distributed to the class for observation of miracidia within eggs and the emergence of the miracidia on hatching. Miracidia hatch as a result of stimulation by light over a period of 2–20 min.

Instruction to students
By referring to Fig. 7.1.1, make a series of drawings of the hatching process of a miracidium from its egg.

Hatching of miracidia

1. Collect some eggs that have been recently transferred to aerated distilled water, pH 6.5, on a cavity slide (or plain slide).
2. Cover with a coverslip and observe by phase-contrast microscopy.
3. Examine and draw the outline of the egg capsule and enclosed miracidium within the egg. You may be able to detect the outline of the operculum, which lies at one end of the egg directly above the viscous cushion.
4. Draw and note the size of the viscous cushion at regular intervals until hatching occurs.

At this stage, whilst the miracidium is relatively immobile within the egg, you should be able to observe tufts of beating flagella in the body of the parasite. There are two of these flame cells situated approximately in the middle of the body. You should also be able to see the paired eye-spots, which are densely pigmented and appear as closely opposed unilaterally concave discs.

By scanning the field you should be able to select for observation those eggs that look as though they are about to hatch. They contain an active miracidium and a discernible viscous cushion at one end of the egg. As hatching becomes imminent the viscous cushion begins to swell to something like three times its original volume, creating internal pressure on the operculum, which then flips open. With some patience and care you should witness the opening of the operculum and emergence of a miracidium. It can be rapid, so be vigilant. Observe the hatching of some miracidia under the binocular microscope for about 15 min.

Remember that hatching is a dynamic process that varies slightly from egg to egg. For each observation/drawing, make a record of the time of hatching in relation to the first exposure to light at the beginning of the exercise.

5. Note and draw the sequence of events of hatching in terms of the shape and movement of the miracidium and emergence of any other material from the egg after the opening of the operculum.
6. Sketch the amount of constriction that the body of the miracidium displays as it squeezes through the opercular opening. Compare the normal width of the miracidium in the egg with the diameter of the operculum after hatching.

Swimming behaviour of the miracidia

NB: This refers to the swimming behaviour immediately post-hatching without stimuli (in diffuse light conditions). Note:

- Speed across a field of view (low power) in a straight line.
- Rotation on the axis of direction.
- Amount of turning, i.e. the number of sharp or right-angled turns.
- *Positive phototaxis.* Place a strong light at one side of your preparation under a dissecting microscope and note the accumulation of miracidia, on that side of the dish.
- *Negative geotaxis.* As a class demonstration, the class supervisor will transfer any excess miracidia to a 10 ml glass measuring cylinder and place it in a dark cupboard for the duration of the exercise. After at least 60 min, samples of miracidia will be drawn off in quick succession starting with the top 2 ml and the remainder in similar aliquots each placed in numbered solid watch-glasses for estimation of numbers of miracidia.

Responses to presence of snails

Make a sketched record of the swimming behaviour of miracidia in response to the presence of snails and snail mucus presented on a glass rod and placed on one side of a solid watch-glass.

- Do the miracidia swim directly towards the snail/glass rod?
- Do they turn more frequently than control miracidia in clean water?

NB: Wilson & Denison (1970) have shown with *F. hepatica* miracidia that the presence of the snail host, *L. truncatula*, or a sample of its mucus, greatly modifies the behaviour of the free-swimming larvae, causing an increase in the rate of turning (klinokinesis). This response is also evoked by ether or water extracts of the mucus, suggesting that relatively small organic molecules are responsible. Analysis of mucus produced by *L. truncatula* (see Wilson, 1968) has revealed concentrations of free glucose, of 16 amino acids (notably arginine, glutamic acid and alanine) and of various lipids and salt.

Effect of snail mucus on miracidial behaviour

By placing a very small slice of snail foot onto a cavity slide before adding 30–40 hatched miracidia and observing at high magnification, you should see the shedding of the ciliated epithelial cells. This is the most significant event in the transformation from miracidium to the sporocyst. NB: approximately one in five preparations are successful.

Sequence of epidermal shedding from the miracidium
Record, at 5 min intervals, the shedding of ciliated cells that balloon from the surface and become detached to swim away. Note that the entire process may take up to 40 min. At this time, the surface is being replaced by membrane-bound cytoplasmic extensions from underlying cells. These extensions merge and fuse to form a new surface layer, the tegument. This is a syncytium that links a number of sub-muscular tegumental cells via a common surface cytoplasm; its establishment as the surface layer is the principal event in transformation from miracidium to sporocyst.

Exposure of live *L. truncatula* to miracidia results in penetration of the head/foot region of the snail. Using its penetration glands, the miracidium dissolves a route through the snail epidermis and, losing its cilated coat, becomes transformed into the sporocyst, which enters the snail to reproduce asexually.

The infections mature through the sporocysts, which give rise to one or more generations of rediae which, in turn, develop cercariae. Intramolluscan development takes approximately 35 days to provide cercariae, which emerge from the snail. Cercariae swim in water by beating their tails, and finally encyst on vegetation or other objects as the metacerarial stage, which is infective to the vertebrate host (sheep/cattle/human).

Expected results

Miracidia swim approximately in a straight line across an evenly lit dish or gently follow the perimeter of a circular dish/ELISA plate multi-well chamber. When freshly hatched, they exhibit corkscrewing of a very small amplitude along the line of movement. This is best observed when miracidia cross a dish; for example, when stimulated by the mucus rod. As miracidia age, their movements slow down and the corkscrewing is of a greater amplitude. Miracidia exhibit both positive phototaxis and negative geotaxis, which would, as explained above, bring them to the perimeter of mud-pools where *Lymnaea truncatula* may normally be found. In the cut-foot preparation, at ×400 magnification, look for the key transformation event that occurs after attachment of miracidia. This is the loss of the miracidial ciliated plates/cells. Each cell forms into a ball on the miracidia and is released to swim away bearing cilia, which continue to beat. The cut-foot preparation used to demonstate the release of cells rarely completes to the penetration stage. In penetrating live *Lymnaea truncatula*, transformation from miracidium to sporocyst also includes digestion of the snail epidermis prior to a very rapid entry into the snail; this usually takes around 40 min.

Ideas for further exploration

- Effect of wavelength of light on hatching.
- Effect of vibration on hatching.
- Effect of light intensity on free-swimming miracidia.
- Effect of temperature on survival/longevity and swimming speed of miracidia.
- Effects of selected chemicals on miracidial behaviour, using agar blocks* impregnated with amino acids or lipids. Record number of contacts; time attached; contacts with return.

   *Preparation of agar blocks: 1% agar with 5 mM chemical attractants, e.g. aspartic, glutamic, lactic, sialic and butyric acids (MacInnis, 1970). Bring to boil, briefly cool and pour into Petri dishes. Covered dishes may be stored for up to a week.

   Blocks may be cut into approximate truncated pyramidal shapes (7 mm × 5 mm × 4 mm) to facilitate viewing of surfaces by a binocular microscope from above. Remember to prepare control blocks. If you have a camera attachment for a microscope then time-exposure will reveal tracks of miracidia in different conditions; this works best with dark field microscopy.

REFERENCES

Dawes, B. (1959). Penetration of the liver fluke *Fasciola hepatica* into the snail *Lymnaea truncatula*. *Nature* **184**, 1334.

MacInnis, A. J. (1970). Experiment 16. Responses of miracidia to chemical attractants. In *Experiments and Techniques in Parasitology* (eds. A. J. MacInnis & M. Voge). San Francisco, W. H. Freeman and Co. (out of print).

Smyth, J. D. (1994) *Introduction to Animal Parasitology*, 2nd edn. Chapter 14. Cambridge, Cambridge University Press.

Smyth, J. D. & Halton, D. W. (1983). *The Physiology of Trematodes*, 2nd edn. Cambridge, Cambridge University Press.

Tyler, S. & Tyler, M. S. (1997). Origin of the epidermis in parasitic Platyhelminthes. *International Journal for Parasitology* **27**, 715–738.

Wilson, R. A. (1968). The hatching mechanism of the egg of *Fasciola hepatica*. *Parasitology* **58**, 79–89.

Wilson, R. A. (1969). The fine structure of the protonephridial system in the miracidium of *Fasciola hepatica*. *Parasitology* **59**, 461–467.

Wilson, R. A. & Denison, J. (1970). Studies on the activity of the miracidium of the common liver fluke *Fasciola hepatica*. *Comparative Biochemistry and Physiology* **32**, 301–314.

7.2 Effects of age and environmental factors on the swimming behaviour of the cercariae of *Cryptocotyle lingua* (Trematoda)

J. G. REA & S. W. B. IRWIN

Aims and objectives

This exercise is designed to demonstrate that:

1. Parasite behaviour alters with age.
2. Swimming behaviour is influenced by environmental factors.
3. Parasites have evolved sensory mechanisms that are receptive to specific stimuli encountered during their active life.

The exercise also aims to:

1. Encourage students to consider the role of the observed responses in parasite dispersal.
2. Stimulate students to reflect on behavioural strategies that may lead to successful transmission.
3. Engage students in a critical debate relating to experimental design and the limitations of laboratory-based experimental models.

Introduction

The experiments described below investigate the swimming responses of the cercariae of the digenetic trematode, *Cryptocotyle lingua*. The swimming behaviour of cercariae may be adapted to ensure dispersal and infection of a host species. Some species show particularly high swimming activity after being shed (Haas *et al.*, 1994). This clearly contributes to escape from the host snail's habitat and assists dispersion. Most cercariae tend to disperse in particular habitats by responding to environmental stimuli, such as gravity, light, water currents, temperature, ionic content and physical boundaries (MacInnis, 1976). Combes *et al.* (1994) suggested that dispersion had both demographic and genetic consequences. It may facilitate an encounter with a host and limit inbreeding. Other non-adult stages, **407** such as eggs, cysts and some larvae, cannot respond to environmental signals by active processes and are transmitted by passive means. Cercarial host-finding behaviour patterns and the environmental cues that stimulate them have been analysed. The evidence indicates that the behaviour of actively host-invading species is adapted to maximise transmission success (Haas, 1994).

The selection of microhabitats frequented by the hosts and responses to specific stimuli emanating from them are also vital for successful host location. Such behaviour has been selected in the evolution of parasite life histories (Rea & Irwin, 1994).

Laboratory equipment and consumables
(per student or group)

Swimming chambers can be constructed from the clear plastic trays used in storage cabinets. They are available from suppliers of electrical components and are inexpensive. The trays come in a range of sizes but those approximately 120 mm in length are ideal. To reduce internal reflections, the interior of the chamber can be painted matt black. A transparent window in the base at one end should be left so that light stimuli can be transmitted through the sea water. If Ilford spectrum filters are used for the wavelength experiments, an aperture 38×13 mm in size is appropriate; blue (450 nm), yellow-green (540 nm), orange (600 nm) and red (700 nm) are recommended. The window must be covered with opaque glass to ensure even illumination. Apertures at either end of the chamber are needed for the colour preference experiment. A lid made of stiff black card or plastic can be used to cover the chamber up to the level of each window. A cold light source, such as a Schott KL 1500 fibre optic, is required. Alternative illumination may be used if the heating effect can be eliminated. A 10 ml syringe fitted with a 0.63 mm bore hypodermic needle is necessary to collect and accurately count the cercariae prior to release into the swimming chamber. The tip should be cut off as a blunt end makes counting easier. The time taken for the cercariae to reach the light can be obtained using a digital stop clock. Although the cercariae can be seen with the naked eye students are likely to require binocular microscopes. Host snails can be maintained in 1 litre beakers and cercariae kept in standard 90 mm diameter Petri dishes. For light intensity experiments, a light meter measuring lux is adequate. Ethanol (75%) can be used to kill the cercariae *in situ* so that they may be counted. The colour preference experiments require a quantum meter to ensure that behavioural differences are due solely to wavelength of light.

Sources of parasite material

The snail host is the marine gastropod *Littorina littorea*, which is common on rocky shores. Although infection levels can be quite low, once a suitable location has been found, infected specimens can be distinguished by the presence of orange pigments in the foot. Snails over 2.5 cm in height frequently yield large numbers of cercariae. Reference to Beedham (1972) will help with the identification of the mollusc, while the article by James (1968) contains a very useful key for larval digeneans from littorinids. See also the Additional information section in Exercise 1.2.

Safety

There are no biological hazards associated with the cercarial stage of this parasite. Care should be taken when removing the tips of hypodermic needles and also with their subsequent use. Students need to be vigilant when using sea water close to electrical equipment.

Instructions to staff

Once collected, approximately 20 snails at a time can be maintained for short periods in beakers of sea water at 5 °C. To ensure sufficient numbers of cercariae are obtained, approximately 100 snails are required. Three days prior to their intended use, the sea water should be removed and the snails kept in darkness at 5 °C. When returned to sea water at room temperature and exposed to bright light, large numbers of cercariae are usually emitted. These will aggregate at the surface and must be removed immediately otherwise they sink to the bottom, making collection difficult. The cercariae should then be transferred to a Petri dish. Light projected from the side of a dish will prevent dispersion. If cercariae are not emitted by this method, snails may be crushed and their rediae macerated. Although this will release some immature individuals, those that are actively swimming and gather at a light source may be used. In order that the cercariae are appropriately aged, it will be necessary to cause their staggered release some hours prior to this exercise. The experiments are based on the photactic response of cercariae. They will swim from the extreme edge of the swimming chamber, the release point, to the darkness/light boundary. Distance and time taken can be measured and therefore the swimming rate calculated. Most experiments can be conducted at room temperature. When the independent variable is temperature, environmental chambers are the best option. If they are not available, either chilling or warming of the sea water will be necessary. This approach provides ample opportunity for students to think about the limitations of laboratory-based experiments. All investigations should be conducted in a darkened room.

Instructions for students

Fig. 7.2.1 illustrates the life cycle of *Cryptocotyle lingua*. This is an example of active transmission. The experiments described in this exercise will examine some factors that influence cercarial dispersion and successful host location.

It should be noted that on release from the snail host the cercariae are strongly photoresponsive. As cercariae age, their tails tend to drop off, thus preventing swimming activity. Physical disturbance may also cause some to become decaudate.

Experiment 1: Effect of age on horizontal swimming rate

You will be provided with cercarial populations 0, 2, 4, and 6 h old. Using the equipment provided, decide how to measure swimming rate to the light source. Identify the independent, dependent and controlled variables for this and subsequent experiments. Released samples should include approximately 100 cercariae. You must determine the number of repeat releases necessary to produce valid results, bearing in mind that between each repetition the cercariae will have aged. Remember that it is inevitable that some cercariae will become decaudate (tailless) due to handling, so not all will arrive at the light. How many arrivals from each release do you time – the first only, the first 10, the first 20? You must make these decisions before you start.

Experiment 2: The effect of temperature on horizontal swimming rate

Using the same method as in Experiment 1, or a modified procedure as a result of your experience, investigate swimming rates

Fig. 7.2.1 Life cycle of *Cryptocotyle lingua*. 1. Parasite eggs released in bird faeces. 2. Eggs ingested by snail. 3. Cercariae shed from snail. 4. Cercarial penetration of fish. 5. Fish eaten by primary host.

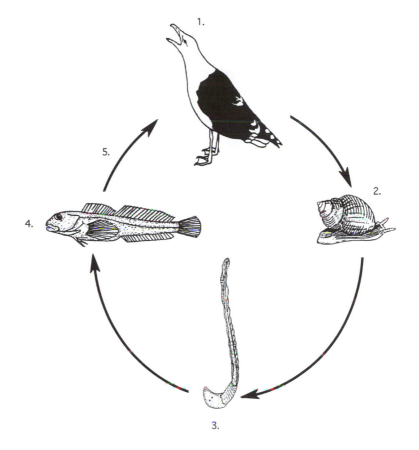

at 5, 15 and 25 and 30 °C. Will the temperature of the sea water change significantly during the experimental period? If changes occur how can they be minimised? What are the limitations of your chosen method?

Experiment 3: The effect of light intensity on horizontal swimming rate

The variable control on the light source adjusts the intensity of light. A meter can then be used to measure the light transmitted through the sea water. Find out the effect of light intensity on the swimming rate of the cercariae. Is there an optimum light intensity? Is there any evidence of an inhibitory effect at the higher intensities?

Experiment 4: The effect of wavelength of light on horizontal swimming rate

Coloured filters can be placed across the window in the base of the chamber. These can produce transmitted light of known wavelength: blue (450 nm), yellow-green (540 nm), orange (600 nm) and red (700 nm) are available. If the light intensity is kept constant the quality of light on swimming behaviour can be investigated. A quantum meter can be used to count the number of quanta (photons) transmitted. If that value is kept constant, differences in swimming rates can be attributed entirely to wavelength.

Experiment 5: Cercarial wavelength preferences

The swimming chamber has been adapted to offer a choice of coloured light. Filters can be placed at either end and the cercariae released in the middle. Five minutes should be allowed for them to disperse. A plastic divider may then be positioned in the middle and ethanol added to the water to kill the parasites. The number of cercariae in each side can then be counted. How many combinations of choice can be offered? Do the cercariae express a colour preference?

Analysis of results

Graphs should be produced illustrating the effect of the independent variable on mean swimming rates. Ninety-five per cent confidence limits should be obtained for each value. The results of the effect of age and wavelength may be linear, in which case regression analysis should be carried out and lines of best fit applied. The results of the two-way colour preference experiments should be graphed to show the percentage of cercariae preferring each colour when presented with a two-way choice.

Expected results

The results have been described by Rea & Irwin (1992). In summary, swimming rates decline rapidly with age. This is probably due to diminishing energy reserves or a change in the response of the cercariae to light. Rates are likely to peak at 15 °C. This may represent the metabolic maximum for contracting caudal muscles. Swimming rates tend to increase with increasing light intensity but very bright light usually has an inhibitory effect. A negative relationship between swimming rate and increasing wavelength is expected. When cercariae are given a choice, they tend to swim towards the shorter wavelengths. This may be due to the transmittance characteristics of sea water. In clear water, maximum transmittance is in the blue region of the spectrum with red light being almost completely absorbed at a depth of 5 m. The cercariae are therefore inclined to be more responsive to wavelengths of light that prevail in their environment.

Ideas for further investigation

- What influence would the decline in swimming rates with age have on parasite dispersal and transmission? Design an experiment that would test if cercarial age affects their ability to infect host tissue.
- If a high light intensity has an inhibitory influence on swimming, what effect would this have on the vertical distribution of cercariae in the water column at sea or in a rock pool?
- Cercariae are known to survive for a longer period if maintained at temperatures equivalent to the open sea. They may experience much higher temperatures however if trapped in rock pools. What, in terms of transmission, are the advantages and disadvantage of host-finding in the open sea compared to small tidal pools?
- If the cercariae demonstrate distinct colour preferences, what is the significance of these preferences? Of what relevance are they to the evolution of cercarial photoreceptors?
- These experiments have not dealt with the vertical distribution of the cercariae in the water column or alternating cercarial activity patterns, i.e. swimming, sinking or resting behaviour. Design an experiment to find out how the proportion of time spent by cercariae in each mode changes with age, light intensity and temperature.
- Cercariae are sensitive to shadow stimuli. What is the adaptive significance of this behaviour? Design an experiment to investigate the effect of shadowing on cercarial transmission success.

Similar exercises

Students could investigate the behavioural responses of other species of cercariae including *Diplostomum* sp. and members of the family Microphallidae, which use the gastropod *Littorina saxatilis* as an intermediate host.

REFERENCES

Beedham, G. E. (1972). *Identification of the British Mollusca*. Amersham, Hulton Educational Publications Ltd.

Combes, C., Fournier, A., Mon, H. & Theron, A. (1994). Behaviours in trematode cercariae that enhance parasite transmission: patterns and processes. *Parasitology* **109** (Suppl), S3–S13.

Haas, W. (1994). Physiological analyses of host-finding behaviour in trematode cercariae: adaptations for transmission success. *Parasitology* **109** (Suppl), S15–S29.

Haas, W., Haberl, B., Schmalfuss, G. & Khayyal, M. T. (1994). *Schistostoma haematobium* cercarial host-finding and host-recognition differs from that of *Schistosoma mansoni*. *Journal of Parasitology* **80**, 345–53.

James, B. L. (1968). The distribution and keys of species in the family Littorinidae and of their digenean parasites, in the region Dale, Pembrokshire. *Field Studies* **2**, 615–650.

MacInnis, A. J. (1976). How parasites find hosts: Some thoughts on the inception of host parasite integration. In *Ecological Aspects of Parasitology* (ed. C. R. Kennedy), pp. 3–20. Amsterdam, North-Holland.

Rea, J. G. & Irwin, S. W. B. (1992). The effects of age, temperature, light quantity and wavelength on the swimming behaviour of the cercariae of *Cryptocotyle lingua*? (Digenea: Heterophyidae). *Parasitology* **105**, 131–137.

Rea, J. G. & Irwin, S. W. B. (1994). The ecology of host-finding behaviour and parasite transmission: past and future perspectives. *Parasitology* **109** (Suppl), 31–39.

7.3 Changes in host behaviour as a consequence of parasite infection

H. HURD

Aims and objectives

This exercise is designed to demonstrate:

1. Parasite-induced behavioural modifications in their hosts.
2. Changes in behaviour at specific stages of infection.
3. The importance of critical evaluation of the experimental observations in terms of possible adaptive manipulation of the host.

Introduction

Many parasites with complex life cycles have adopted strategies that enhance their chances of transmission to the next host. Where no free-living stages exist, these tactics often involve a parasite-induced alteration in host behaviour that increases periods of contact with the next host. Helminths with life cycles that involve passive transfer between hosts make use of predator–prey associations (the food chain) for transmission. In many such cases, changes in host appearance or behaviour that may render them more vulnerable to predation have been described. For example, species of fresh-water shrimps infected with the cystacanth stages of acanthocephalan worms change in colour and behavioural patterns so that they are more conspicuous and spend more time in the vicinity of the precise duck, mammal or fish predator that acts as definitive host for that parasite species (see also exercise 7.5). Additional examples are reviewed by Hurd (1990).

Predation of the intermediate host will only result in successful transmission if the parasite stage is mature and infective. In some cases it has been shown that the intermediate host does not become more vulnerable to predation until the parasite has completed that stage of its development.

The experiments in this exercise are designed to compare **415** aspects of the behaviour of non-infected beetles and beetles infected with the cysticercoid stage of the rat tapeworm, *Hymenolepis diminuta.*

Laboratory equipment and consumables
(per student or group)

1. *Light/dark choice chambers*
   - These can be constructed from plastic Petri dishes. One half is painted black (top and bottom) or covered in black paper. The bottom should be lined with white/black paper to provide a grip for the beetles, making sure that they cannot crawl underneath the paper.
2. *Pheromone choice chambers*
   - Empty Petri dishes.
   - Filter paper circles that fit the Petri dishes.
   - Petri dishes lined with filter paper circles that fit the bottom of the dishes. These dishes should contain 10 non-infected beetles of the same sex and approximate age that have been in the dish, without food, for 12–24 h.
   - Gloves should be worn to handle the filter papers.
3. *Activity arena*
   - A box, tank, etc. positioned in an area of even illumination and having an area of approximately 500 cm^2.
   - Stop watches / timers will be required for all experiments.
4. *Beetles*
   - A supply of *Tenebrio molitor*, the yellow mealworm beetle, of known sex and infected for 4–5 or 11–15 days with cysticercoids of the rat tapeworm, *Hymenolepis diminuta*.
   - Large broad forceps may be used to handle the beetles.
   - A supply of uninfected beetles of the same sex and ages as the infected groups.

Sources of parasite material

Eggs of *Hymenolepis diminuta* can be obtained from several laboratories, including The Centre for Applied Entomology and Parasitology, Keele University (infected beetles can also be supplied). *Tenebrio molitor* can be obtained from commercial suppliers. Beetles should be checked for the presence of cysticercoids at the correct stage prior to the class. See also Additional information section in Exercise 1.14.

Safety

H. diminuta has been categorised as a group 2 infective hazard. Local regulations should be consulted regarding handling and containment. In general, care must be taken to ensure that beetles do not escape and that all infected material is destroyed by autoclaving or soaking in methylated spirits. If ingested, mature cysticercoids of *H. diminuta* (9 days or more post-infection) can infect humans. The bran that is used as beetle food can create a dust hazard. Staff maintaining beetles may wish to wear a mask when feeding, sorting and sexing beetles.

Instructions for staff

T. molitor can be sexed by examining the external genitalia of the pupae (Bhattacharaya *et al.*, 1970). It is preferable to use groups of single sex beetles for these experiments because they produce sex pheromones that can act as attractants that make it difficult to interpret the observations. However, if the light/dark response experiments are performed on single insects, sexing would be unnecessary. The activity experiments are performed on single beetles.

Beetles should be starved for 2 days prior to infection. Tapeworm eggs (collected by a standard salt floatation method) can be mixed with an equal volume of apple pulp to make them more palatable and then presented to the starved beetles for 24–36 h. Infected and control beetles should be maintained at 25–28 °C on a diet of bran, supplemented weekly with small pieces of apple or potato. Care should be taken to keep the bran dry. Beetles 4–6 days post-infection will provide immature infections and those 12 or more days post-infection will contain mature parasites.

Instructions for students

Examine the diagram illustrating the life cycle of *Hymenolepis diminuta*, the rat tapeworm (Fig. 7.3.1). This is an example of passive transmission via a food chain. The following experiments will examine aspects of the behaviour of *Tenebrio molitor* that could affect its vulnerability to predation.

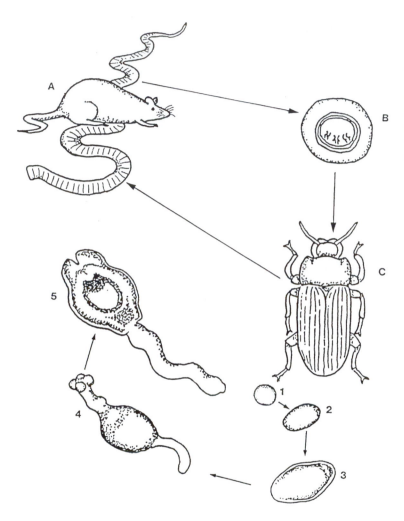

Fig. 7.3.1 The life cycle of *Hymenolepis diminuta*. A. Adult *Hymenolepis diminuta* in the rat gut; B. embryonated eggs are voided with the rat's faeces; C. Maturing cysticercoids pass through 1–5 stages in the haemocoel of the intermediate host, *Tenebrio molitor*. Adapted from original drawing by M. J. Taylor.

Experiment 1: Beetle response to light

1. Take groups of 5 non-infected beetles of the same age and sex and place them in the light/dark choice chamber.
2. After 5 min record the number of beetles in the light section.
3. Decide how to record beetles that are half way between sections before you begin.
4. Express the results as:

    number of beetles in the light / total number tested.

    Values of (or near) 1 indicate a photopositive response, values of 0.5 suggest a random distribution in response to light.
5. How do *T. molitor* respond to light?

6. Does this response vary according to age or sex?
7. Repeat these experiments with beetles infected with mature or immature parasites to determine whether infection affects photokinesis.

Experiment 2: Beetle activity

1. Place a beetle in the arena and watch it for a few minutes.
2. Decide how you will record activity.
3. Using the stop watch provided record the amount of time the beetle spends moving in a 5 min period.
4. Record activity for several beetles from each test group, allowing 2 min for each beetle to acclimatise to the arena before beginning timing.
5. Record the mean time spent moving during a 5 min period for each test group and compare the results from each group.

Experiment 3: Response to aggregation pheromone

1. You are provided with a Petri dish in which beetles have been maintained.
2. Remove the filter paper from the base of the dish and cut it in half. Wear gloves or use forceps to handle the paper. The paper will be impregnated with a pheromone that the beetles have produced.
3. Place one half of the impregnated filter paper in the bottom of a new Petri dish and cover the other half with clean paper.
4. Introduce 5 non-infected beetles of the same sex as those in the original dish into the choice chamber and leave undisturbed in a place with even illumination for 20 min.
5. Record the number of beetles on the non-treated side and express your results as:

Number of beetles on the non-treated side / total number tested.

The experiment can be repeated to increase the sample size and to examine the response of beetles of different age. Values near to zero indicate that beetles are attracted to the pheromone impregnated side and values near 0.5 suggest a random distribution.

6. Do non-infected beetles respond to pheromones from same sex beetles?
7. Does this response differ with age?
8. Investigate the effect of infection upon this behaviour.

Analysis of results

Class results can be combined and subjected to analysis to test the following hypotheses:

1. Parasitised beetles were less active than uninfected beetles.
2. Normal beetles seek a dark environment: parasitised beetles show no such preference.
3. Control beetles aggregate in response to the odour of their own species: this response is lost when infected.
4. Beetles with immature and mature parasites behave in a similar way.

Discuss the appropriate method of analysis. For example, A two-way contingency table analysis (G-test) could be used to test the null hypothesis that parasitisation has no effect upon beetle behaviour and the Student's t-test could be used to compare activity.

Expected results

Results from similar experiments have been described by Hurd and Fogo (1991). In brief, non-infected beetles exhibit negative photokinesis and aggregate in response to same-sex pheromone. Beetles infected with mature metacestodes show a significant decline in photonegative response and aggregation response but this is not observed with immature infections. Similarly, a significant decrease in activity only occurs in beetles with mature infections.

Points of significance include:

1. *T. molitor* displays behaviour patterns that **may** keep them together in secluded areas, less exposed to predators. However, this interpretation is very subjective and could lead to discussion.
2. Parasitised beetles are more likely to move to less dark areas, away from other members of the species and could thus be at greater risk of predation.
3. Behaviour patterns are only disrupted when parasites have matured. Intermediate host predation before this stage would result in parasite death rather than transmission.

Ideas for further exploration

- These host behavioural changes would only be of value to the parasite if the beetle is eaten by a definitive host such as the rat. Design experiments to test whether:
  (i)   infected beetles are more conspicuous, and
  (ii) rats are more likely to catch infected beetles.
- Which other responses to environmental stimuli could be affected by infection? How could you investigate these?
- Changes in host behaviour such as these have been called 'adaptive', i.e. they are not just the result of pathology associated with infection. What criteria do you think would need to be satisfied for behavioural changes to be said to be adaptive?

Similar exercises

Students could try this exercise with other intermediate hosts. *Tribolium castaneum* is thought to suffer greater fitness reduction than *Tribolium confusum* when infected. Aspects such as response to temperature and gravity could be explored.

REFERENCES

Bhattacharya, A. K., Ameel, J. J. & Waldbauer, G. P. (1970). A method for sexing living pupal and adult yellow mealworms. *Annals of the Entomological Society of America* **63B**, 1783.

Hurd, H. (1990). Physiological and behavioural interactions between parasites and invertebrate hosts. In: *Advances in Parasitology* **29** (Eds. J. R. Baker & R. Muller), pp. 271–317. London, Academic Press.

Hurd, H. & Fogo, S. (1991). Changes induced by *Hymenolepis diminuta* (Cestoda) in the behaviour of the intermediate host *Tenebrio molitor* (Coleoptera). *Canadian Journal of Zoology* **69**, 2291–2294.

Poulin, R. (1995). 'Adaptive' changes in the behaviour of parasitized animals: a critical review. *International Journal for Parasitology* **25**, 1371–1383.

7.4 Behaviour of the amphipod *Gammarus pulex*, infected with cystacanths of acanthocephalans

D. B. A. THOMPSON & A. F. BROWN

Aims and objectives

This exercise is designed to demonstrate:

That a parasite can alter the behaviour of its intermediate host to make it more vulnerable to predation by its definitive host.

For students, this exercise aims to:

1. Demonstrate the influence of a parasite on the behaviour of its intermediate host in order to complete its life cycle.
2. Facilitate the design of experiments and formulation of conclusions on the strategies employed by parasites to complete their life cycle.
3. Encourage understanding about studies that might be conducted in the laboratory and in the field to study the manipulation of host behaviour by parasites.
4. Examine a facet of biology embracing the fields of parasitology and animal behaviour.

Introduction

Over the past two decades, exciting developments have been made in understanding the coevolution and coexistence of parasites and their hosts (e.g. Moore, 1995). One of the most intriguing areas of host–parasite ecology is the ability of parasites to change the appearance and/or behaviour of one (intermediate) host to facilitate transmission to another (definitive) host.

This practical exercise describes experiments applicable to an association between a parasite and its two hosts. The experiments are designed to explore morphological and behavioural changes in the intermediate host that facilitate transmission of the parasite to the definitive host. Specifically, we are concerned with the effects of an acanthocephalan parasite on the behaviour of shrimps (gammarids), and then predation by its definitive host-fish (Fig. 7.4.1).

423

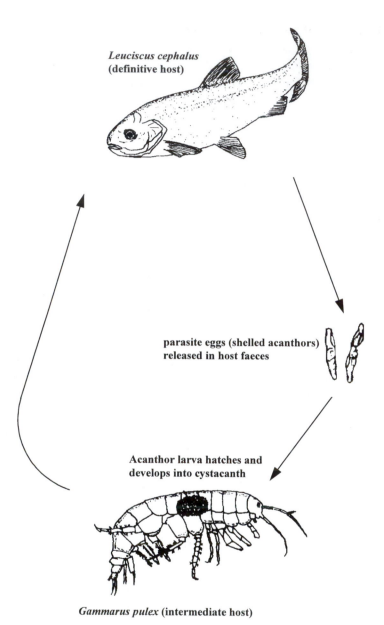

**Leuciscus cephalus
(definitive host)**

**parasite eggs (shelled acanthors)
released in host faeces**

**Acanthor larva hatches and
develops into cystacanth**

Gammarus pulex (intermediate host)

Fig. 7.4.1 Life cycle of *Pomphorhynchus laevis*. The parasite matures and reproduces in the definitive host. Eggs released in this host's faeces are ingested by the intermediate host, where the acanthor larvae hatch and develop into orange-pigmented cystacanths (infective stage) in the haemocoel. The life cycle is completed when the infected intermediate host is eaten by the definitive host.

The Acanthocephalans, popularly known as thorny- (or spiny-) headed worms, as adults are parasites of vertebrates and usually use arthropods as intermediate hosts. Two locally abundant freshwater crustaceans – the amphipod shrimp *Gammarus pulex* and the isopod pond slater *Asellus aquaticus* – are parasitised by a variety of larval acanthocephalans. In Britain and Ireland, *Gammarus pulex* is parasitised by *Polymorphus minutus* in most water bodies and also by *Pomphorhynchus laevis* in the catchments of the rivers Thames, Severn, and Hampshire Avon/Dorset Stour. *Asellus aquaticus*, on the other hand, is parasitised by *Acanthocephalus anguillae* in the River Trent and by *A. lucii* in many ponds and canals where the main definitive host, perch *Perca fluviatilis*, is present (Brown & Thompson, 1986; Marriott *et al.*, 1989). Water-birds such as mallard, *Anas platyrhynchos*, are predators of *G. pulex* and are definitive hosts of *Polymorphus minutus*, whereas certain fish, including barbel *Barbus barbus* and chub *Leuciscus cephalus*, are definitive hosts of *Pomphorhynchus laevis*.

In a series of classic papers, Holmes & Bethel (1972) and Bethel & Holmes (1973, 1974, 1977) described how amphipod shrimps (*Gammarus lacustris* and *Hyalella azteca*) parasitised by the infective larvae of *Polymorphus paradoxus*, *P. marilis*, or *Corynosoma constrictum* exhibited altered evasive behaviour. That is, parasitised individuals were more likely to swim in well-illuminated areas and to cling to surface vegetation. Consequently, infected individuals were rendered more conspicuous to predators, and so were eaten most frequently.

Details of a series of laboratory experiments using *Gammarus pulex* infected with *Pomphorhynchus laevis* are provided. One, but occasionally as many as nine, cystacanths are found within the haemocoel of *Gammarus*, where they are bright orange in colour (see Brown & Thompson, 1986). As infected gammarids are readily recognised, and both intermediate and definitive hosts are easily maintained under laboratory conditions, there is excellent scope for a study of the influence of parasites on the shrimps' evasive behaviour and on predation by fish.

Laboratory equipment, consumables and general methods

Large samples of gammarids are taken in pond nets by kick-sampling river/stream substrate and vegetation. Five kicks can yield over 1000 shrimps, of which at least 10% may be visibly parasitised. Collected samples should be placed in buckets of well-aerated fresh water for transportation to the laboratory. Fish should also be taken if possible. Of these, sticklebacks *Gasterosteus aculeatus* and minnows *Phoxinus phoxinus* may be numerous and should be retained for use as natural predators of *Gammarus* in the laboratory. Also, take some stones, water plants or aquatic macrophytes to be placed in the laboratory tanks!

Fish and gammarids should be held separately in the laboratory at 18–20 °C and a photoperiod of 12L:12D. Infected G. *pulex* will live for up to ten weeks at 18 °C, and longer at lower temperatures. Infected and uninfected individuals are best held together in a well aerated 250–500 cm³ tank. Half a bucket of substrate from the water body of origin should be added to 200–300 cm³ of aerated tap water. Food is provided by a weekly supply of well-soaked leaves (e.g. sycamore, *Acer pseudoplatanus*).

Fish should be kept in 200–300 cm³ tanks with under-gravel filtration and fed on *Calliphora* maggots. Sticklebacks are ideal because they can be accommodated in small tanks. Each group of students will require a maximum of five tanks measuring 0.5m × 0.25m × 0.25m. Unless stated otherwise, reference hereafter is to tanks with these dimensions.

A binocular microscope will aid identification and enumeration of parasites.

Sources of parasite material

Infected gammarids are found mainly in streams. An early survey is recommended in order to ensure that infected gammarids are present; these are easily identified by the bright orange colour of the cystacanth. In the USA, various species of gammarids can be caught locally. Alternatively, the latest Ward's catalogue should be consulted and it may be possible to trace potential suppliers through recent publications in one of the biological publication databases and/or the www.

Safety

There are no known biological hazards associated with handling gammarids or fish infected with the parasites named above. Otherwise, standard laboratory practice should be followed when using glassware and water.

Instructions to staff

Students should, ideally, be divided into groups of four – one to supply infected and uninfected gammarids for each experimental treatment, two to take observations, and the fourth to carry out the statistical analysis. Guidance on three specific experiments is given below.

Gammarids infected with *Pomphorhynchus laevis* or *Polymorphus minutus* are readily identified by the red–orange colour of the contained cystacanth. Since the two parasites are quite different in appearance, they can be distinguished easily. The gammarid should be dissected and the cystacanth placed on a coverslip. *P. minutus* is hard and pink–orange in colour; when the proboscis is everted, the spiny forebody is visible (to promote eversion it will help to warm the coverslip gently over a light bulb). In contrast, *P. laevis* is bright orange and bulbous at the base of the proboscis, has no spiny forebody, and has a wrinkled, soft appearance.

Not all larval stages occurring in the shrimp are infective (or capable of inducing altered behaviour). In fact, three different types of infected shrimps may be recognised. These contain: (i) pale, uninfective developing larvae (acanthellae); (ii) a single infective, coloured cystacanth; (iii) more than one infective, coloured cystacanth. These differences are obvious to the naked eye if gammarids are observed in a small tank with fresh water, a gravel substrate, and some vegetation. Gammarids infected with (i) usually swim horizontally and spend much time burrowing in the substrate; those with (ii) or (iii) swim repeatedly in a spiral towards the water surface and either cling to the vegetation for prolonged periods or sink to the bottom to lie on the substrate surface.

Instructions to students

Experiment 1: The effects of light on the behaviour of infected and uninfected *G. pulex*
Infected *G. pulex* may increase their risk of predation by swimming towards a light source where they are likely to be rendered more conspicuous to surface-feeding and mid-water fish. In this experiment we compare the response of infected and uninfected shrimps to a direct source of light.

Procedure
A tank is filled with water to a depth of 20 cm. The top and sides of one half are covered by a black polythene sheet; the other half is illuminated from 25 cm above by a 60 watt lamp. Twenty uninfected gammarids are placed in the tank and allowed to acclimatise for 30 min. Their position in light, and therefore dark, regions is recorded at 30 sec intervals for 30 min. The experiment is repeated using infected gammarids in a second tank. Replicates should be undertaken, switching light and dark regions in the tanks and using different gammarids.

Questions

- Have you found a difference between parasitised and non-parasitised gammarids in their peferences for light and dark areas? Can you think of explanations for this?

Experiment 2: The effects of habitat on the behaviour of infected and uninfected *G. pulex*

Gammarids, particularly females, spend most of their time in gravel. Infected individuals, however, tend to be observed swimming and attached to surface vegetation in greater numbers than uninfected ones. This experiment tests these observations.

Procedure

Five stones (average diameter 4 cm) are placed in tanks filled to a depth of 20 cm, and two sycamore leaves are placed on the water surface. After 1 h, 20 uninfected gammarids are added to one tank, and 20 infected gammarids to the other. After allowing them to acclimatise for 30 min, record the number of gammarids attached to vegetation, attached to stones, and swimming in open water (the remainder are concealed under stones). Records are taken at 30 sec intervals for 15 min. Replicates should be performed by switching tanks and using fresh gammarids.

Questions

- Have you noticed any differences in the tendencies to swim in the open, to be attached to vegetation, or to keep under the stones? Where are the gammarids most vulnerable to predation by fish?

Experiment 3: Predation by fish on infected and uninfected *G. pulex*

We have assumed that the parasite-induced behaviour of gammarids renders them more prone to predation. We can now test this idea by running experiments that tease apart the effects of changes in appearance and behaviour of infected gammarids.

Procedure

First, 20 infected and 20 uninfected gammarids are added to a tank (with substrate and floating vegetation, as in Experiment 2). After 1 h of acclimatisation, add a single fish that has been starved for 3–5 h. This procedure may require a Home Office licence in the UK. It should soon commence feeding on the shrimps. Record predation during the first hour, and after 6 h count remaining numbers of infected and uninfected gammarids.

To take the experiment further, use cocktail sticks to apply a tiny dot of orange enamel paint to 20 uninfected gammarids. By mimicking the parasitic infection, test whether or not a change in the shrimps' appearance influences the risk of predation. Twenty unmarked (uninfected) and 20 marked (apparently infected) gammarids are placed in a tank, and the procedure above is repeated with different fish.

Questions

- Have you recorded differences in predation by fish on infected versus uninfected gammarids? What about the significance of the appearance of the infected gammarids – is it their 'altered' behaviour and/or their colour that renders them more prone to being predated?

Analysis of results

The data should be presented in the form of tables that illustrate differences between infected and uninfected gammarids for the different treatments. Present your data in the form of means ± standard errors, and analyse the data using *t*-tests. Some of the data are more appropriate for χ^2 tests. Can you plan at the outset how you are going to collect the data and analyse it?

Expected results

The results in this experiment have been described by Brown & Thompson (1986). Further experiments using a similar gammarid–parasite system are described by Marriott *et al.* (1989) and Bakker *et al.* (1997).

You should find that the infected gammarids, compared with the uninfected ones, show a tendency to swim more frequently in lighted regions, and away from situations likely to conceal them. You should also find that the infected gammarids are more likely to be taken by fish, and that this is largely a result of their 'altered bahaviour' rather than their coloured appearance.

Ideas for further investigation

This study should spur you into posing lots of questions! Here are some suggestions for further investigation:

- Are there differences in size between infected and uninfected gammarids?
- Do the infected gammarids appear to be responding to the movements of the fish?
- Can you think of studies that would help you unravel how the 'altered' behaviour of the gammarids has evolved?
- Can you think of ways of researching how the parasite actually exerts its effect on gammarid behaviour, and is it only the infective stage of the parasite that does this?
- Does the altered behaviour of infected gammarids render them more vulnerable to all predators, or is it just the 'definitive' hosts that exert the behavioural response?
- Are the responses likely to change for floating predators, such as ducks, compared with swimming predators such as fish? Can you think of ways of tackling this question?
- Why is the cystacanth coloured orange? Can you design experiments to compare the visibility to fish of other colours?
- Do uninfected gammarids attack infected ones in order to reduce their own risk of predation?
- Are there behavioural differences between gammarids taken from streams and lakes, where the environmental pressures and predator assemblages are likely to be different?
- What other facets of gammarid behaviour might be modified, and what are the likely consequences for the individual shrimps and their populations as well as for the parasites and their populations?

Similar exercises

A range of host–parasite experiments can be undertaken to investigate the altered behaviour of infected animals. The *Gammarus pulex* – *Polymorphus minutus* system is as amenable as the one described above (e.g. Marriott *et al.*, 1989). It is even possible to carry out these sorts of experiments with terrestrial mammals, such as the mouse *Mus musculus* infected with the nematode *Trichinella spiralis*. (Note: in the UK, approval by and licence from the Home Office are required for these experiments). There are endless ways of looking at how parasites alter the behaviour of their intermediate hosts.

REFERENCES

Bakker, T. C. M., Mazzi, D. & Zala, S. (1997). Parasite-induced changes in behavior and color make *Gammarus pulex* more prone to fish predation. *Ecology* **75**, 1098–1104.

Bethel, W. M. & Holmes, J. C. (1973). Altered evasive behaviour and responses to light in amphipods harboring acanthocephalan cystacanths. *Journal of Parasitology* **59**, 945–956.

Bethel, W. M. & Holmes, J. C. (1974). Correlation of development of altered evasive behaviour in *Gammarus lacustris* (Amphipoda) harboring cystacanths of *Polymorphus paradoxus* (Acanthocephala) with the infectivity to the definitive host. *Journal of Parasitology* **60**, 272–274.

Bethel, W. M. & Holmes, J. C. (1977). Increased vulnerability of amphipods to predation owing to altered behaviour induced by larval acanthocephalans. *Canadian Journal of Zoology* **55**, 110–116.

Brown, A. F., Chubb, J. C. & Veltkamp, C. J. (1986). A key to the species of Acanthocephala parasitic in British Freshwater Fishes. *Journal of Fish Biology* **28**, 327–334.

Brown, A. F. & Thompson, D. B. A. (1986). Parasite manipulation of host behaviour. Acanthochephalans and shrimps in the laboratory. *Journal of Biological Education* **20**, 121–127.

Holmes, J. C. & Bethel, W. M. (1972). Modification of intermediate host behaviour by parasites. *Supplement No. 1, Zoological Journal of the Linnean Society* **51**, 123–149.

Marriott, D. R., Collins, M. L., Paris, R. M., Gudgin, D. R., Barnard, C. J., McGregor, P. K., Gilbert, F. S., Hartley, J. C. & Behnke, J. M. (1989). Behavioural modifications and increased predation risk of *Gammarus pulex* infected with *Polymorphus minutus*. *Journal of Biological Education* **23**, 135–141.

Moore, J. (1984). Parasites that change the behaviour of their host. *Scientific American*, June, 82–89.

Moore, J. (1995). The behaviour of parasitized animals. *Biological Science* **45**, 89–96.

Price, P. W. (1980). *Evolutionary Biology of Parasites.* Princeton, New Jersey, Princeton University Press.

Smith-Trail, D. R. (1980). Behavioral interactions between parasites and hosts: host suicide and the evolution of complex life cycles. *American Naturalist* **116**, 77–91.

7.5 Effects of *Schistocephalus solidus* (Cestoda) on stickleback feeding behaviour

I. BARBER

Aims and objectives

This exercise is designed to:

1. Investigate the life history of the cestode, *Schistocephalus solidus*, and its effect on its stickleback host.
2. Collect and analyse quantitative data on parasite loads, stomach contents and body condition in fish.
3. Conduct behavioural experiments, and interpret the data that such experiments generate.
4. Practice summarising and synthesising conclusions from experimental studies into a report.

Introduction

Schistocephalus solidus is a cestode that matures in the intestine of piscivorous (fish-eating) birds, with the three-spined stickleback serving as an intermediate host (see Fig. 7.5.1 for details of the parasite's life cycle). While in the stickleback, the worm is in the form of a plerocercoid metacestode or juvenile stage. Plerocercoids can grow to an enormous size, regularly reaching 40% of the host's mass, and in extreme cases weighing as much as the host itself. *S. solidus* infection is known to alter the behaviour of both the copepod and the fish hosts (for example by suppressing the escape response) in ways that make both intermediate hosts more vulnerable to predation by the next host in the life cycle.

The aim of this exercise is to familiarise you with this common host–parasite system, and to enable you to carry out an investigation into some of the important effects infection can have on the foraging behaviour and nutrient status of hosts. In addition, you will use data from a number of recent studies examining how the parasite influences foraging and body condition in fishes.

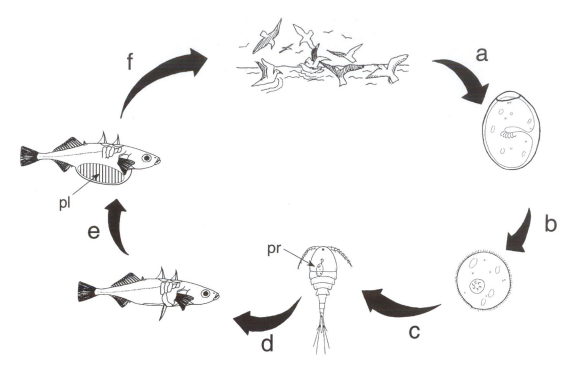

Laboratory equipment and consumables (per student or group)

Equipment

Dissecting microscope with light source
Dissecting tray with a base allowing pins to be inserted
Small Pasteur pipette and bulb
Small (100 ml) glass beaker filled with water
Fine dissecting instruments (reverse dissection scissors, watchmakers' forceps)
Ruler; dissection pins
Animal waste bin
Several balances (accurate to 0.001 g) should be available for the class

Consumables

Paper towel   40% formalin or 95% alcohol
Disposable gloves

Sources of parasite material

The *S. solidus* life cycle requires three hosts (see Fig. 7.5.1): a copepod, a stickleback and a piscivorous bird, so a proportion of

Fig. 7.5.1 The life cycle of the cestode *Schistocephalus solidus*. Eggs pass out with the faeces of infected piscivorous birds (a) into the aquatic environment, and after a period of development, the first stage coracidium larva hatches (b). This is ingested by a copepod (c) and undergoes further development to form a procercoid (pr). When an infected copepod is ingested by a three-spined stickleback (d) the parasite burrows through the gut and into the body cavity of the fish, where it undergoes extensive growth and development to form a plerocercoid (pl), which can cause abdominal swelling (e). If an infected fish is consumed by a piscivorous bird (f), then the plerocercoid rapidly becomes sexually mature, and eggs are produced, completing the life cycle.

three-spined sticklebacks from populations that inhabit lowland still waters, where there is significant predation by piscivorous birds, usually harbour plerocercoids of *S. solidus*. Generally, park ponds frequented by gulls are a good source of infected material. Workers in the USA should consult Exercise 2.3 for advice on sources of sticklebacks.

Safety

There are few safety or health hazards associated with this exercise, but since it involves dissection care must be taken whilst using sharp dissection instruments. As with all parasitology practical classes, laboratory coats must be worn. Although the parasite, *S. solidus*, will probably still be alive when you dissect the fish, this cestode poses little risk to human health unless ingested (*S. solidus* is a recorded parasite of humans in cultures where small fish are eaten raw). However, a small proportion of people may be allergic to secretions produced by the parasite, and so the parasite should not be handled unless gloves are worn.

Instructions for staff

Equal numbers of infected and uninfected stickleback should be maintained together in a large aquarium for one week prior to the exercise and fed on bloodworm (Chironomid larvae). Two days before the exercise, feeding should be stopped and, immediately before the class, bloodworm introduced such that the number available equals $3 \times$ the number of fish. This will ensure that there is competition for available food items, and that some will ingest more than others. Once all prey have been consumed, fish should be left for 10 min or so, then removed and sacrificed by exposure to an overdose of benzocaine (10 ml/l). Students should work in pairs, and each pair should be provided with a freshly killed infected and uninfected stickleback. Fish should be presented in dissecting trays and kept moist.

Instructions for students

Materials
You will be working in pairs and will be provided with a freshly killed infected and uninfected stickleback from a site in which the prevalence of *S. solidus* infection is known to be high.

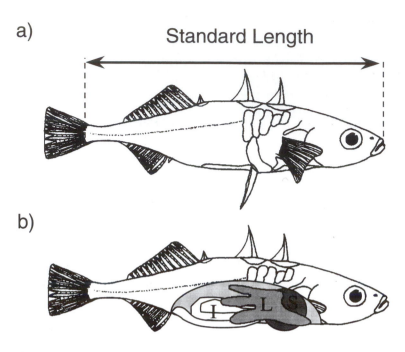

a)

Standard Length

b)

Fig. 7.5.2 (a) Side view of a three-spined stickleback, showing standard length and where it should be cut to expose the body contents. (b) Diagram of the internal organs of an uninfected stickleback. I, intestine; L, liver; S, stomach.

Methods

Measure the two fish (standard length; from the tip of the snout to the end of the caudal peduncle, or tail stalk (see Fig. 7.5.2a) to the nearest mm. Surface dry each fish (dab the fish gently with absorbent tissue to remove surface moisture) and weigh it on one of the balances provided, to the nearest 0.001 g. Carefully dissect the fish, beginning by inserting the point of your dissection scissors into the vent and cutting the body wall as shown in Fig. 7.5.2a. Do not use a scalpel! The infected fish can be recognised by its distended abdomen, and when you dissect this fish you will notice the large white *S. solidus* plerocercoid(s) in the body cavity. Remove these parasite(s), count and weigh them individually (to the nearest 0.001 g).

- **Q1.** What is the position of the *S. solidus* plerocercoids in the body cavity?

Dissect the uninfected fish in the same way. Remove the multi-lobed, pinkish liver (see Fig. 7.5.2b) from each fish, and weigh it to the nearest 1 mg. Removing the liver will expose the stomach (anterior, ball-shaped and swollen, depending on the number of prey items consumed). Remove the stomach by cutting the oesophagus anterior to the stomach and tease out the intestine, which leads from the stomach to the vent.

Remove the stomach and carefully cut it open, ensuring the contents remain in place. Examine the stomach under a binocular microscope and count the number of bloodworms (on which the fish were feeding prior to the start of the exercise) contained therein.

You will calculate two commonly used indices of how well-nourished the fish are, namely, the condition factor and the hepatosomatic index. The condition factor (CF) gives a measure of how heavy the fish is for its length, and you should calculate this as follows:

$$\text{CF (\%)} = [\text{ weight of fish (in g, not including parasite weight) / length (in cm)}^3] \times 100$$

This formula assumes that the weight of a stickleback is predicted from the cube of its length, which is generally true. Make sure that you subtract the weight of the *S. solidus* plerocercoids from the fish weight before you calculate this index for your infected fish.

The liver is an important energy store in fish and the relative weight of the liver, or hepatosomatic index (HSI), gives a useful indication of energy reserves. You should calculate this as follows:

$$\text{HSI (\%)} = \text{liver weight (in g) / [weight of fish (in g, not including parasite weight)]} \times 100$$

In addition, you should calculate how heavily infected your parasitised stickleback is by calculating the parasite index (i.e. the proportion of infected fish weight contributed by *S. solidus*). Parasite index (PI) should be calculated as follows:

$$\text{PI (\%)} = [\text{weight of } S.\ solidus \text{ (in g) / weight of fish} + \text{weight of } S.\ solidus] \times 100$$

When you have calculated these indices, give your data to the class demonstrator, who will set up a class database. As a class, or in smaller groups, you will then use these data sets to answer the following questions:

- **Q2**. What is the distribution of *S. solidus* plerocercoids amongst the infected fish? (Draw a histogram to show the distribution of frequencies of fish with 1 worm, 2 worms, 3 worms, etc.).
- **Q3**. Do infected and uninfected fish differ in condition factor or in hepatosomatic index?

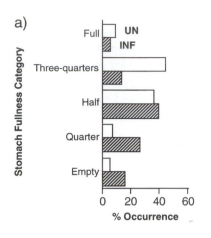
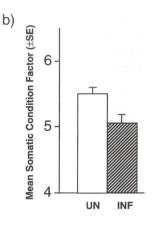
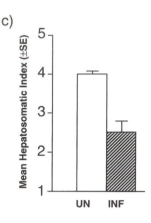

- **Q4**. Is there a relationship between the parasite index of the infected fish (i.e. how heavily it is infected) and *either* the condition factor *or* hepatosomatic index?
- **Q5**. What is the distribution of the number of worm consumed by (i) uninfected and (ii) infected sticklebacks, and are these distributions different from each other?

While the class results are being compiled and analysed, examine Figs. 7.5.3 a–c, which show the results of similar analyses using more extensive samples, and provide answers to the following questions:

- **Q6**. What conclusions can you draw from this larger data set about the effects of *S. solidus* infection on stomach fullness and body condition in sticklebacks?
- **Q7**. What consequences might your answer to Q6 have for the biology of parasitised fish in natural populations in the wild?

One possible explanation for the relatively empty stomachs and poorer body condition of *S. solidus*-infected fish is that the presence of the worm in the body cavity physically constrains the host from taking in food. In a study designed to test this possibility, uninfected and *S. solidus*-infected sticklebacks were deprived of food for 24 h and then fed live bloodworms one at a time until they were satiated. Fig. 7.5.4. shows the number of bloodworms eaten in relation to fish length for the two categories of sticklebacks.

- **Q8**. What do these data suggest about the effect of *S. solidus* infection on food intake in sticklebacks?

Fig. 7.5.3 (a) Distribution of stomach fullness categories (re-drawn from Tierney, 1994), (b) Mean ± SE somatic condition factor, and (c) Mean ± SE hepatosomatic index in sticklebacks with (hatched bars) and without (open bars) *S. solidus* infection (re-drawn from Tierney *et al.*, 1996.)

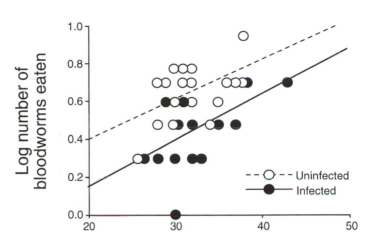

Fig. 7.5.4 The relationship between the number of bloodworm eaten to satiation and length of fish, for sticklebacks with and without *S. solidus* infection.

- **Q9**. Do they support the suggestion that *S. solidus* infection physically constrains food intake in sticklebacks?
- **Q10**. Do your class results agree with these published results, and if not why might this be?

In a separate experiment, video tapes of infected and uninfected fish feeding were used to determine the time taken to handle and swallow large and small bloodworms under two conditions, namely when the fish had empty stomachs ('Empty') or when the fish had eaten one prey item ('Fed'). The sequence in which fish experienced these two conditions was randomised, to allow for any effects of experience of the first exposure. From the known energetic content of prey of both sizes, the profitabilities (rate of energy intake in Joules sec^{-1}) of large and small prey were calculated for infected and uninfected sticklebacks, depending on whether they were empty or full. These values are shown in Fig. 7.5.5.

- **Q11**. What is the more profitable prey size (large or small) for infected and uninfected sticklebacks when they have empty stomachs and when they have just fed?

Finally, in a third experiment different groups of infected and uninfected sticklebacks were given a simultaneous choice of a large and a small bloodworm presented behind glass and a record taken of which worm they bit at first; previous studies have shown that this is a good measure of food preference in sticklebacks. This test was repeated twice for each fish, once

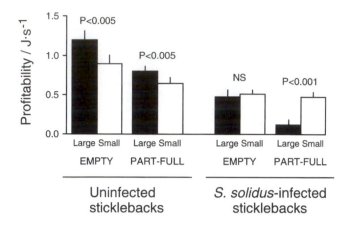

Fig. 7.5.5 Median (and interquartile range) profitability of large and small bloodworms for uninfected and infected sticklebacks with empty stomachs ('Empty') and after having eaten one blood worm ('Part-full').

when they had an empty stomach and once when they had just been fed a single worm, again randomising the two conditions. Table 7.5.1 shows the number of fish in each condition that bit at the larger and the smaller worm.

- **Q12.** Which is the preferred prey size for each category of fish?
- **Q13.** Combining your answers to Q11 and Q12, how does profitability influence prey preference in sticklebacks?

Using the class data and the results presented in this schedule, prepare a report describing the effects of *S. solidus* infection on feeding in sticklebacks, how these might come about and their probable consequences for body condition, for assessment. A short discussion with the whole class based on these reports may then follow.

Ideas for further exploration

1. Is the effect of the parasite on host feeding behaviour (i.e. how many bloodworms it ingests) related to the extent of the infection, measured as the parasite index (PI) ? Test your hypothesis with appropriate statistical analysis. Explain why fish that have higher PIs may do worse in competition for food with other less heavily infected, or uninfected, fish.
2. How does infection with the parasite affect the growth of the fish? The standard length measurement is an approximate indicator of somatic growth, so compare the lengths of infected and uninfected sticklebacks in the class sample? Test your hypothesis with appropriate statistical analysis. If there are differences in the mean lengths of infected and

Table 7.5.1 *Number of sticklebacks biting first at the large or the small*
bloodworms in relation to infected state and stomach fullness

| | No. of first bites to: | | |
Condition	Large worms	Small worms	χ^2 test
Uninfected fish, empty stomachs	13	2	$P<0.01$
Uninfected fish, partially full	12	3	$P<0.05$
Infected fish, empty stomachs	6	5	NS
Infected fish, partially full	2	11	$P<0.05$

Notes:
Data for Cunningham *et al.*, 1994
NS, not significant.

uninfected fish in the sample, explain what may cause these
differences.
3. In fish with multiple infections (i.e., harbouring more than
one *S. solidus* plerocercoid), are all of the worms the same
weight, or does, for instance, one large worm dominate the
infection? Explain how the observed pattern of weights of
parasites in multiple infections may arise – is it more likely
that fish are infected successively or simultaneously with
these parasites?
4. How does infection alter the outline of the fish when seen
from above, and from the side? How might this alter the
extent to which infected fish are predated by piscivorous
birds (which are definitive hosts of the parasite) and piscivorous fish (which are not). How could morphometric measurements be used to estimate the PI of *living* fish ?

Expected results

Because food was limited in the feeding period prior to the
class, thus generating competition for individual prey items,
students should notice that infected fish consume fewer of the
available prey than uninfected fish. This is a result of the
reduced competitiveness of infected fish, which has been well-
documented (see References). In addition, students should find
that infected fish have lower body condition (as measured by
the condition factor) and lower energy reserves (measured as
hepatosomatic index). This is a result of both their reduced
competitiveness and feeding on less nutritional prey in the wild

and the high energetic demands of the growing parasite. Students should notice that the worm show an overdispersed distribution (i.e. most infected fish have one or two worms, and a smaller number with higher burdens) and that the worms live in the body cavity, intertwined with the viscera, and often close to the body wall so that they can be seen moving under the surface of the skin. Answers to the data interpretation exercise should be self-explanatory, given this information and that presented in the schedule.

Information on similar exercises

Three-spined sticklebacks have a wide distribution in the Northern hemisphere, and the distribution of the parasite closely matches that of the host. However, if sticklebacks are not available, a similar host–parasite system – that of small cyprinid, catostomid or galaxid fish infected with plerocercoids of a related pseudophyllidean cestode, *Ligula intestinalis*, could be used equally well as a substitute, and a similar exercise could be carried out using these fish. The *L. intestinalis* life cycle is analagous to that of *S. solidus*, and the parasite has a similar suite of effects on its host; however, *L. intestinalis* has the advantage of having a world-wide distribution. Again, infected fish are most likely to be caught in still waters (lakes and ponds) rather than in streams and rivers. In the UK and Europe, common host fish are minnows (*Phoxinus phoxinus*), roach (*Rutilus rutilus*) and gudgeon (*Gobio gobio*); in North America, commonly infected fish are shiners (*Notropis* spp.), suckers (*Catostomus* spp.) and minnows (*Pimephales* spp.); in Southern Africa, *Barbus* spp.; in India, *Danio*, *Labeo* and *Rasbora* spp. and in Australasia, *Galaxias* spp.

REFERENCES

Barber, I. & Huntingford, F. A. (1995). The effect of *Schistocephalus solidus* (Cestoda: Pseudophyllidea) on the foraging and shoaling behaviour of three-spined sticklebacks, *Gasterosteus aculeatus*. *Behaviour* **132**, 1223–1240.

Cunningham, E. J., Tierney, J. F. & Huntingford, F. A. (1994). Effects of the cestode *Schistocephalus solidus* on food intake and foraging decisions in the three-spined stickleback, *Gasterosteus aculeatus*. *Ethology* **97**, 65–75.

Milinski, M. (1990). Parasites and host decision-making. In: *Parasitism and Host Behaviour* (eds. C. J. Barnard & J. M. Behnke), pp. 95–116. London, Taylor & Francis.

Tierney, J. F. (1994). Effects of *Schistocephalus solidus* (Cestoda) on the diet and food intake of the 3-spined stickleback *Gasterosteus aculeatus*. *Journal of Fish Biology* **44**, 731–735.

Tierney, J. F., Huntingford, F. A. & Crompton, D. W. T. (1996). Body condition and reproductive status in sticklebacks exposed to a single wave of *Schistocephalus solidus* infection. *Journal of Fish Biology* **49**, 483–493.

APPENDIX 1: REAGENT INDEX

A range of salines, media, buffers and other reagents are used in the exercises detailed in this book. In each case, their composition can be found in the consumables section of each exercise, or in an appendix specific to an exercise, where reagents are complex. For convenience of the user, this index identifies the exercise where details of reagents can be found. Note that the formulation of commercially available reagents (e.g. Hanks' saline) is not normally detailed.

	Exercise
Buffers	
Acetate buffer	6.7
Bacterial cell lysis buffer	6.1
DNA wash buffer	6.5
EcoRI reaction buffer	6.2
Gel loading buffer	6.2, 6.4
Lysis buffer	6.5
Sodium cacodylate buffer	6.7
TBE gel buffer	6.2, 6.4
TE buffer	6.1, 6.2. 6.4, 6.5
Tris/HCl buffer	6.5, 6.7
Cotton blue-lactophenol	1.13
Cunningham's medium	6.6
ELISA reagents	4.3, 4.6
Lugol's iodine	1.8, 1.9, 5.2, 5.3
Plasmid preparation solutions	6.4
SDS PAGE reagents	4.7
Western blotting reagents	4.7
Salines	
Ascaris Ringers Solution	3.5
Ascaris saline	3.6
Citrate-saline	1.13
Formol saline, buffered	4.4
Hanks' balanced salt solution (or saline)	1.5, 1.13, 1.14, 1.15, 3.2
Hedon Fleig solution	3.4, 3.5
Insect Ringer	2.1, 2.5, 3.2
Invertebrate saline	1.1, 4.1
Krebs Ringer tris	3.3

Locke's saline	3.6
Mammalian Ringer solution	2.4
Phosphate buffered saline (PBS)	4.6, 5.1, 6.5, 6.6, 6.7
Physiological saline (normal or mammalian)	1.1, 1.8, 1.9, 1.12, 3.4, 3.5, 4.1, 4.4, 4.5, 5.3, 5.4
Tyrode's salt solution	3.1, 3.2

APPENDIX 2: UK SUPPLIERS

Amersham Pharmacia Biotech UK
23 Grosvenor Road
St Albans
Herts AL1 3AW
Tel: 01727-814000
Fax: 01727-814001

Anachem Ltd
Anachem House
20 Charles Street
Luton
Bedfordshire LU2 0EB
Tel: 01582-745000
Fax: 01582-488815
E-mail: sales@anachem-ltd.com
www.anachem-ltd.com

BDH Chemicals (see Merck
Eurolab)

Becton & Dickinson (UK) Ltd
21 Between Towns Road
Cowley
Oxford OX4 3LY
Fax: 01865-781627

Boehringer (see Roche Molecular
Biochemicals)

Clark Electromedical Industries
PO Box 8
Pangbourne
Reading
Berkshire RG8 7HU

Clontech Laboratories UK Ltd
Unit 2, Intec 2
Wade Road
Basingstoke
Hampshire RG24 8NE
Tel: 01256-476-500
Fax: 01256-476-499
E-mail: orders@clontech.co.uk

DAKO Ltd
Denmark House
Angel Drove
Ely
Cambridgeshire CB7 4ET
Tel: 01353-669911
Fax: 01353-668989
www.dakoltd.co.uk

Difco
(see Becton
& Dickinson (UK)
Ltd)

Fluka Chemicals (see Sigma-
Aldrich Company Ltd)

GIBCO/BRL (see Life Technologies
Ltd)

Hendley Ltd
Oakwood Industrial Estate
Loughton
Essex

ICN Pharmaceuticals Ltd
1 Elmwood
Chineham Business Park
Basingstoke
Hampshire RG24 8WG

Tel: 01256-374-620
Fax: 01256-374-621

Invitrogen BV/NOVEX
PO Box 2312
9704 CH Groningen
The Netherlands

Tel: 0800-966-193
Fax: +31(0)-50-5299-281
E-mail:
tech_service@invitrogen.nl

Jencons (Scientific) Ltd (see also
Merck Eurolab)
Cherry Way Industrial Estate
Stanbridge Road
Leighton Buzzard
Bedfordshire LU7 8UA

Life Technologies Ltd
3 Fountain Drive
Inchinnan Business Park
Paisley PA4 9RF

Tel: 0141-814-6100
Fax: 0141-814-6287
E-mail: euroinfo@lifetech.com.

Merck Eurolab
Tel: 01202-664778
Fax: 01202-664735

www.merckeurolab.ltd.uk

NUNC (see Life Technologies Ltd)

Peninsula Laboratories (Europe)
Ltd
PO Box 62
17K Westside Industrial Estate
Jackson Street
St Helens
Merseyside WA9 3AJ

Tel: 01744-612108
Fax: 01744-730064

Philip Harris Education (England,
Northern Ireland, Wales)
Lynn Lane
Shenstone
Lichfield
Staffordshire WS14 0EE

Tel: 01543-482202
Fax: 01543-483056
E-mail:
sales@education.philipharris.co.
uk

Philip Harris Education
(Scotland)
E6 North Caldeen Road
Calder Street
Coatbridge
Lanarkshire ML5 4EF

Tel: 01236-437716
Fax: 01236-435183
E-mail:
sales@education.philipharris.co.
uk

Promega Life Science
Southampton

Tel: 01703-760-225
Fax: 01703-767-014
www.promega.com

Qiagen Ltd
Boundary Court
Gatwick Road
Crawley
West Sussex RH10 2AX

Tel: 01293-422911
Fax: 01293-422922
www.qiagen.com

Roche Molecular Biochemicals
Roche Diagnostics Ltd
Bell Lane
Lewes
East Sussex BN7 1LG

Tel: 0808-100-99-98
Fax: 0808-100-80-60

Scotlab Ltd
Kirkshaws Road
Coatbridge
Lanarkshire ML5 8AD

Sigma-Aldrich Company Ltd
Fancy Road
Poole
Dorset BH12 4QH

Tel: 0800-717181
Fax: 0800-378538
E-mail:
ukorder@euronotes.sial.com
www.sigma-aldrich.com

APPENDIX 3: US SUPPLIERS

American Type Tissue Collection
(ATTC)
10801 University Boulevard
Manassas
VA 20110-2209

Tel: 703-365-2700
www.atcc.org/

Biomedical Research Institute
12111 Parklawn Drive
Rockville
MD 20852

Tel: 301-881-3300
Fax: 301-881-7640
E-mail:
SNAILSRUS@compuserve.com

Carolina Biological Supply Co.
Burlington, P.O. Box 6010
NC 27215

Tel: 336-584-0381
Fax: 336-584-7686
E-mail: carolina@carolina.com

GIBCO BRL/Life Technologies
International Orders
9800 Medical Center Drive
P.O. Box 6482
Rockville
MD 20849-6482

Tel: 301-610-8709
Fax: 301-610-8724
E-mail: laorders@lifetech.co

Jones Biomedicals & Laboratory,
Inc.
5911 East Spring Street
Suite 368, Long Beach
CA 90808

Tel: 562-234-0053
Fax: 413-828-0553
E-mail: service@jonesbiomed.cjb.net
http://jonesbiomed.cjb.net

Presque Isle Culture
PO Box 8191
Presque Isle
PA 16505

Tel: 814-833-6262
Fax: 814-871-7634
E-mail: PIC@velocity.net

Sea Life Supply
740 Tioga Avenue
Sand City
CA 93955-3034

Tel: 831-394-0828
Fax: 831-899-3399
E-mail: info@sealifesupply.com

Ward's Natural Science
Establishment, Inc.
5100 West Henrietta Road
PO Box 92912
Rochester
NY 14692-9012

Tel: 800-962-2660
Fax: 800-635-8439
www.wardsci.com

Marine Biological Laboratory
Aquatic Resources Division
7 MBL Street
Woods Hole
MD 02543

Tel: 508-548-7375
Fax: 508-289-7900
www.mbl.edu

INDEX

Note: page numbers in *italics* refer to figures and tables

452